RELATIVITY

RELATIVITY

Proceedings of the Relativity Conference in the Midwest,
held at Cincinnati, Ohio, June 2-6, 1969

Edited by

Moshe Carmeli

and

Stuart I. Fickler

Aerospace Research Laboratories
Wright-Patterson Air Force Base, Ohio

and

Louis Witten

Department of Physics
University of Cincinnati
Cincinnati, Ohio

PLENUM PRESS • NEW YORK-LONDON • 1970

Library of Congress Catalog Card Number 74-112865

ISBN-13: 978-1-4684-0723-5 e-ISBN-13: 978-1-4684-0721-1
DOI: 10.1007/978-1-4684-0721-1

Softcover reprint of the hardcover 1st edition 1970
A Division of Plenum Publishing Corporation
227 West 17th Street, New York, N. Y. 10011

United Kingdom edition published by Plenum Press, London
A Division of Plenum Publishing Corporation, Ltd.
Donington House, 30 Norfolk Street, London W.C. 2, England

PREFACE

The Relativity Conference in the Midwest was held in Cincinnati from June 2-6, 1969 and was sponsored jointly by the Aerospace Research Laboratories and the University of Cincinnati. During 1969, the Aerospace Research Laboratories celebrated the twentieth year of its existence and the University of Cincinnati celebrated its sesquicentennial year. Because of the extended interest of the Aerospace Research Laboratories in the Theory of Relativity and the recent arrival at the University of Cincinnati of one of the conference organizers it was felt that sponsoring a National Conference on Relativity designed primarily for participation by American scientists would represent a worthwhile contribution to the joint celebrations.

This was the first formal national conference specifically devoted to relativity held in this country since the 1957 conference on The Role of Gravitation in Physics held in Chapel Hill. The format of the week-long conference consisted primarily of invited papers and extensive informal discussions. Contributed papers were generally discouraged except as their subject matter fit into the discussions. At no time was a clock used to terminate an invited talk; it was frequently necessary, however, to terminate the discussions after some hours because of the onset of weariness in mind and stomache. By general demand a special informal session was organized and held on Tuesday night to discuss "Superspace." The Superspace Session started at 8:00 P.M. and faded away beginning at 2:00 A.M.

This volume contains a summary of most of the invited papers. Because of the great interest in Superspace generated at the Conference it also contains an extended introduction and discussion of the subject. Each day of the meeting contained its own highlights and no general summary of these could be given without a complete description of the proceedings. However it is worth recording the universal acclaim given to the extended informal discussions conducted.

We want to acknowledge gratefully the assistance we had in organizing the conference and in typing the proceedings. Especially we are grateful to Robert B. Adams for his general control of all the arrangements. We thank Christine Adams and Mary Beckwith for their help with the preliminary typing and registration and Michelle Brumfield and Suzanne Louis for typing many of the final manuscripts. We appreciate the efforts of Diana Fenichel, Margrit Franke, and Lorraine Witten, who planned the entertainment of the ladies. Our final and great acknowledgement goes to the invited speakers, to the chairman of the sessions all of whom acted as discussion leaders, and to all the participants at the conference.

Moshe Carmeli, Stuart I. Fickler, Louis Witten Cincinnati, 1969

THE PROGRAM OF THE

RELATIVITY CONFERENCE IN THE MIDWEST

JUNE 2-6, 1969

Sponsored by the Aerospace Research Laboratories and the University of Cincinnati

Organizing Committee: Moshe Carmeli, Stuart I. Fickler, Louis Witten

Sunday, June 1, 1969

7:30 P.M. - 12:00 P.M. - Registration and Reception

Monday, June 2, 1969

Morning Session: Stuart I. Fickler, Chairman

9:00 A.M. Thomas N. Bonner, Provost for Academic Affairs; University of Cincinnati
Welcoming Remarks

9:15 A.M. John R. Klauder - Bell Telephone Laboratories
Soluble Models of Quantum Gravity

11:00 A.M. Arthur Komar - Yeshiva University
The Quantization Program for General Relativity

Afternoon Session: Irwin Goldberg, Chairman

2:00 P.M. Bryce S. DeWitt - University of North Carolina
Quantum Theories of Gravity

3:30 P.M. Discussion Period

Tuesday, June 3, 1969

Morning Session: Dieter R. Brill, Chairman

9:15 A.M. John A. Wheeler - Princeton University
Particles and Geometry

11:00 A.M. Peter G. Bergmann - Syracuse University
The Sandwich Theorem

Afternoon Session: Kip S. Thorne, Chairman

2:00 P.M. Charles Misner - University of Maryland
Horizons and the Singularity in Closed Anisotropic Homogeneous Cosmological Models

3:30 P.M. Discussion Period

Wednesday, June 4, 1969

Morning Session: John Stachel, Chairman

9:15 A.M. S. Chandrasekhar - University of Chicago
Post-Newtonian Methods and
Conservation Laws

11:00 A.M. James L. Anderson - Stevens Inst. of Technology
Relativistic Transport Theory via the
Grad Method of Moments

Afternoon Session: James L. Anderson, Chairman

2:00 P.M. Rainer K. Sachs - University of California
General Relativistic Kinetic Theory

3:30 P.M. Discussion Period

Thursday, June 5, 1969

Morning Session: Allen I. Janis, Chairman

9:15 A.M. Joe Weber - University of Maryland
Gravitational Radiation Experiments

11:00 A.M. William M. Fairbank - Stanford University
Low Temperature Experiments Relating
to General Relativity

Afternoon Session: Joshua N. Goldberg, Chairman

2:00 P.M. Eugene Guth - Oak Ridge National Laboratory
Historical Development of Einstein's
Ideas of Gravitation

3:15 P.M. Kip S. Thorne - California Inst. of Technology
Gravitational Radiation Damping

4:15 P.M. Discussion

7:30 P.M. Dinner - Losantiville Room, The Donald Core
Tangeman University Center
After Dinner Speaker: Alvin J. Reines
Professor of Jewish Philosophy
Hebrew Union College
Relativistic Concept of Liberal Religion

Friday, June 6, 1969

Morning Session: Fred J. Belinfante, Chairman

9:15 A.M. Nathan Rosen - Israel Institute of Technology and Weizmann Institute of Science
The Nature of the Schwarzschild Singularity

11:00 A.M. Robert Geroch - Birkbeck College
Singularities

Afternoon Session: Ivor Robinson, Chairman

2:00 P.M. Jeffrey Winicour - Aerospace Research Laboratories
Energy - Momentum of Radiating Systems

3:15 P.M. Ivor Robinson - Southwest Center for Advanced Studies
Vacuum Metrics Without Symmetry

4:15 P.M. Discussion

CONTENTS

SOLUBLE MODELS OF QUANTUM GRAVITATION

John R. Klauder

Bell Telephone Laboratories, Incorporated

Murray Hill, New Jersey

1. INTRODUCTION: MODELS TO BE DISCUSSED

The quantum theory of gravitation has essentially all the difficulties of conventional field theories, as well as several of its own. Among the latter are the technical difficulties of gauge aspects, constraints and positivity requirements on the metric, and the conceptual difficulties of interpretation. No lesser theory can describe the gravitational field. Nevertheless, considerable insight may frequently be gained by studying models exhibiting some of the distinctive features. The prime example of this is the electromagnetic field which is frequently used to gain insight into the properties of gauge freedoms. Other theories or models can shed light onto the difficulties associated with the positivity requirements on the metric tensor; for example, the space-like metric $g_{rs}(\underset{\sim}{x})$, when viewed as a 3×3 matrix, is positive definite.

We have studied in some detail three model theories in order to gain insight into the quantum nature of the positivity restrictions on the metric.[1] Since our models focus on these positivity restrictions, we have for convenience adopted models algebraically similar to the gravitational theory but which have been sufficiently simplified so that constraints do not appear. This is only a temporary shortcoming, for we hope to study more complicated models having constraints as well.

In principle, our approach to generate models is to adopt the classical Einstein action functional, possibly in the presence of external sources, and to impose certain restrictions on the allowed variations of the variables. In practice, modifications of this approach have been adopted to facilitate the discussion of existence and solubility questions. In all cases we adopt a Palatini-like approach treating the metric and connection as independent variables. To motivate our first and simplest model imagine that the metric is conformally flat and is a function of time alone (long wavelength limit). In particular, we restrict the contravariant metric density so that $\mathfrak{g}^{\mu\nu}(x) \to p(t)\, \mathfrak{g}^{\mu\nu}_{\text{LORENTZ}}$, where $p(t) > 0$ is a physical requirement. In like manner, we restrict the connection to a single function $q(t)$ and adopt for our first model the classical system defined by the action functional

$$I = \int \{p(t)\dot{q}(t) - p(t)q^2(t) - V[p(t)]\}\, dt\ , \tag{1}$$

where $p(t) > 0$ and V symbolizes, say, the interaction with an external source. The algebraic similarity of the initial terms to the Einstein action functional should be apparent. Section 2 is devoted to the quantum theory of such a model with special emphasis on the restriction $p(t) > 0$.

For our second model we promote the first example to a true field theory; namely, we set $\mathfrak{g}^{\mu\nu}(x) \equiv \mathfrak{g}(x)\mathfrak{g}^{\mu\nu}_{\text{LORENTZ}}$, but neglect spatial derivatives in forming the action functional. Specifically, the second model is based on the classical action functional

$$I = \int \{\mathfrak{g}(x)\dot{\Gamma}(x) - \mathfrak{g}(x)\Gamma^2(x) - V[\mathfrak{g}(x)]\}\, d^4x\ . \tag{2}$$

Here $\mathfrak{g}(x) > 0$ is a scalar analog of the metric, while $\Gamma(x)$ is a scalar analog of the connection. Finite energy requires that $V[1] = 0$. Note that in the second model there is no coupling of different points (as with derivatives), a kinematic simplification which largely accounts for the solubility of the quantum theory.

These second models are, of course, bona fide field theories. If carried out in conventional fashion their quantization is beset with all the traditional difficulties. There are divergences and renormalizations that would be required in a perturbation approach. However, far superior methods of approach are available which may be applied without ambiguity, and which lead to the existence of solutions by techniques which avoid "boxes," momentum space cutoffs or any divergences whatsoever. Among the notable consequences we find is the fact that distinct classical theories (different V's) require inequivalent representations of the basic field operators. We determine the form of the solution on the basis of very general arguments. Subsequent to this determination we present an operator form of solution which is especially noteworthy for its simplicity. Section 3 is devoted to a discussion of the foregoing model.

The last model we shall treat generalizes the second model by representing the metric as a positive 3×3 matrix with elements $g_{rs}(x)$. With the connection represented by $\Gamma_{rs}(x)$ we adopt as our classical action functional

$$I = \int \{g_{rs}\dot{\Gamma}_{rs}-\Gamma_{rt}g_{rs}\Gamma_{st}-V[g_{rs}]\} \, d^4x \; . \tag{3}$$

Here g_{rs} and Γ_{rs} are both symmetric in their index pair and summation from 1 to 3 is understood for repeated indices. We emphasize that the 3×3 matrix $g(x)$ with elements $g_{rs}(x)$ is required to be positive, $g(x) > 0$. Here we meet face-to-face the complicated positivity requirements that a true gravitational theory must encounter. Again the action functional has been chosen to be algebraically similar to relativity, but kinematically simplified so that existence questions can be settled. Our discussion of this last model appears in Sec. 4.

It is our belief that the analysis of these models has shed considerable light on the quantum treatment of the positivity restrictions for the gravitational field. We have found especially useful an alternate kinematical group to the Heisenberg group. In brief, rather than adopt the usual relation $[P,Q] = -i$, we study the Lie algebra relation of the affine group, $[P,B] = -iP$, which consistently incorporates the positivity requirements. This commutation relation and its generalization to a field theory are basic in our study of quantum models of gravitation.

2. BASIC IDEAS, AND THE SINGLE DEGREE OF FREEDOM MODEL

Let us consider the simplest of the models, that pertaining to a single degree of freedom. The classical action for this model is given in Eq. (1). Following standard procedures, we would introduce a canonical pair of operators Q and P for which

$$[Q,P] = i \; .$$

The physics of the problem demands that the spectrum of the operator P be positive, $P > 0$. In the usual momentum representation P is represented by multiplication by k, and to ensure the spectral condition the allowed wave functions $\psi(k)$ must all vanish if $k < 0$. In other words, the requisite Hilbert space is $L^2(0,\infty)$, the space of square integrable wave functions on the interval 0 to ∞. As usual $Q = i\partial/\partial k$, but it is of fundamental importance to note that Q is not the generator of unitary transformations on $L^2(0,\infty)$. Specifically, the operator $U(s) = \exp(isQ) = \exp(-s\partial/\partial k)$ is not norm preserving for $s < 0$ since it wants to "slide" wave functions below $k = 0$ which is outside $L^2(0,\infty)$. This lack of unitarity implies that Q <u>can not</u> be diagonalized. That is, there is no "Q-representation" compatible with the spectral condition $P > 0$.

Technically, one says Q is symmetric, but not self-adjoint. Observe also that the function $\exp(-i\beta k)$, for any β such that $\mathrm{Im}\beta < 0$, is a normalizable eigenstate of Q with a complex eigenvalue β.

To circumvent these problems it is expedient to introduce

$$B = \tfrac{1}{2}(PQ+QP) = \frac{i}{2}\left(k\,\frac{\partial}{\partial k} + \frac{\partial}{\partial k}\,k\right)$$

which satisfies the commutation relation

$$[B,P] = iP \ . \tag{4}$$

This relation is seen to characterize a two-parameter Lie algebra, which belongs to the so-called <u>affine group</u>, the group of translations and dilations (without reflections) of the real line: $x \to p^{-1}x - q$. A detailed investigation shows that there are two unitarily inequivalent, faithful, irreducible representations of (4), one with $P > 0$ and the other with $P < 0$.[2] We confine our attention to the former and consider the unitary operators

$$U[p,q] = e^{-iqP}\, e^{i\,\ell np\, B}$$

where $p > 0$ and q are group parameters. Basic relations of importance are

$$U[p,q]^{\dagger} = U[p^{-1},-pq] \ , \tag{5a}$$

$$U[p,q]^{\dagger}\{\alpha P+\beta Q+\gamma B\}U[p,q] = \alpha pP + \beta(p^{-1}Q+q) + \gamma(B+pqP) \ . \tag{5b}$$

The quantum Hamiltonian suggested for the model of Eq. (1) is of the form

$$\mathcal{H} = QPQ + \mathcal{V}(P) = -\frac{\partial}{\partial k}\,k\,\frac{\partial}{\partial k} + \mathcal{V}(k) \ . \tag{6}$$

Although Q is not self adjoint, the operator $\mathcal{H}$ is self adjoint on $L^2(0,\infty)$. For any potential $\mathcal{V}$ we find that

$$[\mathcal{H},P] = 2i\ B \ . \tag{7}$$

Since $\mathcal{V}$ must be real for $\mathcal{H}$ to be Hermitian, any eigenstate $\psi(k)$ may be chosen real, which is a consequence related to "time-reversal" invariance.

It is convenient to have a functional description of this system generated by the relevant kinematic group, the affine group. To this end let $|0\rangle$ denote a normalized fiducial vector in the Hilbert space which satisfies the relations

$$\langle 0 | 0 \rangle = \langle 0 | P | 0 \rangle = 1 \,, \tag{8a}$$

$$\langle 0 | B | 0 \rangle = \langle 0 | Q | 0 \rangle = 0 \,. \tag{8b}$$

The set of vectors

$$| p,q \rangle \equiv U[p,q] \, | 0 \rangle$$

constitutes an overcomplete family of states, and is of fundamental importance in our approach. Of special interest are diagonal matrix elements such as

$$\begin{aligned} \langle p,q \,|\, \{\alpha P + \beta Q + \gamma B\} \,|\, p,q \rangle &= \langle 0 \,|\, \{\alpha p P + \beta (p^{-1} Q + q) + \gamma (B + pqP)\} \,|\, 0 \rangle \\ &= \alpha p + \beta q + \gamma (pq) \end{aligned}$$

which holds because of (5b) and (8). Let us also consider the relation

$$H(p,q) \equiv \langle p,q \,|\, \mathcal{H} \,|\, p,q \rangle \tag{9}$$

in the light of Eq. (7). It follows that

$$\begin{aligned} \frac{\partial}{\partial q} H(p,q) &= - \langle p,q \,|\, i[\mathcal{H},P] \,|\, p,q \rangle \\ &= 2 \langle p,q \,|\, B \,|\, p,q \rangle = 2pq \,. \end{aligned}$$

Hence

$$H(p,q) = pq^2 + V(p) \,,$$

where

$$\begin{aligned} V(p) = \langle p,0 \,|\, \mathcal{H} \,|\, p,0 \rangle &= \langle 0 \,|\, p^{-1} QPQ + \mathcal{V}(pP) \,|\, 0 \rangle \\ &\equiv p^{-1} \Lambda + v(p) \,. \end{aligned}$$

Here $\Lambda = \langle 0 \,|\, QPQ \,|\, 0 \rangle$ is a positive constant which vanishes as $\hbar \to 0$. Moreover as $\hbar \to 0$, P becomes dispersionless with unit mean and $\mathcal{V}$, V and v all coincide. Thus it is reasonable to regard H(p,q) defined by (9) as a suitable candidate for the classical Hamiltonian.[3] Additional evidence for this point of view will emerge in our discussion of the field theoretic models.

Let the vector $|0\rangle$ be represented by the wave function $\varphi_0(k)$. Then the function

$$\begin{aligned} \varphi_0(p,q) &\equiv \langle p,q \,|\, 0 \rangle \\ &= \int_0^\infty p^{-\frac{1}{2}} \, \varphi_0^*(p^{-1}k) \, e^{ikq} \, \varphi_0(k) \, dk \end{aligned}$$

uniquely determines $\varphi_o(k)$ up to an overall phase factor. Further, if $\varphi_o(k)$ is real and nonvanishing--as would be the case for the ground state of a Hamiltonian of the form (6)--then the function

$$E(q) \equiv \varphi_o(1,q) = \langle 1,q \mid 0\rangle$$

$$= \int_0^\infty e^{ikq}\, \varphi_o^2(k)\, dk$$

uniquely determines $\varphi_o(k)$.

The unitarity condition (5a) leads to the general relation

$$\varphi_o^*(p,q) = \varphi_o(p^{-1},-pq) ,$$

while reality of $\varphi_o(k)$ (time reversal invariance) further implies that

$$\varphi_o(p,q) = \varphi_o(p^{-1},pq) .$$

This relation leads to consequences which are especially important in the field theoretic models which we shall discuss next.

3. LOCAL AFFINE FIELD THEORY: SCALAR MODEL OF THE GRAVITATIONAL FIELD

We now study the model characterized by the action functional of Eq. (2). The condition $\mathcal{G}(x) > 0$ leads to complications in the Heisenberg commutation relation, and we adopt instead the affine field commutation relation

$$[\kappa(\underset{\sim}{x}),\pi(\underset{\sim}{y})] = i\delta(\underset{\sim}{x}-\underset{\sim}{y})\ \pi(\underset{\sim}{x}) , \tag{10}$$

where $\pi(\underset{\sim}{x}) > 0$ is the analog of $P > 0$ and $\kappa(\underset{\sim}{x})$ is the analog of B of Sec. 2. Equation (10) has an infinite number of unitarily inequivalent, irreducible solutions (just as is the case for the usual canonical field operators). To study these operators it is especially convenient to study the unitary operators

$$U[\mathcal{G},\Gamma] \equiv e^{-i\int \Gamma(\underset{\sim}{x})\pi(\underset{\sim}{x})d\underset{\sim}{x}}\ e^{i\int \ln \mathcal{G}(\underset{\sim}{x})\kappa(\underset{\sim}{x})d\underset{\sim}{x}} .$$

Let $|0\rangle$ denote the ground state of the dynamical system under study, and consider the states

$$|\mathcal{G},\Gamma\rangle \equiv U[\mathcal{G},\Gamma]\ |0\rangle .$$

The "ultralocal" nature of the second model suggests that

$$\langle \mathcal{G},\Gamma \mid 0\rangle = \exp\{-\int d\underset{\sim}{x}\; W[\mathcal{G}(\underset{\sim}{x}),\Gamma(\underset{\sim}{x})]\} .$$

A careful analysis shows that only special forms for W are compatible with the group structure of U.[4] In particular, we find that

$$\langle \mathcal{G},\Gamma \mid 0\rangle = \exp\{-\tfrac{1}{2}\iint d\underset{\sim}{x}dk\; [\,|\varphi_{\mathcal{G},\Gamma}(k)|^2 - \varphi^*_{\mathcal{G},\Gamma}(k)\varphi(k) + \varphi^*(k)\varphi_{\mathcal{G},\Gamma}(k)]\} , \tag{11a}$$

where $\varphi(k)$ is a fixed function and

$$\varphi_{\mathcal{G},\Gamma}(k) \equiv \mathcal{G}(\underset{\sim}{x})^{-\frac{1}{2}} e^{-ik\Gamma(\underset{\sim}{x})} \varphi(\mathcal{G}(\underset{\sim}{x})^{-1}k) - \varphi(k) . \tag{11b}$$

Note that $0 < k < \infty$ and all k integrations are from 0 to ∞. Valid group representations are obtained if $\varphi(k) \in L^2(0,\infty)$ and also for certain $\varphi(k) \notin L^2(0,\infty)$. In particular, it suffices if $\varphi(k) \propto k^{-\frac{1}{2}}$ near $k = 0$, as in the example $\varphi(k) = k^{-\frac{1}{2}} \exp(-k/2)$. If $\varphi \in L^2$, the above expression may be re-expressed in the simpler form

$$\langle \mathcal{G},\Gamma \mid 0\rangle = \exp\{\iint d\underset{\sim}{x}dk\; \varphi^*_{\mathcal{G},\Gamma}(k)\varphi(k)\} .$$

The properties of the group representation are rather different in the two cases. If $\varphi \in L^2$, the group $U[\mathcal{G},\Gamma]$ is highly reducible and contains a seemingly unphysical subrepresentation in which $\pi(\underset{\sim}{x}) = \kappa(\underset{\sim}{x}) = 0$. On the other hand, if $\varphi \notin L^2$ the representation of $U[\mathcal{G},\Gamma]$ is irreducible and no such unphysical subrepresentation appears. It is our present feeling that the latter case is the physical one.[5] However, much of our subsequent discussion applies in either case. In particular, it is noteworthy that each distinct $\varphi(k)$, apart from an overall phase, leads to an inequivalent representation of the operators $U[\mathcal{G},\Gamma]$.

For purposes of interpretation we impose the requirements

$$\langle 0 \mid \pi(\underset{\sim}{x}) \mid 0\rangle = 1 , \tag{12a}$$

$$\langle 0 \mid \kappa(\underset{\sim}{x}) \mid 0\rangle = 0 , \tag{12b}$$

which, respectively, are ensured if

$$\int \varphi^*(k)\; k\; \varphi(k)dk = 1 ,$$

$$\tfrac{i}{2}\int k[\varphi^*(k)\varphi'(k) - \varphi^{*\prime}(k)\varphi(k)]\; dk = 0 .$$

(A real function $\varphi(k)$ automatically fulfills the second relation.) The conditions (12) imply that

$$\langle g,\Gamma \mid \{\alpha\pi(\underset{\sim}{x})+\beta\kappa(\underset{\sim}{x})\} \mid g,\Gamma\rangle = \langle 0 \mid \{\alpha\, g(\underset{\sim}{x})\pi(\underset{\sim}{x})+\beta[\kappa(\underset{\sim}{x})+g(\underset{\sim}{x})\Gamma(\underset{\sim}{x})\pi(\underset{\sim}{x})]\} \mid 0\rangle$$

$$= \alpha\, g(\underset{\sim}{x}) + \beta\, g(\underset{\sim}{x})\Gamma(\underset{\sim}{x}) \ .$$

There are other diagonal matrix elements of basic importance. Let $\underset{\sim}{\wp}$ denote the space translation generator, and consider

$$\underset{\sim}{P}(g,\Gamma) \equiv \langle g,\Gamma \mid \underset{\sim}{\wp} \mid g,\Gamma\rangle \ .$$

It follows that

$$\frac{\delta}{\delta\Gamma(\underset{\sim}{x})}\underset{\sim}{P}(g,\Gamma) = \langle g,\Gamma \mid i[\pi(\underset{\sim}{x}),\underset{\sim}{\wp}] \mid g,\Gamma\rangle$$

$$= -\langle g,\Gamma \mid \underset{\sim}{\nabla}\,\pi(\underset{\sim}{x}) \mid g,\Gamma\rangle$$

$$= -\underset{\sim}{\nabla}\langle 0 \mid g(\underset{\sim}{x})\pi(\underset{\sim}{x}) \mid 0\rangle = -\underset{\sim}{\nabla}\, g(\underset{\sim}{x}) \ .$$

A similar but more detailed calculation shows that

$$\frac{\delta}{\delta g(\underset{\sim}{x})}\underset{\sim}{P}(g,\Gamma) = \underset{\sim}{\nabla}\Gamma(\underset{\sim}{x}) \ ,$$

while translation invariance of $\mid 0\rangle$ implies that $\underset{\sim}{P}(0,0) = \langle 0 \mid \underset{\sim}{\wp} \mid 0\rangle = 0$. The unique solution to these relations is given by

$$\underset{\sim}{P}(g,\Gamma) = \int g(\underset{\sim}{x})\,\underset{\sim}{\nabla}\Gamma(\underset{\sim}{x})\,d\underset{\sim}{x} \ ,$$

which is recognized as the classical generator of space translations.[3]

Let us also consider the diagonal matrix elements of the Hamiltonian $\mathcal{H}$ under the two hypotheses that

$$\mathcal{H} \mid 0\rangle = 0 \ ,$$

$$[\mathcal{H}, \pi(\underset{\sim}{x})] = 2i\kappa(\underset{\sim}{x}) \ , \qquad (13)$$

[cf., Eq. (7)]. If we set

$$H(g,\Gamma) \equiv \langle g,\Gamma \mid \mathcal{H} \mid g,\Gamma\rangle \ ,$$

then

$$\frac{\delta}{\delta\Gamma(\underset{\sim}{x})} H(g,\Gamma) = \langle g,\Gamma \mid i[\pi(\underset{\sim}{x}),H] \mid g,\Gamma\rangle$$

$$= 2\langle g,\Gamma \mid \kappa(\underset{\sim}{x}) \mid g,\Gamma\rangle$$

$$= 2\, g(\underset{\sim}{x})\Gamma(\underset{\sim}{x}) \ .$$

Hence

$$H(g,\Gamma) = \int g(\underset{\sim}{x})\Gamma^2(\underset{\sim}{x})d\underset{\sim}{x} + V\{g\}$$

$$= \int \{g(\underset{\sim}{x})\Gamma^2(\underset{\sim}{x})+V[g(\underset{\sim}{x})]\}\, d\underset{\sim}{x} \ ,$$

the last form following from the special symmetries involved in the present model.

The time reversal nature of the relation $H(g,-\Gamma) = H(g,\Gamma)$ leads, much as in the single degree of freedom case, to the relation

$$\langle g,\Gamma \mid 0\rangle = \langle g^{-1}, g\Gamma \mid 0\rangle \ ,$$

which implies that the basic function $\varphi(k)$ is real. Let us assume that $\varphi(k)$ is real and nonvanishing and derive some of the consequences. When such is the case it is clear that the complete function $\langle g,\Gamma \mid 0\rangle$ is determined by the partial information contained in

$$E(\Gamma) \equiv \langle 1,\Gamma \mid 0\rangle \equiv \langle \Gamma \mid 0\rangle$$

$$= \exp\{-\iint d\underset{\sim}{x}dk[1-e^{ik\Gamma(\underset{\sim}{x})}]\,\varphi^2(k)\} \ ,$$

where we introduce the notation $|\Gamma\rangle \equiv |1,\Gamma\rangle$. Equation (14) leads, on expansion to first order in g-1, to the relation

$$\langle \Gamma \mid \kappa(\underset{\sim}{x}) \mid 0\rangle = \tfrac{1}{2}\Gamma(\underset{\sim}{x})\langle \Gamma \mid \pi(\underset{\sim}{x}) \mid 0\rangle$$

$$= -\frac{i}{2}\Gamma(\underset{\sim}{x})\frac{\delta}{\delta\Gamma(\underset{\sim}{x})}E(\Gamma) \ .$$

From (10) we can deduce that

$$\langle \Gamma \mid \kappa(\underset{\sim}{x}) \mid \Gamma'\rangle = \tfrac{1}{2}\{\Gamma(\underset{\sim}{x})+\Gamma'(\underset{\sim}{x})\}\langle \Gamma \mid \pi(\underset{\sim}{x}) \mid \Gamma'\rangle \ .$$

This last relation permits us to determine certain matrix elements of the Hamiltonian. Since

$$\left\{\frac{\delta}{\delta\Gamma(\underset{\sim}{x})} + \frac{\delta}{\delta\Gamma'(\underset{\sim}{x})}\right\}\langle\Gamma\,|\,\mathcal{H}\,|\,\Gamma'\rangle = \langle\Gamma\,|\,i[\pi(\underset{\sim}{x}),\mathcal{H}]\,|\,\Gamma'\rangle = 2\,\langle\Gamma\,|\,\kappa(\underset{\sim}{x})\,|\,\Gamma'\rangle\ ,$$

it follows that

$$\langle\Gamma|\,\mathcal{H}\,|\Gamma'\rangle = \int \Gamma(\underset{\sim}{x})\Gamma'(\underset{\sim}{x})\,\langle\Gamma\,|\,\pi(\underset{\sim}{x})\,|\,\Gamma'\rangle\, d\underset{\sim}{x}\ . \tag{15}$$

This derivation closely follows that of Araki for canonical field theories.[6]

Under the present assumptions [real, nonvanishing $\varphi(k)$] the matrix elements $\langle\Gamma\,|\,\mathcal{H}\,|\,\Gamma'\rangle$ actually determine the Hamiltonian $\mathcal{H}$. To spell out what the Hamiltonian $\mathcal{H}$ is like let us first introduce the operator

$$\begin{aligned} h &= -\frac{\partial}{\partial k}\,k\,\frac{\partial}{\partial k} + \left(\frac{\partial}{\partial k}\,k\,\frac{\partial}{\partial k}\right)\varphi(k)/\varphi(k) \\ &\equiv a^{\dagger}a\ , \end{aligned} \tag{16}$$

where the operator a is given by

$$a \equiv k^{\frac{1}{2}}\,\varphi(k)\,\frac{\partial}{\partial k}\,\varphi(k)^{-1}$$

which is well defined whenever $\varphi(k)$ is nonvanishing. Clearly h is positive and $h\,\varphi(k) = 0$. In terms of h we find that

$$\begin{aligned} \langle g,\Gamma\,|\,e^{-i\mathcal{H}t}\,|\,g',\Gamma'\rangle &= \langle g,\Gamma\,|\,0\rangle\,\langle 0\,|\,g',\Gamma'\rangle \\ &\times \exp\,\{\textstyle\iint d\underset{\sim}{x}dk[\varphi^{*}_{g,\Gamma}(k)e^{-iht}\varphi_{g',\Gamma'}(k)]\} \end{aligned} \tag{17}$$

which expresses a complete set of matrix elements of the evolution operator. This is a relation of fundamental importance. It holds true in both cases, $\varphi \in L^2$ and $\varphi \notin L^2$. Another distinction which lends further credence to the latter being the physical case can be drawn from (17). Suppose that $\varphi \in L^2$, then as $t \to \infty$ the limit of (matrix elements of) $\exp(-iht)$--the so-called weak operator limit--converges to a projection operator onto the ground state $\varphi(k)$. As such the left-hand-side, (the matrix element of) $\exp(-i\mathcal{H}t)$, also converges as $t \to \infty$ to a projection operator but <u>not</u> a one-dimensional projection operator. On the other hand, if $\varphi \notin L^2$, then h has a continuous spectrum and matrix elements of $\exp(-iht)$ converge to zero as $t \to \infty$. As a consequence

$$\lim_{t\to\infty}\ \langle g,\Gamma\,|\,e^{-i\mathcal{H}t}\,|\,g',\Gamma'\rangle = \langle g,\Gamma\,|\,0\rangle\,\langle 0\,|\,g',\Gamma'\rangle\ ,$$

which corresponds to the projection operator onto the single state $|0\rangle$. In other words, if $\varphi \in L^2$ the ground state of $\mathcal{H}$ is degenerate, while if $\varphi \notin L^2$ the ground state of $\mathcal{H}$ is nondegenerate. Obviously the latter is a physically more desirable situation.

The first time derivative of (17) evaluated at t = 0 leads to matrix elements of $\mathcal{H}$. In particular, let us consider the diagonal matrix elements

$$\langle g,0 \,|\, \mathcal{H} \,|\, g,0\rangle = \iint d\underset{\sim}{x}dk \,\left| a[g(\underset{\sim}{x})^{-\frac{1}{2}}\varphi(g(\underset{\sim}{x})^{-1}k)-\varphi(k)] \right|^2$$

$$= \iint d\underset{\sim}{x}\, k\, dk\; \varphi^2(k) g(\underset{\sim}{x})^{-1} \left| \frac{\partial}{\partial k} \frac{\varphi(g(\underset{\sim}{x})^{-1}k)}{\varphi(k)} \right|^2$$

$$\equiv \int d\underset{\sim}{x}\; V[g(\underset{\sim}{x})] \; .$$

Since distinct φ most certainly lead to distinct V, which in turn lead to distinct $H(g,\Gamma)$, we find that the quantum theory of every distinct classical Hamiltonian involves an inequivalent representation of the basic field operators $\pi(\underset{\sim}{x})$ and $\kappa(\underset{\sim}{x})$.

Finally, let us consider a specific example, namely where

$$\varphi(k) = k^{-\frac{1}{2}} \exp(-k/2) \; ,$$

which has been normalized so as to ensure (12a). With this choice it follows that

$$V[g(\underset{\sim}{x})] = \tfrac{1}{4}\,[g(\underset{\sim}{x})^{\frac{1}{2}} - g(\underset{\sim}{x})^{-\frac{1}{2}}]^2 \; ,$$

$$E(\Gamma) = \exp\{-\int d\underset{\sim}{x}\; \ln[1-i\Gamma(\underset{\sim}{x})]\} \; ,$$

$$h = -\frac{\partial}{\partial k}\, k\, \frac{\partial}{\partial k} + \tfrac{1}{4}(k+\tfrac{1}{k}) \; .$$

Operator Formulation

The solutions given above may be re-expressed in an exceptionally simple alternate form. While the formulas we present constitute a direct reformulation of the preceding results,[7] one would be extremely hard pressed to come to this simple formulation ab initio.

We begin with an introduction of conventional boson annihilation and creation operators, $A(\underset{\sim}{x},k)$ and $A^\dagger(\underset{\sim}{x},k)$, defined for configuration space points $\underset{\sim}{x}$ and an extra variable k, $0 < k < \infty$.

These operators obey the commutation relations

$$[A(\underset{\sim}{x},k),A^\dagger(\underset{\sim}{x}',k')] = \delta(\underset{\sim}{x}-\underset{\sim}{x}')\delta(k-k')$$

and also satisfy

$$A(\underset{\sim}{x},k)\,|0\rangle = 0 \ .$$

Thus these operators are the usual Fock representation operators. The fundamental states $|\mathcal{G},\Gamma\rangle$ have the property of being eigenstates of the annihilation operators,

$$A(\underset{\sim}{x},k)\,|\mathcal{G},\Gamma\rangle = \varphi_{\mathcal{G},\Gamma}(k)\,|\mathcal{G},\Gamma\rangle \ , \tag{18}$$

where $\varphi_{\mathcal{G},\Gamma}(k)$ is given in Eq. (11b). Thus the states $|\mathcal{G},\Gamma\rangle$ have some of the properties of coherent states of the radiation field.[8] In particular, Eq. (18) enables matrix elements of normally ordered expressions in A and $A^\dagger$ to be readily evaluated.

Along with the operators A and $A^\dagger$, let us define the "translated-Fock operators"

$$B(\underset{\sim}{x},k) \equiv A(\underset{\sim}{x},k) + \varphi(k) \ ,$$

$$B^\dagger(\underset{\sim}{x},k) \equiv A^\dagger(\underset{\sim}{x},k) + \varphi^*(\underset{\sim}{k}) \ .$$

The B operators also satisfy the canonical commutation relations, but they constitute an inequivalent representation--indeed, if $\varphi \notin L^2$ the B representation is not even <u>locally</u> equivalent to the Fock representation. We now introduce

$$\pi(\underset{\sim}{x}) \equiv \int B^\dagger(\underset{\sim}{x},k)kB(\underset{\sim}{x},k)\,dk \ ,$$

$$\kappa(\underset{\sim}{x}) \equiv -\frac{i}{2}\int B^\dagger(\underset{\sim}{x},k)\left\{\frac{\overleftarrow{\partial}}{\partial k}\,k - k\,\frac{\overrightarrow{\partial}}{\partial k}\right\}B(\underset{\sim}{x},k)\,dk \ ,$$

$$\underset{\sim}{\wp} \equiv -\frac{i}{2}\iint A^\dagger(\underset{\sim}{x},k)\,\overset{\leftrightarrow}{\underset{\sim}{\nabla}}\,A(\underset{\sim}{x},k)\,d\underset{\sim}{x}dk \ ,$$

$$\mathcal{H} \equiv \iint A^\dagger(\underset{\sim}{x},k)\,h(k)\,A(\underset{\sim}{x},k)\,d\underset{\sim}{x}dk \ ,$$

$$= \iint B^\dagger(\underset{\sim}{x},k)\,h(k)\,B(\underset{\sim}{x},k)\,d\underset{\sim}{x}dk \ ,$$

where $h(k)$ is given by the expression (16). From well known properties of the annihilation and creation operators it is seen that (i) π has positive spectrum, (ii) π and κ obey the proper commutation relations (10), (iii) $\mathcal{H}$ and π obey the proper relation

(13), (iv) $\mathcal{H}|0\rangle = 0$ and (v) $\mathcal{H} \geq 0$. Degeneracy of the ground state would arise if $h(k)$ had a discrete state with eigenvalue zero, which occurs only if $\varphi \in L^2$. We can also readily see if $\varphi \in L^2$ that the operators $\pi(\underset{\sim}{x})$ and $\kappa(\underset{\sim}{x})$ would be represented by zero in a certain subspace [that subspace annihilated by $B(\underset{\sim}{x},k)$]. Moreover, if $\varphi \in L^2$ the operator $N(y) = \int B^\dagger(\underset{\sim}{y},k)B(\underset{\sim}{y},k)\, dk$ commutes with both $\pi(\underset{\sim}{x})$ and $\kappa(\underset{\sim}{x})$ demonstrating that the representation is highly reducible. If $\varphi \notin L^2$, $N(\underset{\sim}{y})$ is not defined and the representation of $\pi(\underset{\sim}{x})$ and $\kappa(\underset{\sim}{x})$ is irreducible; also $\pi(\underset{\sim}{x}) > 0$ as desired.

The Hamiltonian $\mathcal{H}$ is a bilinear expression in B and $B^\dagger$, but it can be constructed formally as a nonlinear function of π and κ. We first note that to make sense of the square of $\pi(\underset{\sim}{x})$ (at a point) an infinite renormalization is required. Specifically, if $e_n(\underset{\sim}{x})$ is a sequence of smooth functions such that $e_n^2(\underset{\sim}{x}) \to \delta(\underset{\sim}{x})$, then the sequence of operators

$$c_n(\underset{\sim}{z}) = \iint \pi(\underset{\sim}{x})\pi(\underset{\sim}{y})e_n(\underset{\sim}{x}-\underset{\sim}{z})e_n(\underset{\sim}{y}-\underset{\sim}{z})\, d\underset{\sim}{x}d\underset{\sim}{y}$$

converges to

$$\lim_{n\to\infty} c_n(\underset{\sim}{z}) = \int B^\dagger(\underset{\sim}{z},k)k^2B(\underset{\sim}{z},k)\, dk \, .$$

Note that such a product leads to multiplication of the factor inside the k-integral. We may formally denote this kind of product by $Z\pi^2(\underset{\sim}{z})$, where (symbolically) $Z^{-1} = \delta(0)$. With this notation it follows that

$$\mathcal{H} = \int Z^{-1}\, h[Z\pi(\underset{\sim}{z}),Z\kappa(\underset{\sim}{z})]\, d\underset{\sim}{z}$$

$$= \iint B^\dagger(\underset{\sim}{z},k)\, h\left[k, \frac{i}{2}\left(\frac{\partial}{\partial k}k + k\frac{\partial}{\partial k}\right)\right] B(\underset{\sim}{z},k)\, dkd\underset{\sim}{z}$$

which is a formal construction of $\mathcal{H}$ from π and κ.

Finally let us remark on the form of the field operator $\varphi(\underset{\sim}{x})$ in this model. This operator, just as in the single degree of freedom case can only be symmetric and never self adjoint. The formal statement

$$\kappa(\underset{\sim}{x}) = \tfrac{1}{2}\{\varphi(\underset{\sim}{x})\pi(\underset{\sim}{x})+\pi(\underset{\sim}{x})\varphi(\underset{\sim}{x})\}$$

leads to the formal relation

$$\varphi(\underset{\sim}{x}) = \tfrac{1}{2}\{\pi(\underset{\sim}{x})^{-1}\kappa(\underset{\sim}{x})+\kappa(\underset{\sim}{x})\pi(\underset{\sim}{x})^{-1}\}$$

which in fact is essentially correct. Thus $\varphi(\underset{\sim}{x})$ is not bilinear in $B^\dagger$ and B but, roughly speaking, the ratio of two such expressions. This complicated relation makes it especially clear how convenient the formulation in terms of π and κ really is.

4. LOCAL AFFINE FIELD THEORY: MATRIX MODEL OF THE GRAVITATIONAL FIELD

The matrix model of the gravitational field can be studied in much the same manner as the scalar model. We have found it expedient to introduce the following generalization of the basic commutation relations expressed in terms of a symmetric, positive matrix operator $\pi_{rs}(\underset{\sim}{x})$ and a nonsymmetric matrix operator $\kappa_{rs}(\underset{\sim}{x})$:

$$[\kappa_{rs}(\underset{\sim}{x}),\kappa_{tu}(\underset{\sim}{y})] = \tfrac{1}{2}\, i\delta(\underset{\sim}{x}-\underset{\sim}{y})[\delta_{st}\kappa_{ru}(\underset{\sim}{x})-\delta_{ru}\kappa_{ts}(\underset{\sim}{x})] \;,$$

$$[\kappa_{tu}(\underset{\sim}{x}),\pi_{rs}(\underset{\sim}{y})] = \tfrac{1}{2}\, i\delta(\underset{\sim}{x}-\underset{\sim}{y})[\delta_{su}\pi_{rt}(\underset{\sim}{x})+\delta_{ru}\pi_{ts}(\underset{\sim}{x})] \;.$$

These relations are suggested by the heuristic equation (summation understood)

$$\kappa_{tp} = \tfrac{1}{2}\,\{\pi_{t\ell}\varphi_{\ell p}+\varphi_{p\ell}\pi_{\ell t}\}$$

expressed in terms of a symmetric field operator $\varphi_{\ell p}$ which fulfills

$$[\pi_{tu}(\underset{\sim}{x}),\varphi_{rs}(\underset{\sim}{y})] = -\tfrac{1}{2}\, i\delta(\underset{\sim}{x}-\underset{\sim}{y})[\delta_{rt}\delta_{su}+\delta_{ru}\delta_{st}] \;.$$

The basic unitary operators of interest are given by

$$U[\gamma,\Gamma] \equiv e^{-i\int \Gamma_{rs}\pi_{rs}d\underset{\sim}{x}}\, e^{i\int \gamma_{ba}\kappa_{ab}d\underset{\sim}{x}} \;,$$

where $\Gamma_{rs}(\underset{\sim}{x})$ is a symmetric matrix function and $\gamma_{ba}(\underset{\sim}{x})$ is a nonsymmetric matrix function related to the metric in a way to be determined. Suppose we let $\underline{\pi}$ and $\underline{\kappa}$ denote the 3×3 matrices, and introduce the matrix

$$\underline{S} \equiv \exp \tfrac{1}{2}\,\underline{\gamma} \;.$$

Then, it follows that

$$U[\gamma,\Gamma]^{\dagger}(\alpha\,\underline{\pi}+\beta\underline{\kappa})U[\gamma,\Gamma] = \alpha\;\underline{S}\,\underline{\pi}\,\underline{S}^{T} + \beta(\underline{S\kappa S}^{-1}+\underline{S\pi S}^{T}\underline{\Gamma}) \;,$$

where ordinary matrix multiplication is meant. We require that

$$\langle 0\,|\,\alpha\pi_{rs}(\underset{\sim}{x})+\beta\kappa_{rs}(\underset{\sim}{x})\,|\,0\rangle = \alpha\,\delta_{rs} \;,$$

which leads to

$$\langle \gamma,\Gamma \mid \pi_{rs}(\underset{\sim}{x}) \mid \gamma,\Gamma\rangle = s_{rt}(\underset{\sim}{x})s_{st}(\underset{\sim}{x}) \equiv g_{rs}(\underset{\sim}{x}) ,$$

which is positive and symmetric as desired. We also find that

$$\langle \gamma,\Gamma \mid \kappa_{rs}(\underset{\sim}{x}) \mid \gamma,\Gamma\rangle = s_{rt}(\underset{\sim}{x})s_{ut}(\underset{\sim}{x})\Gamma_{us}(\underset{\sim}{x})$$

$$= g_{ru}(\underset{\sim}{x})\Gamma_{us}(\underset{\sim}{x}) .$$

Many of the basic relations for the matrix field are analogs of those that hold for the scalar field. General properties of the space translation generator $\underset{\sim}{\wp}$ lead to the relation

$$\underset{\sim}{P}(g,\Gamma) \equiv \langle \gamma,\Gamma \mid \underset{\sim}{\wp} \mid \gamma,\Gamma\rangle = \int g_{rs}(\underset{\sim}{x})\underset{\sim}{\nabla}\Gamma_{rs}(\underset{\sim}{x})d\underset{\sim}{x} ,$$

while the equation

$$[\mathcal{H}, \pi_{rs}(\underset{\sim}{x})] = i[\kappa_{rs}(\underset{\sim}{x})+\kappa_{sr}(\underset{\sim}{x})]$$

leads to

$$H(g,\Gamma) \equiv \langle \gamma,\Gamma \mid \mathcal{H} \mid \gamma,\Gamma\rangle$$

$$= \int \{g_{rs}(\underset{\sim}{x})\Gamma_{rt}(\underset{\sim}{x})\Gamma_{st}(\underset{\sim}{x})$$

$$+ V[g_{rs}(\underset{\sim}{x})]\} \, d\underset{\sim}{x} .$$

If we set $|\Gamma\rangle \equiv |0,\Gamma\rangle$ (for $\gamma_{ab} \equiv 0$) and exploit time-reversal invariance, then we can derive an analog to Eq. (15) which reads

$$\langle\Gamma \mid \mathcal{H} \mid \Gamma'\rangle = \int \Gamma_{rt}(\underset{\sim}{x})\Gamma'_{st}(\underset{\sim}{x}) \langle\Gamma \mid \pi_{rs}(\underset{\sim}{x}) \mid \Gamma'\rangle \, d\underset{\sim}{x} .$$

This important relation permits a determination of the Hamiltonian to be made and links the dynamics to the representation of the field operators. To present that relation we adopt an operator formulation analogous to that for the scalar field. In particular, consider operators such that

$$[A(\underset{\sim}{x},\underline{k}),A^\dagger(\underset{\sim}{x}',\underline{k}')] = \delta(\underset{\sim}{x}-\underset{\sim}{x}')\delta(\underline{k}-\underline{k}')$$

where $\underline{k} = \{k_{rs}\}$ denotes a point in a six-dimensional space restricted so that the matrix $\underline{k} > 0$. In addition, we let

$$A(\underset{\sim}{x},\underline{k}) \mid 0\rangle = 0 ,$$

and we define

$$B(\underset{\sim}{x},\underline{k}) = A(\underset{\sim}{x},\underline{k}) + \varphi(\underline{k}) \ .$$

The function $\varphi(\underline{k})$--the basic function which characterizes the operator representation--should not be square integrable in accord with arguments in Sec. 3. The field operators read

$$\pi_{rs}(\underset{\sim}{x}) = \int B^{\dagger}(\underset{\sim}{x},\underline{k})\ k_{rs}\ B(\underset{\sim}{x},\underline{k})\ d\underline{k}$$

$$\kappa_{rs}(\underset{\sim}{x}) = \int B^{\dagger}(\underset{\sim}{x},\underline{k})\ b_{rs}\ B(\underset{\sim}{x},\underline{k})\ d\underline{k} \ ,$$

where

$$b_{rs} = \frac{i}{2}\left(k_{rt}\frac{\partial}{\partial k_{ts}} + \frac{\partial}{\partial k_{st}}k_{tr}\right) .$$

The dynamics is determined by

$$\mathcal{H} = \iint A^{\dagger}(\underset{\sim}{x},\underline{k})\ h\ A(\underset{\sim}{x},\underline{k})\ d\underset{\sim}{x}d\underline{k}$$

$$= \iint B^{\dagger}(\underset{\sim}{x},\underline{k})\ h\ B(\underset{\sim}{x},\underline{k})\ d\underset{\sim}{x}d\underline{k} \ ,$$

where

$$h = -\frac{\partial}{\partial k_{rt}}k_{rs}\frac{\partial}{\partial k_{st}} + \upsilon(k_{rs})$$

and $h\,\varphi(\underline{k}) = 0$. With this brief outline it can be seen how the matrix model may be treated along the lines which proved successful for the scalar model.

The present discussion has focussed on problems associated with the positivity requirements. We feel that the use of the affine group and its generalizations has been--and will be--of considerable value in this regard. Further research along these lines should include models possessing constraints as well, and work in this direction is in progress.

REFERENCES

1. E. W. Aslaksen, thesis, Lehigh University (1968); J. R. Klauder, 5th International Conference on Gravitation and the Theory of Relativity, Tbilisi (1968), to be published.
2. I. M. Gel'fand and M. A. Naimark, Dokl. Akad. Nauk SSSR 55, 570 (1947); M. A. Naimark, Normed Rings, translated by L. F. Boron (P. Noordhoff, Ltd., Groningen, 1964), p. 381; E. W. Aslaksen and J. R. Klauder, J. Math. Phys. 9, 206 (1968).

3. The identification of diagonal matrix elements of quantum generators with their classical counterparts is the essence of the weak correspondence principle. See, for example, J. R. Klauder, J. Math. Phys. 8, 2392 (1967).
4. J. R. Klauder, "Exponential Hilbert Space: Fock Space Revisited," to be published; E. W. Aslaksen, thesis, Lehigh University (1968).
5. Reducible representations can not be excluded per se as has been demonstrated by their explicit relevance in canonical field theories. Earlier analyses (Ref. 1) were unaware of all the consequences that followed when $\varphi \notin L^2$. We discuss the distinctions here in some detail.
6. H. Araki, J. Math. Phys. 1, 492 (1960), Sec. 8.
7. J. R. Klauder, to be published.
8. J. R. Klauder and E. C. G. Sudarshan, Fundamentals of Quantum Optics (W. A. Benjamin, 1968), Chap. 7.

THE QUANTIZATION PROGRAM FOR GENERAL RELATIVITY*

Arthur Komar

Belfer Graduate School of Science

Yeshiva University

Abstract

The problem of the construction of a quantum theory of gravitation is attacked by a variety of methods. The fundamental epistemological difficulties are ellucidated and certain novel qualitative features of the sought-for quantum theory are described.

I Introduction

Alone among the classical theories, gravitation has eluded quantization for forty years. In view of the fact that the classical field equation permit the existence of homogeneous solutions which may be identified with gravitational radiation, it is hard to believe that the gravitational radiation, when eventually observed, would not transfer its energy, momentum and angular momentum in quanta. Thus we have every reason to expect the existence of a quantum theory of gravitation which would account for such phenomena. Even should it eventually be found that gravitational fields do not transfer energy in quanta, the import of such a discovery would not be fully intelligible without a quantum theory of gravitation against which to compare it.

*This work is supported in part by the United States Air Force under Grant No. AF-AFOSR 68-1524

The classical theory of gravitation is an exceedingly intricate theory. In addition to the ten field equations for the ten components of the gravitational potential being non-linear, they are subject to a four-function gauge group which entails that four of the equations are constraints upon the initial data. It has therefore proven very difficult to exhibit a complete, local, non-redundent set of canonical variables which one could use for the application of the procedure of canonical quantization. Over the decades, numerous approaches directed toward resolving or circumventing these difficulties have met with only middling success, and have come to naught when applied to the development of a quantum theory of gravitation.

Though the technical mathematical difficulties are admittedly formidable, after forty years in the desert one should suspect that perhaps, in addition to such difficulties, fundamental questions of principle are barring our route to the promised land. For this reason, before proceeding any further, it is necessary to examine the foundations of the procedure customarily employed to construct a quantization of a given classical theory. The second section of this paper will be devoted to such an analysis. In the third section we shall apply the insights obtained to a quantization of the space solutions which are asymptotically flat. The proposed procedure will be applicable to the quantization of any set of coupled classical field theories of massless particles. (The extention of our techniques to fields of massive particles is currently under active investigation.) The fourth section will be devoted to a consideration of the Hamilton-Jacobi version of classical relativity. In view of the fact that we have every reason to expect that the Hamilton-Jacobi functional to correspond to the WKB approximation to the (as yet to be determined) Schroedinger state vector, it will enable us to draw some interesting conclusions concerning the novel qualitative features of the quantized theory.

II The Quantization Program

The dynamical variables of a classical theory play a dual role. On the one hand they are used attributatively, that is they lable attributes of the physical system under consideration, such as location, momentum, energy, etc. On the other hand they are used as operators, that is, they generate infinitesimal canonical transformations. The attributative usage is the most immediate and natural, and consequently was employed long before the operator aspect of the dynamical variables was recognized. However, when one considers that by employing canonical transformations we can map theories into one another (e.g. the simple harmonic oscillator into the free particle), it becomes evident that the operator aspect of a dynamical variable is vital to its epistemological identification. In particular, the operator aspect is characterized by the algebra of the Poisson brackets, which in turn provides a realization of the Lie Algebra of the Canonical Group of the theory in question.

With the transition to quantum theory the dual role of the dynamical variables bifurcates. The operator aspect of the variables is realized by a Hermitian operator on a linear vector space, while the attribute aspect is realized by the eigenvalues of that operator. The procedure of canonical quantization generally consists of relating the classical variables with Hermitian operators in such a way that a subset of the algebra of the Poisson brackets is preserved as a commutator algebra of the corresponding linear operators. It is possible to represent the entire algebra of the canonical group of the classical theory by unitary mappings of a linear vector space[1,2] (the square-integrable functions over the phase space). The theory so obtained is isomorphic to classical mechanics. It is evident that in addition to realizing observables by linear mappings of a vector space in such a fashion that the algebra of some subgroup of the canonical group is preserved, the procedure of quantization requires an additional hypothesis which materially alters the physical implication of the resulting theory.

The essential novel ingredient in the procedure of quantization is the requirement that if the Hermitian operation, A, is to correspond the classical observable, α, then corresponding to the classical observable, $\beta \equiv f(\alpha)$ is the linear operator $B \equiv f(A)$. An equivalent statement, frequently employed, is that the image of the Poisson bracket is to retain the property of being a differentiation. That is $[A,BC] = B[A,C] + [A,B]C$. It is evident that this is not a statement within the albebra of the canonical group, but rather a statement within the enveloping algebra. This requirement, which is closely related to the probability interpretation of the theory, seems quite natural for the attribute interpretation of the dynamical variables. For any classical mechanical system the attribute β is a trivial consequence of the attribute α. We should therefore expect that the corresponding attributes of the quantum mechanical system, realized by the eigenvalues of B and A respecively, be related in the identical trivial fashion. (The same algebraic relation will, of course also persist for expectation values of B and A.) From the operator point of view, however, although the significance of maintaining the form of the algebraic relationship between α and β are trivial consequences of each other, the mappings in phase space which they generated are by no means trivially related. (For example consider $\alpha = p$, $\beta = \alpha^2$. Although α generates an infinitesimal translation along a configuration space axis, β generates a proper canonical transformation which can readily be expressed only in phase space.)

With the imposition of the above requirement upon the quantum image of functions of classical dynamical variables it no longer becomes possible to preserve the algebra of the entire canonical group in terms of commutators of Hermitian operators. The central question in proceeding with the construction of a quantum theory is the determination of the subgroup of the classical group which is to be preserved. An inspection of existing quantum theories reveals that they were obtained by the preservation of the commutator algebra of the symmetry group of the space-time, the arena of physics: the Galilean group for non-relativistic physics, the Poincaré group for special relativistic theories. It is for this reason that the quantization proceeds most easily in rectangular coordinates, for these coordinates are adapted to the Killing directions of the space-time manifold. (It is perfectly feasible to quantize a classical system for example in polar coordinates. In that event however one must express the Killing directions in terms of polar coordinates, and then require that algebra of the motions in the Killing directions be represented by linear operators in the Hilbert space of the quantum theory.)

The central role played by the space-time symmetry group in the quantization of a classical theory is a quantitative application of Bohr's Correspondence Principle. The epistemological significance of the fundamental dynamical quantities such as energy, momentum, angular momentum and center of mass stem (in their operator aspect) from their identification with elements of the Lie algebra of the space-time group. This becomes particularly evident when we for example consider in detail the angular momentum. The angular momentum was originally singled out because it provided a first integral for mechanical systems which were invariant under rotations. It provided that first integral of motion due to the fact that, in its operator aspect, it generated infinitesimal rotations of the mechanical system. Thus the angular momentum necessarily provides a realization of the Lie algebra of the rotation group. In view of the fact that the rotation group is an abstract mathematical group baring no relevance to the question of whether the physical theory under consideration is classical or quantum mechanical, it is evident that to the extent that a measuring device designed to observe angular momentum functions under the conditions that the quantities being measured are conserved provided the total physical system is rotationally invariant, in every theory which has symbols which are to correspond to the angular momentum so measured, those symbols must provide an exact realization of the Lie algebra of the rotation group. We can even conclude that should quantum mechanics eventually be replaced by some superior theory, if such theory will contain quantities corresponding to angular momentum they must necessarily provide a realization of precisely the same Lie algebra.

Thus far the correspondance between the classical and quantum theories which we have indicated has been primarily kinematical. The dynamics of the classical theory is determined by the Hamiltonian constraint, that is, by requiring that the generator of infinitesimal time translations be constrained to equal some particular, arbitrarily given element of the Lie algebra of the canonical group. The dynamical correspondance between a classical theory and its quantized version is obtained by preserving the functional form of the Hamiltonian constraint as a function of the corresponding kinematical operators (modulo ambiguities of factor-ordering).

The procedure of quantization which we have indicated above singles out the configuration variables associated with the rectilinear Killing directions and their canonically conjugate momenta. It preserves the Lie algebra of functions which are at most quadratic polynomials of these prefered coordinates and momenta. Thus should the Hamiltonian constraint be a more complicated function of the prefered variables, that is, should the system under consideration be essentially more ellaborate than a free particle or a harmonic oscillator, then the dynamical correspondence is at best approximate. The net effect is that a dynamical correspondence can at best be maintained between the classical trajectory and the mean position of a wave packet (Ehrenfest's principle). For the harmonic oscillator it is well known that a more detailed dynamical correspondence can be established.

III Asymptotic Quantization

In the previous section we emphasized the critical role for quantization played by the assumed preferred symmetries of the space-time. Upon turning to the general theory of relativity, we find that the distinguished space-time group is the Einstein group, that is, the group of curvilinear coordinate transformations in four dimensions. This group, being a function group, necessitates that every theory deriveable from an invariant action principle must necessarily have four field equations which are constraints upon the initial data. Furthermore, the four constraints generate the canonical mappings which correspond to the coordinate transformations of the Einstein group. In view of the fact that the observables of a theory, when viewed in their operator aspect, must generate canonical transformations compatible with the equations of motion, which in the present case include the four constraints, it follows that the observables of the general theory of relativity must commute with the generators of the Einstein group! It therefore appears to be impossible to employ the observable dynamical variables of a general relativistic theory to provide a realization of the Einstein group. There being no other distinguished space-time symmetry group readily available, it would appear that the possibilities of constructing a unique quantum theory of gravitation is placed in jeopardy.

One possible recourse is to abandon the relationship between the algebraic structure to be preserved under quantization and a distinguished space-time group, and merely inquire whether some sub-algebra of the classical observables can be preserved consistently upon establishing a function preserving correspondence between the observables and Hermitian operators. One could surely call such a construction a quantum theory of gravitation. However, there is every reason to believe that such a construction would not be unique, for without the correspondence principle to guide in the selection of the preferred sub-algebra many inequivalent choices would become available, none of them appearing to have any simple classical interpretation. Rather than exploring this point further, in this section we propose to impose a preferred space-time symmetry.

When constructing a quantum theory it is convenient first to select a preferred independent set of classical dynamical variables whose canonical algebra is to be preserved. Customarily, this is done by selecting a space-like Cauchy surface adapted to the symmetry of the space-time (as determined by the Killing directions) and on this surface selecting the canonical Cauchy data adapted to the symmetry of the surface. However, for the purpose of having a suitable complete independent set of dynamical variables, it is sufficient that the surface employed be Cauchy—it need not be space-like. The field equations of Einstein's theory are non-linear. We are therefore not at liberty to form linear superpositions which could assure solutions which behave reasonably at infinity. In order to specify fully the theory under consideration we shall addend to the Einstein field equations the condition that the solutions are to be assymptotically flat, or more rigorously, conformally regular[3]. By this one stroke we accomplish all of the following:

(a) There exists a preferred (null) Cauchy surface

(b) The preferred Cauchy surface is at infinity, where the independent radiation modes of the gravitational field are readily identifiable, and where the non-linear terms become less effective since they fall off more rapidly with distance than do the linear terms.

(c) The null-Cauchy surface at infinity has a preferred space-time group, the Bondi-Metzner-Sachs group[4], which contains the Poincaré group as a sub-group.

(d) The algebra of the null Cauchy data at infinity can be arranged to provide a realization of the Bondi-Metzer-Sachs group[4].

It is evident that the independent dynamical variables defined on the infinite null Cauchy surface are ideally suited for quantization. The canonical commutation relations or their equivalent are no longer assigned on a space-like hypersurface where the self-interactions (due to the non-linearity) being effective subject the validity of the commutation relations to question. At null infinity, where linearized terms predominate we can have more confidence in imposing upon the independent modes of gravitational radiation the equivalent of the free field commutation relations as operator relations in a Hilbert space[5]. One can easily employ the resulting operators to construct a Fock space for gravitons, essential for interpreting graviton-scattering problems. The gravitational field at finite space-time points is obtained as a functional of the null Cauchy data by integrating the field equations of the full non-linear theory in from infinity. In view of the fact that the null-Cauchy data are now operators, it would appear that we have effectively succeeded in quantizing this gravitational theory in a fashion uniquely consistent with all of our requirements. The principle draw back is that although for the classical theory the relationship between the field at finite points and its null Cauchy data is essentially unique, when we attempt to integrate non-linear operator equations we immediately obtain ambiguities due to inequivalent factor orderings.

In order to clarify these problems we are currently exploring applying these techniques of quantization to simpler interacting field theories modelled after Lorentz-covarient electrodynamics. The virtue of such an approach is that at null-infinity we must put in the true dressed particles including the correct coupling constant. The theory as presently stated should not permit mass, charge, or wave function renormalization, for at infinity the interacting terms are not effective. The principle vice of this approach is that we are currently able to apply it to theories which are assymptotically conformally flat—that is, at this stage of the art we can only handle interacting fields of rest mass zero!

IV Hamilton-Jacobi Theory

The assymptotic approach to quantization is particularly appropriate for the treatment of graviton-graviton scattering problems. However, it is not very likely that our technology will develop in the near future to the point where predictable cross sections will be observed. Our interest in the quantum theory of gravitation is primarily motivated by the hope that such a theory will provide new and deeper insights into the structure of physical reality. In this section we shall review some recent results obtained in the investigation of properties of the Hamilton-Jacobi functional of general relativity. These investigations were primarily conducted with the expectation that should one succeed in constructing a viable quantum theory of gravitation preferably without having to refer to explicit boundary conditions at infinity, the resulting Schroedinger state vector would reduce in the WKB approximation to the Hamilton-Jacobi functional. Thus some of the qualitative features of the quantum theory could already be inferred by a purely classical investigation.

The Einstein field equations have been stated in Hamiltonian form by Dirac[6]. In brief, the Einstein theory is usually defined by ten partial differential equations $G_{\mu\nu} = 0$ for the determination of the ten components of the metric tensor $g_{\mu\nu}$. However, not all ten equations are propagation equations. Four of the equation, $G_\mu{}^4 = 0$, place constraints on the Cauchy data within the initial space-like surface which must be satisfied before the propagation can proceed. These four constraints, when viewed as generators of infinitesimal canonical transformations, generate the space-time subgroup of the canonical group of the theory, which we have called the Einstein group. Choosing an essentially arbitrary family of space-like hypersurfaces in order to parametrize the time t, the spatial metric tensor of these surfaces, $g_{mn}(x^s)$, may be taken as our configuration variables. The canonically conjugate set of variables, $p^{mn}(x^s)$, is the second fundamental form of the space-like hypersurface. With the canonical field variables thus characterized, the canonical version of the vacuum field equations is specified by giving the "Hamiltonians", that is, by expressing the four constraint equations, $G_\mu{}^4 = 0$, in terms of the canonical variables: thus $H_\mu(g_{ab}, p^{cd}) = 0$.

The Hamilton-Jacobi functional, $S(g_{ab})$, may be defined as a functional whose domain is the set of tensor functions $g_{ab}(x^s)$ which are everywhere positive definite and characterize three-dimensional Riemannian manifolds which are either closed, assymptotically flat, or satisfy some other well-stated boundary condition of interest, whose range is the real numbers, and which satisfies the four functional equations $H_\mu(g_{mn}, \frac{\delta S}{\delta g_{mn}}) = 0$, obtained by substituting $p^{mn} = \frac{\delta S}{\delta g_{mn}}$ into the Hamiltonian constraints.

The following properties of Hamilton-Jacobi functionals have been proven:

(1) $S(g_{mn})$ is not an explicit function of the space-time coordinates, but depends exclusively on the configuration variables, g_{mn}.[7]

(2) Under the action of the space-time group generated by the constraints, the functional form of $S(g_{mn})$ is invariant, although its value will in general alter under the action of the time-like translations[7].

(3) As a corollary of (2) it follows that within a family of classical trajectories determined by a given $S(g_{mn})$ (via employing g_{ab}, $p^{mn} = \frac{\delta S}{\delta g_{mn}}(g_{ab})$ as Cauchy data) occur all canonical pairs g_{ab}, p^{cd}, which can be obtained from a given one by means of a space-time transformation[7].

(4) The complete family of classical trajectories determined by a given $S(g_{mn})$ is fully characterized by $2 \times \infty^3$ commuting constants of the motion $\alpha_A(x^s)$, which do not commute with $g_{mn}(x^s)$ [8]. To elaborate, the family of Ricci flat four dimensional Riemannian manifolds determined by a given $S(g_{mn})$ have in common a complete, commuting set of observables, and conversely, given the set of all Ricci-flat 4-dimensional Riemannian manifolds having in common such a complete commuting set of observables $\alpha_A(x^s)$, there exists a unique Hamilton-Jacobi functional, which we shall denote by $S(g_{mn},\alpha_A)$, which yields back precisely this family of manifolds.

(5) The invariance group of the Hamilton-Jacobi theory is isomorphic to the proper canonical group of general relativity (i.e. the canonical group modulo the space-time subgroup generated by the constraints); or, equivalently, the functionals, $S(g_{mn},\alpha_A)$, provide a space for the realization of the proper canonical group as a transformation group[9].

(6) The observables $\beta^A(x^s)$ which are canonically conjugate to $\alpha_A(x^s)$ satisfy the familiar relation $\beta^A(x^s) = \frac{\delta S}{\delta \alpha_A(x^s)}$ [9].

(7) The $4 \times \infty^3$ observables $\alpha_A(x^s)$, $\beta^A(x^s)$ forms a complete, independent (but not commuting) set of invariants which determine a unique Ricci-flat four dimensional Riemannian manifold[9].

To the extent that the functional $S(g_{mn},\alpha_A)$ does in fact correspond to the WKB limit of the Schoedinger state vector of the quantized theory we can conclude that to a Schroedinger state will correspond not a single four-dimensional geometry, but rather an infinitely large family of four-geometries which have in common a complete commuting set of observables, but disagree on the canonically conjugate set of observables. In particular this treatment suggests that an appropriate procedure for constructing a quantum theory would be to select an appropriate set of classical observables $\alpha_A(x^s)$, and their canonically conjugate $\beta^A(x^s)$, and represent their algebra in a function-preserving way by means of Hermitian operators in a Hilbert space. We could then expect that wave packets could be constructed peaked about the expectation values $\langle\alpha_A(x^s)\rangle$, $\langle\beta^A(x^s)\rangle$. Since the assignment of the complete set of classical observables, $\langle\alpha_A(x^s)\rangle$, $\langle\beta^A(x^s)\rangle$ uniquely determines a classical four-geometry, we can in this way obtain quantum state which can be interpreted as being peaked about a mean Ricci-flat four-geometry. In this sense such a quantum theory satisfies an Ehrenfest principle and provides an understanding of the extent to which we can retain a classical four-dimensional space-time continuum.

Rather than eliminating all four constraints to obtain the classical observables before proceeding with the construction of the quantum theory, some investigators prefer to retain the fourth constraint and convert it into a Schroedinger equation by means of the substitution $p^{ab} \rightarrow \frac{1}{i}\frac{\delta}{\delta g_{ab}}$. It is not at all clear whether such an approach is consistent with the one we have just presented, nor is it clear whether this latter approach can provide wave packets which satisfies our Ehrenfest principle. On the other hand, the principle deficiency of our observable approach to quantization is the lack of a convincing criterion for the selection of the preferred set of classical observables which are subject to the distinguished treatment of having their canonical commutation relations preserved. (Recall that we no longer have a space-time group at our disposal to guide in the selection.)

V Qualitative Conclusions

Should one obtain a quantum theory of gravitation, the most startling feature it will have is the disolution of the space-time manifold. Thus, in the description of an experimental situation the theory will establish a contingency relationship between the out-come of an experiment and the geometry of the space-time in which the experiment was performed! This is a most encouraging result for there is reason to believe that it might provide a resolution to the age-old paradox of "the reduction of the wave-packet"[10].

In brief, half of the classically required canonical variables must be employed to specify the frame of reference, leaving the remaining half available for unequivocal observeration. If the observer chooses to alter his procedure for the specification of the frame of reference of his apparatus he is at liberty to do so and obtain thereby a consistent, but complementary description of the empirical situation. The penalty, however is that he must conclude that the space-time structure was also altered by the experimental rearrangement! However such consideration are far too speculative to warrant pursuing at this time.

VI References

1. B.O. Koopman, Proc. Nat. Acad. Sci. 17, 315 (1931)
2. J. von Newmann, Ann. of Math. 33, 587 (1932)
3. R. Penrose, Phys. Rev. Lett. 10, 66 (1963)
4. R. Sachs, Phys. Rev. 128, 2851 (1962)
5. A. Komar, Phys. Rev. 134, B1430 (1964)
6. P.A.M. Dirac, Phys. Rev. 114, 924 (1959)
7. P.G. Bergmann, Phys. Rev. 144, 1078 (1966)
8. A. Komar, Phys. Rev. 153, 1385 (1967)
9. A. Komar, Phys. Rev. 170, 1195 (1968)
10. A. Komar, Inter. Jour. Theor. Phys. (to be published)

PARTICLES AND GEOMETRY

John Archibald Wheeler

Joseph Henry Laboratories, Princeton University

Princeton, New Jersey

A drop is hanging on the ceiling as one comes out of the shower. The size of the drop is governed by the balance between the pull of gravity -- which is proportional to the density, the acceleration of gravity, and the cube of the radius -- and the upward pull of surface tension, which is proportional to the product of the surface tension and the circumference. For the drop to be supported it must not exceed a critical size. The critical radius is given in order of magnitude by the condition of equilibrium,

$$\rho\, g\, a^3 \sim a$$

or

$$a \sim (\sigma/\rho g)^{\frac{1}{2}} \sim (\sigma/\rho c^2)^{\frac{1}{2}}(c^2/g)^{\frac{1}{2}}\sigma$$

One can rewrite this expression as the geometric mean

$$a \sim (L_{micro}\; L_{macro})^{\frac{1}{2}}$$

between a quantity which is of microscopic dimensions,

$$L_{micro} = \sigma/\rho c^2$$

$$\sim \frac{0.1\text{eV} \times 10^{-8}\text{cm}}{(16+1+1) \times 10^{9}\text{eV}} \sim 10^{-19}\text{cm}$$

and one of macroscopic magnitude,

$$L_{macro} = c^2/g$$

$$\sim \frac{9 \times 10^{20} cm^2/sec^2}{10^3 cm/sec^2} \sim 10^{18} cm \text{ or 1 light year.}$$

The one quantity measures the thinness of that layer of water which, upon annihilation, would give up enough energy to overcome the surface tension forces. The other distance is the characteristic distance that one must travel with the acceleration g to attain something like the speed of light (a "velocity parameter" θ equal to unity). The radius is of the well known order

$$a \sim (10^{-19} cm \times 10^{18} cm) \sim 0.3 cm$$

The difference in order of magnitude between the size of a water drop and the length M_{micro} is fantastic. However, it is no more fantastic than the difference between the size of an elementary particle, $\sim 10^{-13}$cm, and the basic Planck length,[1]

$$L^* = (\hbar G/c^3)^{\frac{1}{2}} = 1.6 \times 10^{-33} cm$$

The size and mass of the water droplet are not constants of nature. Is it possible that also the size and mass of the elementary particle are not constants of nature?

The size of the water droplet depends upon an accident: the value of the acceleration of gravity at the place where one happens to take his shower. Does the mass of the elementary particle also depend upon an accident: the particular initial conditions which happened to characterize this particular cycle of expansion and recontraction of the universe?

The molecular constitution of my pencil was fixed by the chemistry that went on in a tree a few years ago. The nuclear constitution of the carbon of the pencil was built in by thermonuclear combustion in a star some billions of years ago. Both the molecules and the nuclei are fossils from the past. Are the particles and the universe also fossils[2] from a still greater violence?

More concretely, we state the issues as follows. First, we'll accept Einstein's general relativity or "geometrodynamics" in its standard 1915 form, translated of course into the appropriate quantum version. Second, we accept as tentative working hypothesis the picture of Clifford[3] and Einstein[4] that particles originate from geometry; that there is no such thing as a particle immersed in geometry, but only a particle built out of geometry. Third, we recognize that we have not come far enough along the road to trace out the consequences of this working hypothesis. We don't know how to describe particles as built out of geometry. Therefore we ask, assuming that particles are built out of geometry, what ideas

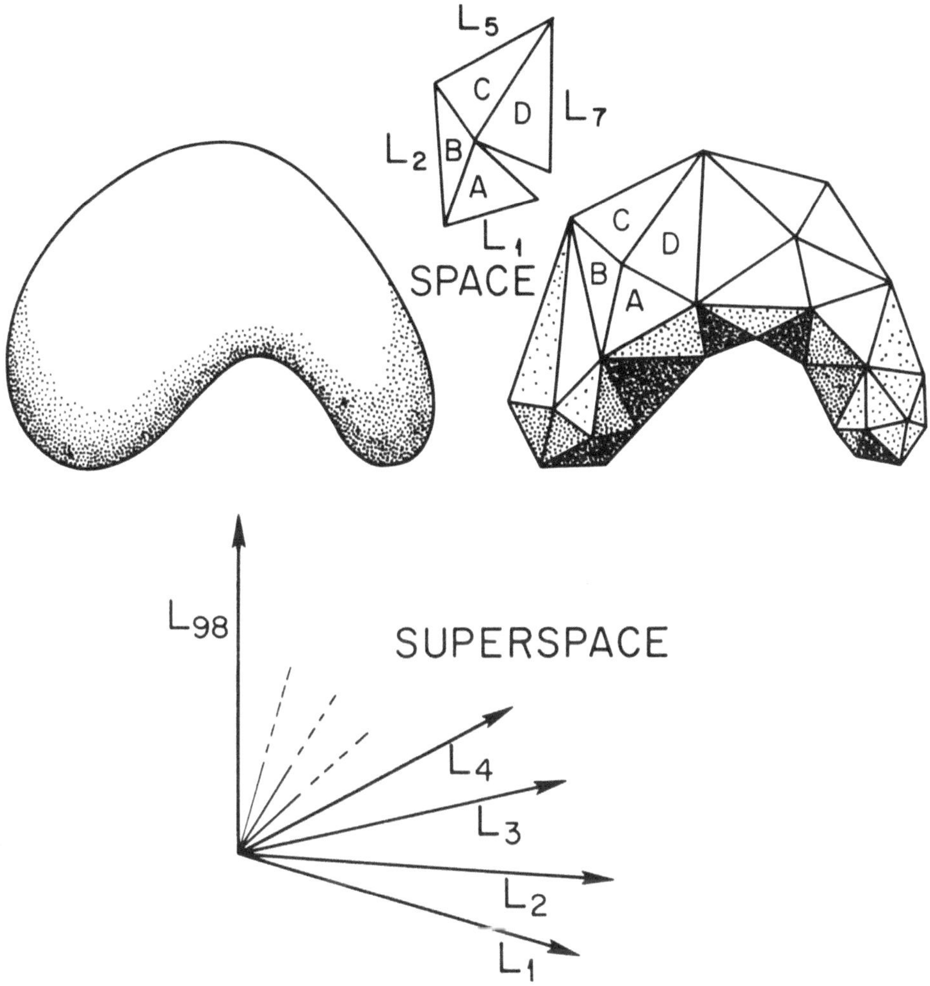

Fig. 1. Simplified version of superspace. Upper left: a 2-geometry (a stand-in for the 3-geometry of the Einstein universe). Upper right: approximation of this 2-geometry by decomposition into triangles (these 2-simplexes replaced by 3-simplexes (tetrahedrons) in the actual analysis!) The curvature at any vertex (cf. insert) is governed by the nearest neighbor lengths, and by nothing more. The geometry in the simplicial approximation is completely fixed by the specification of the 98 edge lengths. Equivalently (lower diagram) the geometry is completely specified by a single point in a space of 98 dimensions. To go to superspace proper (replacement of simplicial decomposition by actual geometry) one has to go from 98 dimensional space to ∞-dimensional space.

can we get as to the building process from the very existence of particles.

As we start, we recognize that the arena of Einstein's geometrodynamics is not space, not even spacetime, but superspace. Space, to be sure, is the dynamical object. And a spacetime is one classical history of that dynamical object. But the arena in which the dynamics unrolls is superspace.

Fig. 1 presents a simplified version of superspace. One sees how the specification of a single point in superspace fixes an entire 3-geometry. Here, by 3-geometry, we mean one equivalence class of all those 3-metrics which are equivalent to each other under diffeomorphism ("dividing out the gauge group of coordinate transformations". There are advantages to augmenting "3-geometry" as just stated to "3-geometry with an identified point" to make superspace into a proper manifold[5]; "six slices of pie united to make a complete pie").

Fig. 2 shows the relation between space, spacetime and superspace. Spacetime is an assembly, a nesting together, with many a crisscrossing of all those 3-geometries which occur in one given classical history of space (compare B and B'; freedom to move ahead further in time in some regions than others in the exploration of spacetime; the "many fingered time" of general relativity). The 3-geometries which occur in this classical history may be called "YES" 3-geometries to distinguish them from the infinitely more numerous "NO" 3-geometries which make up the rest of superspace. They are encountered on other histories, on other leaves slicing through superspace. Einstein's classical field equation, supplemented by initial conditions, makes the most familiar tool to determine a spacetime; or, in the terminology of superspace, to distinguish "YES" 3-geometries from "NO" 3-geometries. An alternative and equally good tool for making this distinction is the Einstein-Hamilton-Jacobi equation,

$$g^{-1}(\tfrac{1}{2}g_{pq}g_{rs}-g_{pr}q_{qs})(\delta S/\delta g_{pq})(\delta S/\delta g_{rs})+ {}^{(3)}R = 0$$

Here the Hamilton-Jacobi functional $S = S(g_{rs})$ ostensibly depends upon the six components g_{rs} of the space-metric. In actuality it depends only on the 3-geometry. This is seen in the familiar zero effect of a mere (infinitesimal) coordinate transformation,[7]

$$x^i \to x^i - \xi^i$$

$$g_{rs} \to g_{rs} + \xi_{r|s} + \xi_{s|r}$$

$$S \to S + \int (\delta S/\delta g_{rs})(\xi_{r|s} + \xi_{s|r})\, d^3x$$

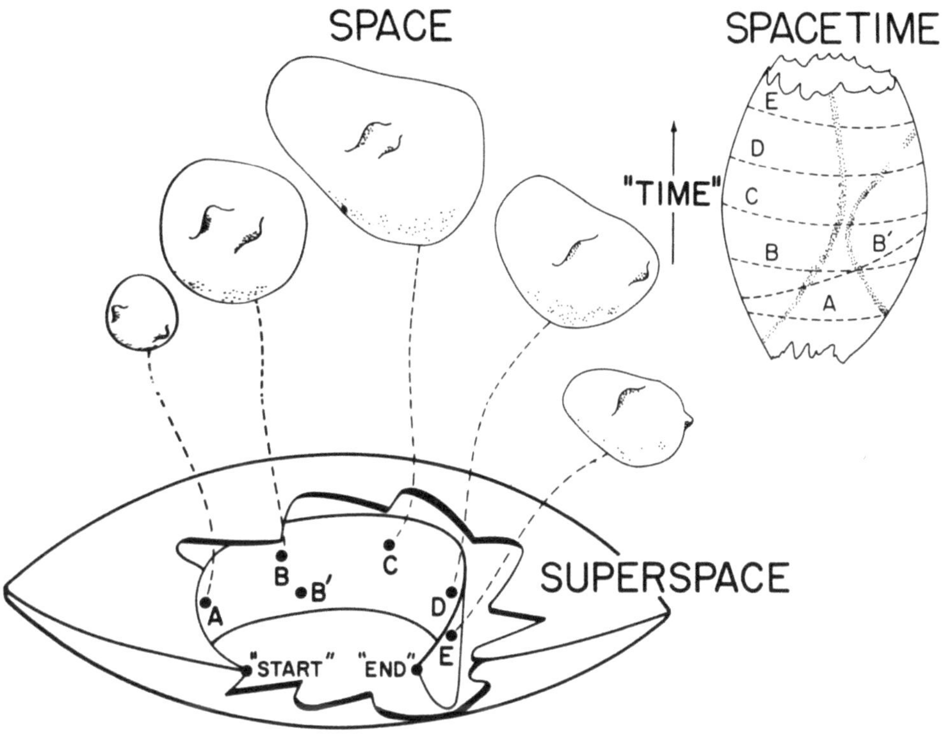

Fig. 2. Space, spacetime and superspace. Upper left: Five sample configurations attained by space in the course of its expansion and recontraction. Upper right: Spacetime. A spacelike cut, like A, through spacetime, gives a momentary configuration of space (Upper left; the early 3-geometry A). Below: Superspace. The one bent "leaf" in superspace (running from "START" to "END") comprises all those 3-geometries which are obtainable as spacelike slices through the given spacetime.

$$= S - 2 \int \underbrace{(\delta S/\delta g_{rs})|_S}_{\text{zero}} \xi_r \, d^3x$$

There is no spacetime in the classical theory. There is no sharp distinction between "YES" and "NO" 3-geometries. There is only a probability amplitude $\psi({}^{(3)}\mathcal{G})$ for this, that and the other 3-geometry. The 3-geometries that occur with appreciable probability amplitude are enormously more numerous than can be accommodated in any one spacetime, any one classical history of space. We have to speak in a different language. Spacetime is not a part of that language.

In quantum geometrodynamics we have to do with a probability amplitude $\psi = \psi({}^{(3)}\mathcal{G})$ that propagates in superspace. One wave with one set of initial conditions (on an "initial value supersurface" cutting through superspace) propagates at one amplitude through superspace. A second wave has another amplitude. Superpose a sufficient number of such waves, with suitable multiplying coefficients, and with suitable phases. The resulting wave packet will closely reproduce the predictions of classical theory for a universe running through its dynamics. For example, the wave packet can be arranged to reproduce in a certain approximation the classically predicted behavior of Misner's mixmaster model of the universe.

Closed space, according to classical geometrodynamics, inevitably undergoes gravitational collapse. It (or geodesics running through it; cf. Taub-Misner-NUT space[8]) arrives in a finite proper time at a condition of infinite compaction and infinitely high curvature. Similar conditions characterize the earliest days of the universe in classical theory.

It is not new for classical theory to predict trouble. The electron headed straight for a center of charge arrives in a finite time at an infinite kinetic energy, according to classical mechanics. But a prediction that is infinity is not a prediction. Quantum theory has another story to tell. The electron does not follow a deterministic track. It has a finite probability to be scattered in any one of a variety of directions.

We expect a similar quantum indeterminism in the earliest days of the expanding universe and in the final days of recontraction. At the nearly singular conditions which prevail near the beginning and end of each cycle of the universe, we anticipate, not catastrophe, but coupling, as in a multichannel wave guide, between the given cycle of the universe and many alternative cycles ("histories"; "leaves in superspace").

The universe will not come out of the collapse, we can believe, with the same conditions (number of particles; size at phase of maximum expansion) that it had when it fell in. It will not even come out with any unique set of properties. Instead, there will be a probability amplitude to come out in one cycle, with one size at the phase of maximum expansion, and one spectrum of coupling constants and particle masses; and probability amplitudes to come out in other cycles, with other dimensions for the universe, and other particle masses.

On this view a particle mass is not a basic quantity of nature. It has as little claim to that title as does the mass of the water droplet that hangs from the ceiling of the shower. Ask why it has its mass, and find oneself asking why one takes a shower where the value of g happens to be 980cm/sec^2. Ask why the particle has its mass, and end up asking why we happen to be living in this particular cycle of the universe.[10] One cycle, one set of masses. Another cycle, another set of masses. That is the picture.

Having followed our line of reasoning so far to see where it will take us, we find ourselves confronted with a central issue. Why are all electrons identical, all protons identical, and so on? If particles are to be viewed as having another set of masses in another history of the universe, why don't electrons, for example, have a variety of masses in this history?

The observed identity of particles in the universe[11] seems an overwhelming objection to the view that particles are constructed out of geometry. How can particles at one side of the universe even get word how to act to look like particles on the other side of the universe? There isn't time for information to get around from one side to the other until late in the cycle of expansion and recontraction! This problem of uniformity is not new. It is met in another context. How can the universe be as homogeneous as we find it to be, Misner asks,[12] when there is no apparent way for matter at one side of the universe to have heard about what the matter is doing at the other side of the universe.

Misner gives an answer -- and gives reasons to believe it is the only workable way for establishing homogeneity. The universe is more and more anisotropic as one goes back into the past. The change of the anisotropy with time follows directly from Einstein's equations. The universe undergoes an anisotropy vibration. First one principal direction of curvature dominates, then another. The path round the universe is shortened so effectively that every point communicates with every other point. Moreover, this communication is established early in the history of the expansion.

Think of a pulse of light emitted from a point at an early time. The wave front starts out spherical. But it rounds a universe foreshortened by anisotropy vibration first in one direction, then in another. In consequence waves come back to the original point from many different directions. The return of the wave by rounding the universe is as effective as if there were complete reflection. Therefore consider by analogy an acoustic pulse emitted in an auditorium with walls of high reflectance and complicated configuration. Wait for many reflections, and make an instantaneous map of the wave front. It will show a fantastic degree of complication, with multiple crossings at many different angles: a "statistical wave front"! This is what goes on in Misner's mixmaster model of the universe.

As for a pulse of light, so for a pulse of pressure, and so for a shock front, and so for an initial vortex tube: practically complete mixing can be achieved.

Question: Can we think of an initial structure -- shock front, vortex line, or other feature of geometry -- that can be so stirred up and ground down to a very fine scale that it gives the basis for making identical particles out of geometry?

In a bread factory one sees dough filled up with bubbles by the action of the yeast, and watches the dough stirred long and vigorously. The big bubbles are broken down into smaller bubbles. More. They are homogenized. Can we think of a similar way to stir up a feature of the geometry so thoroughly and on such a fine scale that it satisfies the following three conditions:

PERSISTENT
UNIFORM
SUBMERGED
STRUCTURE

(1) The structure should be persistent. It can itself be equivalent to virtual particles. Or superposed upon this structure the particles can be "excitons". In either case this structure must stay unchanged in scale, or equivalent mass, throughout the dynamical history of the universe.
(2) It has to be uniform.
(3) The structure has to be on such a fine scale that as far as everyday purposes are concerned, it doesn't show up.

These features asked of geometry recall some of the patterns of winds, clouds and waves (Table 1). Moreover, there are many illuminating analogies between geometrodynamics[13] and hydrodynamics. Therefore it is natural to ask sooner than almost any other question: What dimensionality must an initial feature of the geometry possess if it is to Misner-mix to give "persistent uniform submerged structure"? One-dimensional or 2-dimensional? In the absence of any immediate answer, it is natural to look at three models for the initial feature, 1. Line (closed loop: "vortex core"?)

Table 1. Features of hydrodynamics and their dimensionality.

Feature	Dimensionality of "active part"
Shock	2-dimensional front
Inversion layer	2-dimensional
Water wave	No sharp distinction between active part and non-active part
Turbulence	No sharp distinction between active part and non-active part
Cavitation	0-dimensional (point)
Mach triple point	1-dimensional
Vortex (concentrated)	1-dimensional

2. Surface expanding with the speed of light (light pulse or shock)

3. Surface expanding with less than the speed of light.

A program naturally poses itself: Follow each of these three features as it evolves with time, as it is stretched out, reverberates around the universe, and is mixed to finer and finer scale -- working ahead in time, at first by detailed classical calculations, then by statistical analysis. Program for the future!

Nothing would seem vainer than to hope to see final uniformity of turbulence, of mixing, or of whatever, emerging out of any such classical calculation, carried however far. At every stage percentagewise significant statistical variations will be apparent. And nothing could be more certain than this, that particles, like atoms, derive their exact identity, one with another, from the quantum principle. Classical physics will not and cannot give this identity. But everyday experience holds out the synchronous motor. The driving field runs by at exactly 60 cycles. It might therefore seem necessary to get the rotor turning at exact frequency to have a successful start. Not so. Some ways short of the exact frequency the rotor finds itself caught up and brought to full speed. "Lock-in" occurs. Is there a way for the classical structure ("small scale turbulence of geometry"), brought almost into uniformity by Misner-mixing, to be locked into exact uniformity by the quantum principle? And in that event, the characteristic scale of the "turbulence" must be brought down, by mixing, within how many powers of ten of the Planck length before lock-in occurs? More questions for the future!

If there is any such structure in geometry as we are talking about, it has to describe virtual particles as well as real particles. Therefore it must pervade all space. It shares this all-pervasiveness with the wormholes that are continually coming into being and disappearing everywhere at the Planck scale of distances ("quantum fluctuations in geometry and topology"; "foam-like structure of space"). However, the two kinds of structure are quite different in scale and in other properties, if we are to judge on these matters from the discordant scales of elementary particles (10^{-13}cm) and quantum fluctuations (10^{-33}cm).

Turning from particles now to geometry, what can we say about the structure of geometry at the very smallest distances? Of all the arguments why geometry must undergo quantum fluctuations in topology as well as in curvature at the smallest distances, none is more persuasive than the existence in nature of electric charge. No other natural explanation has ever been put forward except "lines of force trapped in the topology of space". So we accept changes in topology. With that, we accept the idea of a handle "breaking". Viewed classically, the handle becomes thinner and thinner until at last only a single point remains to connect two fingers of space. Then even that contact breaks. Two points that were neighbors have parted company. Put in these terms, the process is discontinuous: from "yes, connected" to "no, disconnected". In quantum physics, however, nothing is discontinuous. The two points that have separated in reality must retain some residual connection. And if they have some connection, then by the principle of democracy among points, every point must have some connection with every other point.

These points are relevant when we think of the direction of advance of physics. In the 1850's the density of lead was an empirical number which could have had as well one value as a number. Today its value follows out of atomic and nuclear physics. But today 3 dimensions is still something which has to be taken as empirical. Some day it must be derived via the quantum principle from deeper considerations, from "pregeometry". There the "points" have a new relationship. What considerations of what is uniquely right, beautiful and true will guide us into the world of "pregeometry"?

REFERENCES

1. M. Planck, Sitzber. preuss. Akad. Wiss. Berlin, Math.-phys. Kl. 1899, 440; J. A. Wheeler, *Geometrodynamics*, Academic Press, New York, 1962.

2. J. A. Wheeler, "Maria Sklodowska-Curie as Copernicus of the World of the Small", in *Maria Sklodowska-Curie: Centenary Lectures; Proceedings of a Symposium, Warsaw, 17-20 October 1967,*

IAEA, Vienna, 1968, pp. 25-32; and J. A. Wheeler, "Strange Matter" in H. Mark and S. Fernbach, eds., *Properties of Matter under Unusual Conditions*, Interscience Publishers, New York·London·Sydney·Toronto, 1969, pp. 365-379.

3. W. K. Clifford, "On the Space-Theory of Matter", a lecture before the Cambridge Philosophical Society 21 February 1870, reprinted in abstract form in his *Lectures and Essays* (L. Stephen and F. Pollock, eds.), Vol. 1, London, 1879, also in his *Mathematical Papers* (R. Tucker, ed.), London, 1882; also in *The World of Mathematics* (J. R. Newman, ed.), Vol. 1, Simon and Schuster, New York, 1956, pp.568-569.

4. J. A. Wheeler, *Einsteins Vision*, Springer-Verlag, Berlin·Heidelberg·New York, 1968; or, slightly abbreviated, but in English, "Superspace and the Nature of Quantum Geometrodynamics", a chapter in C. De Witt and J. A. Wheeler, eds., *Battelle Rencontres; 1967 Lectures in Mathematics and Physics*, W. Benjamin & Company, New York, 1968, pp. 242-307.

5. Point raised by C. W. Misner in discussion of contribution of A. Fischer, these proceedings. For the concept of "geometry-plus-an-identified-point" see also the discussion, and references on Teichmüller space, in the articles of L. Bers and J. A. Wheeler in G. Newton, ed., *Analytic Methods in Mathematical Physics*, Gordon and Breach, New York, 1969.

6. A. Peres, *Nuovo Cimento*, 26, 53 (1962)

7. P. W. Higgs, *Phys. Rev. Letters*, 1, 373 (1958) and 3, 66 (1959)

8. C. W. Misner and A. H. Taub, "A Singularity-Free Empty Universe", *Zh. Eksperim. i Teor. Fiz.*, 55, 233 (1968); [English text: *Soviet Phys.-JETP*, 28, 122 (1969)]

9. Cf. illustration, Fig. 22, p. 143, in Harrison, Thorne, Wakano and Wheeler, *Gravitation Theory and Gravitational Collapse*, University of Chicago Press, 1965.

10. G. W. Leibniz, *Essais de Théodicée sur la bonté de Dieu, la liberté de l'homme, et l'origine du mal*, Amsterdam, 1747; Samuel Clarke, *A Collection of Papers which passed between the late Learned Mr. Leibniz, and Dr. Clarke, In the Years 1715 and 1716. Relating to the Principles of Natural Philosophy and Religion*, London, 1717; L. D. Landau and E. M. Lifshitz, *Statistical Physics*, trans. E. Peierls and R. F. Peierls, Addison-Wesley, Reading, Massachusetts and Pergamon, London, 1958; R. H. Dicke, "Dirac's Cosmology and Mach's Principle", *Nature*, November 1961; B. Carter, "Large Numbers in Astrophysics and Cosmology", preprint, Institute of Theoretical Astronomy, Cambridge, England, September 1968; J. A. Wheeler, "Man's Place in Cosmology", Léon Lecture, University of Pennsylvania, 11 Nov. 1969 (to be published)

11. For one measure of the fantastic precision with which we know electrons to be identical (stability of core of earth against collapse), see the chapter by R. Marzke and J. A. Wheeler, "Gravitation as Geometry, I: The Geometry of Space-time and the Geometrodynamical Standard Meter", in H.-Y. Chiu and W. F. Hoffmann, eds., *Gravitation and Relativity*, W. A. Benjamin & Company, New York, 1964.

12. C. W. Misner, "Relativistic Fluids in Cosmology", p. 155 in *Fluides et Champ Gravitationnel en Relativité Générale*, Centre National de la Recherche Scientifique, Paris, 1969; *Phys. Rev.*, 186, December 1969; see also the article of Misner in these proceedings.

13. For a table of these analogies, see for example E. A. Power and J. A. Wheeler, *Rev. Mod. Phys.*, 29, 480 (1957), reprinted in J. A. Wheeler, *Geometrodynamics*, Academic Press, New York, 1962.

ACKNOWLEDGEMENT

Preparation of this report was assisted in part by the National Science Foundation Grant GD 7669 to Princeton University.

THE SANDWICH CONJECTURE*

Peter G. Bergmann

Syracuse University

Syracuse, New York

The "Sandwich Conjecture" (SC) was originally formulated by Wheeler and his coworkers.[1] In the course of several years both the original author(s) and others have proposed the statement in several different forms, and with different qualifications. Roughly, the SC asserts that in a purely gravitational field the internal geometries of two distinct three-dimensional space-like hypersurfaces determine uniquely the geometry of a four-dimensional space-time (i.e. pseudo-Riemannian) manifold that is required to obey Einstein's field equations (i.e. to be Ricci-flat). In a careful formulation one will have to specify further whether the two three-surfaces are to be "close together", whether the distance between them is to be given as additional data, and what sort of conditions are to be imposed at space-like infinity.

The SC is suggested by similar theorems that hold in classical mechanics, where knowledge of the configuration of a system at two distinct times may determine its trajectory throughout the period of time bounded by the two instants. If the SC were to hold, one would hope that the metric field can be quantized by Feynman integral methods. Quite apart from its bearing on quantization, the SC, and its possible limitations and qualifications, possesses considerable interest for the structure of Einstein manifolds *per se*.

---

*Research supported in part by ARL and by AFOSR.

[1] See, for instance, R. F. Baierlein, D. H. Sharp, and J. A. Wheeler, Phys. Rev. 126, 1864-1865 (1962).

In approaching the SC, I shall be guided by a certain formulation of the laws of the gravitational field itself. This formulation is in turn based on Dirac's Hamiltonian versions of Einstein's equations: In a function space that is described by six pairs of canonical field variables, $g_{mn}(\vec{x})$, $p^{mn}(\vec{x})$, defined on a three-dimensional manifold, and with the proviso that the metric g_{mn} is everywhere positive-definite, these twelve field variables are subject to four first-class constraints at every point $\vec{x}$. Any linear combination of these constraints,

$$H = \int d^3x(\xi^L \mathcal{H}_L + \xi^n \mathcal{H}_n) \qquad (1)$$

generates an infinitesimal transformation of the field variables, which is equivalent to the transition from one space-like hypersurface to another one nearby, both of which lie in the same Einstein manifold. By including all finite canonical transformations which may be constructed with arbitrary choices of the ξ-fields, one can obtain the Cauchy data appropriate to all conceivable space-like cuts through one Einstein manifold. The data on any one of these cuts determines the Einstein manifold uniquely, but obviously highly redundantly, in that the huge manifold of all fields of canonical variables that can be obtained from each other by successions of infinitesimal transformations generated by the Hamiltonians (1) belongs but to one Einstein manifold. It is intuitive to think of the function space of the g's and p's that obey Dirac's constraints, and which I shall call henceforth the phase space (of the gravitational field) as being composed of equivalence classes; each equivalence class corresponds to one distinct Einstein manifold. The equivalence classes cover the phase space without overlap.

It is well known that the equivalence classes have different dimensionalities. For instance, in the presence of a Killing field a certain class of Hamiltonians will map in that equivalence class a point of phase space on itself, whereas the same class produces non-trivial trajectories in other equivalence classes.

The SC is concerned with properties of the configuration space described by the g's alone, without the p's. A fixed three-geometry represents a subspace of the phase space described by the g's and p's together: It is the set of all g-fields that can be transformed into each other by three-dimensional coordinate transformations. The SC proposes that any two of these subspaces together are contained precisely in one equivalence class, that is to say, in one Einstein manifold.

As a preliminary, I shall construct an example of a phase space with constraints and equivalence classes in which the SC holds. Consider an n-dimensional Euclidean space with coordinates x^k. The straight lines of that space are the solutions of the variational problem

$$\delta S = 0, \quad S = \int L d\theta \tag{2}$$

where

$$L = \sqrt{x^{k'} x^{k'}}, \quad x^{k'} \equiv \frac{dx^k}{d\theta} \tag{3}$$

and θ is an arbitrary parameter. By defining "momentum" components p_k in the usual manner,

$$p_k = \frac{\partial L}{\partial x^{k'}} = \frac{dx^k}{ds}, \tag{4}$$

$$ds^2 \equiv dx^k dx^k,$$

I obtain the "Hamiltonian constraint",

$$C = \frac{1}{2}(p_k p_k - 1) = 0 \tag{5}$$

which generates the propagation of a set of Cauchy data x^k, p_k (the latter chosen so as to be consistent with the constraint itself) along the straight line defined uniquely by the Cauchy data. In this example we have canonical variables, a constraint, and equivalence classes of points in phase space. Each distinct equivalence class corresponds to one distinct straight line.

A point in configuration space is a set of x^k. The same numerics define, of course, a point in the original Euclidean space, which is thus homeomorphic with the configuration space. That two distinct points in a Euclidean space have in common exactly one straight line need not be proven here. Nevertheless, a formal handling of this "problem" is perhaps instructive.

It is possible to introduce (n-1) pairs of canonical coordinates which are in one-to-one correspondence with the set of all straight lines. For instance, one may introduce as the coordinates X^k (k=1, ..., n-1) the Cartesian coordinates of the point on the straight line which lies in the hyperplane $x^n = 0$, and as the momenta P_k(k=1, ..., n-1) the original momenta p_k, with p_n omitted. These new canonical coordinates, which describe an 2(n-1) - dimensional <u>reduced phase space</u>, are related to the original coordinates by the equations

$$X^k = x^k - \frac{p_k}{p_n} x^n, \; P_k = p_k, \tag{6}$$

These relations may also be obtained from a solution of the Hamilton-Jacobi equation corresponding to the constraint (5),

$$\frac{\partial S}{\partial x^k} \frac{\partial S}{\partial x^k} -1 = 0, \; S = S(x', \ldots, x^n; P_1, \ldots P_{n-1}), \tag{7}$$

The particular solution leading to Eqs. (6) is

$$S = P_k x^k + (1-P_\ell P_\ell)^{1/2} x^n, \qquad k,\ell = 1, \ldots, n-1. \tag{8}$$

The transformation equations are, as usual,

$$p_m = \frac{\partial S}{\partial x^m}, \; m = 1, \ldots, n, \qquad X^k = \frac{\partial S}{\partial P_k}, \; k = 1, \ldots, n-1. \tag{9}$$

It is trivial to verify that the X's and P's, understood to be the expressions (6), indeed satisfy standard Poisson bracket relations.

These new coordinates, "constants of the motion", remain unchanged throughout each equivalence class. An equivalence class known to contain the point $(\bar{x}^1, \ldots, \bar{x}^n)$ of the configuration space has its canonical coordinates restricted by the (n-1) conditions

$$(1 - P_\ell P_\ell)^{1/2} (X^k - \bar{x}^k) + \bar{x}^n P_k = 0, \qquad k,\ell = 1, \ldots, (n-1) \tag{10}$$

If the equivalence class is then required to contain a second point, say with the coordinate values $\bar{\bar{x}}^1, \ldots, \bar{\bar{x}}^n$, the resulting two sets of (n-1) conditions each determine the equivalence completely, with the exception of the case $\bar{x}^n = \bar{\bar{x}}^n$, which leads to infinite values of the coordinates X^k.

An apparently different situation is presented by the following problem from classical mechanics. Let a force-free mass point occupy consecutively the positions (in physical space) $\bar{x}^k$ and $\bar{\bar{x}}^k$

(k=1, ..., 3). Determine the trajectories in phase space passing through these two points. Obviously, this problem has many substantively different solutions unless the lapse of time between the two passages is also given. Only with this additional piece of information is the magnitude of the linear momentum determined.

Finally, let me remind you of a "pathological" situation. Consider a one-dimensional harmonic oscillator and give the values of its x-coordinate at two different times $\bar{t}$ and $\bar{\bar{t}}$. For most choices of the time lapse this is a well-set problem leading to unique trajectories. The problem degenerates, however, if the lapse of time equals an integral multiple of the half-period of the oscillator. In those cases the problem has either no solutions, or infinitely many solutions, depending on the given values of $\bar{x}$ and $\bar{\bar{x}}$.

I am reminding you of all these contingencies only in order to persuade you that the task of "filling the sandwich" between sets of data pertaining to two distinct stages in the history of a physical system is not a trivial one, and that the existence, and uniqueness, of a solution cannot be taken for granted.

I shall now formulate the general problem, of which the SC is, of course, the example of interest to us. Given a phase space to be described by the canonical coordinates q_k, p_k, let there be a number of first-class constraints $C_a(q,p) = 0$, with the index $\underline{a}$ having a smaller range than the index k,

$$[C_a, C_b] = 0. \tag{11}$$

Assume further that the constraints do not restrict the domain of the configuration variables q_k. Define an <u>equivalence class</u> of points in phase space all those points that may be connected with each other by curves obeying the conditions:

$$C_a = 0,$$

$$\frac{dq_k}{d\lambda} = \frac{\partial H}{\partial p_k}, \quad \frac{dp_k}{d\lambda} = -\frac{\partial H}{\partial q_k} \tag{12}$$

$$H = \xi^a C_a.$$

λ serves as a parameter for the curve to be constructed from the set of ordinary differential equations; the ξ^a are arbitrary functions of the canonical coordinates. Given any two sets of numbers, $\bar{q}_k$ and $\bar{\bar{q}}_k$, (i.e. two points in configuration space), does there exist an equivalence class containing two points (in phase space) whose q-coordinates take the specified values, and if so, is this equivalence class uniquely determined?

If the two points are separated only by an infinitesimal displacement, i.e. if we are given two sets of numbers q_k, $(dq_k/d\lambda)$, the problem may be reduced to the question whether there exists a set of p_k, and a set of ξ^a, so that

$$\overline{\frac{dq_k}{d\lambda}} = \xi^a \frac{\partial C_a(\bar{q},p)}{\partial p_k}, \quad C_a(\bar{q},p) = 0. \tag{13}$$

The number of conditions (13) equals the number of unknowns. Generally, the conditions are highly non-linear. Even in its infinitesimal form the problem does not lend itself readily to analysis.

For the case of the Einstein theory in Dirac's version, A. Komar[2] has utilized the fact that the canonical momentum densities appear in the constraints at most quadratically. Thus Eqs. (13) reduce to the form:

$$\begin{gathered} p^{mr}{}_{|r} = 0, \quad p^{rs}p_{rs} - \frac{1}{2}(p^s_s)^2 + {}^3g\,{}^3R = 0, \\ \dot{g}_{mn} = -\frac{1}{2}(\xi_{m|n} + \xi_{n|m}) + \frac{\xi^L}{\sqrt{{}^3g}}\left(p_{mn} - \frac{1}{2} g_{mn} p^s{}_s\right). \end{gathered} \tag{14}$$

In these equations, the unknowns are the functions p^{mn}, ξ^m, ξ^L on three-space, all other variables being given. The second line Eqs. (14) enable us to express the p^{mn} in terms of the remaining unknowns:

$$p_{mn} = \frac{\sqrt{{}^3g}}{\xi^L}\left[(\dot{g}_{mn} - g_{mn}g^{rs}\dot{g}_{rs}) + \frac{1}{2}(\xi_{m|n} + \xi_{n|m} - g_{mn}\xi^r{}_{|r})\right], \tag{15}$$

---

[2] Private communication. I am indebted to Prof. Komar for a number of stimulating discussions.

and to substitute these expressions directly into the constraints, the first two sets of conditions (14). The constraints $H^m = 0$ lead to three second-order partial differential equations for the ξ^m, which are, with respect to these variables, inhomogeneous and linear. The fourth constraint results in an algebraic equation for ξ^L,

$$g^{mn} g^{rs}[(\dot{g}_{mr} \dot{g}_{ns} - \dot{g}_{mn} \dot{g}_{rs}) + 2(\dot{g}_{mr} \xi_{n|s} - \dot{g}_{mn} \xi_{r|s})$$
$$+ (\xi_{m|r} \xi_{n|s} - \xi_{m|n} \xi_{r|s})] + (\xi^L)^2 \, {}^3R = 0. \qquad (16)$$

As noted by Komar, difficulties arise in the event of two contingencies: (1) if the original data given result in ${}^3R = 0$, or (2) if $(\xi^L)^2$ should come out negative. If ${}^3R = 0$, then the first term of Eq. (16), the square bracket, should vanish, resulting in a mixed-quadratic partial differential equation for the three variables ξ^m. If the expressions (15) for p_{mn} are substituted into the first set of three Eqs. (14), three more equations result, which, however, contain, in addition to the variables ξ^m, reference to the fourth variable ξ^L. Hence the set is not overdetermined. To avoid the second contingency one will have to make the ξ^m satisfy an inequality in addition to the partial differential equations $H_s=0$.

In the absence of these special situations, when ξ^L can be determined from the constraint $H_L = 0$, the remaining constraints represent three second-order partial differential equations for the three variables ξ^m. Because of the form of Eq. (16). once the appropriate expression for ξ^L is submitted, there will be highly nonlinear terms in the three equations $H_s = 0$, and their discussion is enormously complicated, to say the least. Nevertheless, it should be realized that the problem can be carried forward in closed form quite a distance, and that its systematic study is not entirely precluded. In what follows I shall present a much more limited approach, which, however, leads to equations that are linear in the unknowns.

Let me return to the general problem summarized in Eqs. (11), (12), and (13). The unknowns in this problem were the variables p_k, ξ^a, which of course enter in a nonlinear fashion. Suppose now that we had obtained a particular solution of that problem, leaving aside, for the moment, the question of its uniqueness. We can then ask two questions, both of which lead to linear problems: (1) Given the same data as before, do there exist solutions in the infinitesimal vicinity of the solution already in hand? And (2), suppose the given data were modified infinitesimally, does a solution still exist, and how does it differ from the original solution? I shall narrow down the second question by restricting the modification of the given data to the $\dot{q}_k$, the left-hand sides of the first set of Eqs. (13), because the data on a sandwich can, without substantive restriction, be changed in two steps, each step involving the data on one three-surface only. To insist on modifying both three-geometries at the same time merely makes life difficult, without achieving any greater substantive generality.

Obviously, the first question formulated above is a special case of the second, in that the given data are changed by vanishing amounts. Hence I shall take on the second question immediately. To simplify the notation, I shall designate the given data and the particular solution chosen as the point of departure by the symbols q_k, $\dot{q}_k$, p_k, and ξ^a, and the infinitesimal modifications as follows

$$\delta\dot{q}_k = a_k, \quad \delta p_k = \pi_k, \quad \delta\xi^a = \eta^a. \tag{17}$$

With this notation the conditions to be satisfied may be cast in the following forms. The constraints must remain satisfied:

$$\frac{\partial C_a}{\partial p_k}\,\pi_k = 0. \tag{18}$$

The propagation equations (13) must continue to hold:

$$a_k = \frac{\partial C_a}{\partial p_k}\,\eta^a + \xi^a\,\frac{\partial^2 C_a}{\partial p_k \partial p_\ell}\,\pi_\ell = 0. \tag{19}$$

The unknowns of Eqs. (18) and (19) are the variables η^a, π_k. The number of equations equals the number of unknowns, and the problem is linear and inhomogeneous. The system of equations corresponding to the first problem enunciated above is obtained by setting the inhomogeneous terms on the left of Eqs. (19) zero.

Generally, if arbitrarily altered $\dot{q}_k$ do not preclude the existence of a solution, the solution associated with the given data is unique; and , conversely, if the solution associated with given data is non-unique, then at least some changes in the data will preclude the existence of a solution altogether. Both of these statements hold, of course, only locally. That is to say, they refer to the existence and uniqueness of solutions infinitesimally different from a given solution.

I shall now transfer the conditions (18), (19) to the gravitational case. The constraint conditions become:

$$\pi^{ms}{}_{|s} = 0, \tag{20}$$

$$(p^{rs} - \frac{1}{2} g^{rs} p)\, \pi_{rs} = 0. \tag{21}$$

The propagation condition turns into the relation:

$$\begin{aligned} a_{mn} + \frac{1}{2}(\eta_{m|n} + \eta_{n|m}) &= g^{-1/2}(p_{mn} - \frac{1}{2} g_{mn} p)\, \eta^{L} \\ &+ g^{-1/2}\, \xi^{L}(\pi_{mn} - \frac{1}{2} g_{mn}\, \pi). \end{aligned} \tag{22}$$

If this relationship is solved algebraically with respect to π_{mn},

$$\begin{aligned} \pi_{mn} &= \frac{1}{\xi^{L}}\{g^{1/2}[(a_{mn} - g_{mn} a) + (\eta_{(m|n)} - g_{mn}\, \eta^{s}{}_{|s})] - p_{mn}\eta^{L}\}, \\ \pi &= -\frac{2}{\xi^{L}}\, [g^{1/2}(a + \eta^{s}{}_{|s}) + \frac{1}{2} p\, \eta^{L}], \end{aligned} \tag{23}$$

and substituted into (21), the result is:

$$g^{1/2} p^{mn} (a_{mn} + \eta_{(m|n)}) + {}^{3}R\eta^{L} = 0. \tag{24}$$

Again, the case ${}^{3}R = 0$ is special, in that it eliminates from Eq. (24) any reference to the variable η^{L}. The three equations (20) will now contain references to the four variables η^{m}, η^{L}. The coefficients of the resulting system of equations involve both the background metric and the background momentum densities, i.e. the extrinsic characteristics of the imbedded three-surface.

In the more general case, when ${}^{3}R \neq 0$, Eq. (24) can be solved algebraically for the unknown η^{L}, and the resulting expression substituted into the expressions (23), which in turn are subject to the three linear differential equations (20). The resulting system of three second-order partial differential equations for the three variables η^{m} is inhomogeneous-linear, the coefficients depending both on the three-metric g_{mn} and on the p^{mn}, all given. Nothing general can be said about the elliptic, hyperbolic, or parabolic character of the system of equations. In fact, the character of the equation coefficients may well change from one finite region of the three-dimensional domain of the problem to another.

Regardless of this question, I believe that one should not attempt to render the solutions of the problem unique by imposing standard boundary conditions at spatial infinity (or at other "natural" boundaries of the three-dimensional domain). Given two arbitrary three-surfaces, whose geometries are asymptotically flat (or satisfy some other reasonable condition), there is no reason to assume that corresponding coordinate values must belong to points at a normal orientation to each other. Even in a Minkowski universe two hyperplanes may be shifted and rotated relative to each other.

There remains the question whether lack of uniqueness may not simply reflect a degree of arbitrariness in the choice of coordinates in the four-dimensional problem. We have not completed this investigation in its entirety; at present we have good reason for believing that this is not the case. To summarize, there is definite evidence, both by Komar's method and ours, that ${}^3R = 0$ represents a singular case. Even when ${}^3R \neq 0$, solutions are probably far from unique, and they cannot be made unique by the setting of universal and physically appealing boundary conditions.

The SC is intimately associated with the Feynman approach to quantization. If the "thin" sandwich could be filled classically without ambiguity, then one could associate with such a filling a non-zero action (though the classical Einstein action is always zero!) and proceed to decompose every conceivable four-dimensional manifold (even if not Ricci-flat) into Ricci-flat slices between consecutive three-geometries. The Feynman integral over histories would then, presumably, lead to a transition matrix element between any two given three-geometries. The consideractions presented here would seem to lead to the conclusion that this, otherwise attractive, scheme cannot work. Probably it is too early to decide this question definitively. Additional work would appear highly desirable.

Given two three-geometries that differ from each other only infinitesimally, there may be infinite ambiguity in filling the sandwich, but there must be a non-zero distance between them somewhere, no matter how hard one tries to close the gap between them by providing a connecting Ricci-flat four-geometry. Presumably, when one scans all possible sandwich fillings, there exists a lower bound for this distance, and the magnitude of this lower bound may be taken to represent an irreducible "distance" between the two three-geometries in superspace. Here appears to lie the possibility for another topology of superspace, and one that is adapted to the physical inquiry which arouses our interest in superspace in the first place. Again, much more thinking needs to be done before one feels sure of one's ground.

CLASSICAL AND QUANTUM DYNAMICS OF A CLOSED UNIVERSE*†

Charles W. Misner

University of Maryland

College Park, Maryland

This report will attempt to describe briefly a new technique in cosmological theory, and some applications, interpretations, and conjectures that can utilize it. The technique is the use of the Arnowitt, Deser, and Misner[1] (ADM) Hamiltonian methods for formulating and solving the Einstein equations for homogeneous cosmological models.[2] Among the applications are a simplified and more detailed presentation[3] of the Mix-Master Universe.[4] This is a closed universe in which it is possible[5] for light rays to travel completely around the universe in certain epochs near the initial singularity. Some numerical examples of the evolution of these models have been given by Okerson.[6] As another application, these techniques have allowed Ryan[7,8] to give a concise description of a closed expanding universe model which has rotation as well as shear, but does not employ a cosmological constant. This Hamiltonian technique also leads to models of quantized geometry[2] which should be useful as a testing ground for many ideas concerning the full quantum theory of curved space-time. The main application developed so far is a study of quantum effects on the initial singularity in closed universes which I have pursued in collaboration with Dr. Kenneth Jacobs and Professor Harold Zapolsky.[9] One interpretation of some of the results of these more general cosmological models is a new attitude toward the initial singularity. I have suggested[10,11] that the initial singularity is a useful, valid, and physical prediction of relativistic cosmology. At the

---

*Supported in part by NSF Grant GP8560 and NASA Grant NSG-21-002-010.

†Invited talk given June 3, 1969, at the Relativity Conference in the Midwest, Cincinnati, Ohio.

same time, I would argue that proper time is not the relevant parameter for discussing the early history of the Universe, and that in terms of a more appropriate and significant time parameter, the singularity occurred an infinite time in the past. Finally, these developments in cosmological theory suggest to me (and have been partly suggested by) a conjecture about the broad outlines of the history of the Universe. In its current form this conjectured history distinquishes an early quantum phase for the Universe containing the singularity, and a later classical phase including the present. According to this conjecture, the Universe could have begun in an essentially arbitrary quantum state near the singularity, but should evolve into a nearly homogeneous state by the time the radius expanded to about 10^{-33}cm. Thereafter the gravitational field would be classical but any remaining anisotropy could be eliminated prior to attainment of a radius of 1 sec = 3 x 10^{10}cm, giving rise to a Robertson-Walker Universe during the later epochs of Helium formation, galaxy formation, and the present. To substantiate this conjecture one must find, through the theoretical analyses of the quantum epochs, a natural way to give rise to the spectrum of irregularities which will be needed to develop later into galaxies, etc. Also one would like to explain some of the dimensionless 10^{40} numbers which relate cosmological and microphysical constants.

ADM HAMILTONIAN METHODS

The ADM[1] methods for restating Einstein's equations start by making some geometrically motivated notational changes to rewrite the Einstein variational principle $\delta I = 0$, where*

$$16\pi I = \int {}^4R(-{}^4g)^{1/2}\, d^4x. \tag{1}$$

The result of these changes is the formula

$$16\pi I = \int \{\pi^{ij} \frac{\partial g_{ij}}{\partial t} + N\sqrt{g}\ [{}^3R + g^{-1}(\frac{1}{2}\ \pi^2 - \pi^{ij}\ \pi_{ij})] + 2\, N_i\ \pi^{ij}{}_{|j}\}\ dt d^3x \tag{2}$$

where latin indices are spacial only $(i,j = 1,2,3)$ and $g = \det g_{ij}$. By varying π^{ij}, g_{ij}, N, and N_i independently one obtains both the ten Einstein equations and the definition of the π^{ij}. The N's serve both as metric components ($g_{oi} = N_i$, $g^{oo} = -N^{-2}$) and as Lagrange multipliers to give upon variation the constraints

$$g\ {}^3R + \frac{1}{2}\,(\pi^k{}_k)^2 - \pi^{ik}\ \pi_{ik} = 0 \tag{3}$$

$$-\,2\ \pi^{ik}{}_{|k} = 0 \tag{4}$$

*Units are chosen so $G = \hbar = c = 1$.

Because the N's can be thought of as Lagrange multipliers, the variational principle can be put in a much simpler appearing form

$$16\pi I = \int \pi^{ik} \dot{g}_{ik} \, d^4x \quad , \tag{5}$$

but here of course the π^{ik} and g_{ik} are not independent, but related by the constraints (3) and (4). For homogeneous cosmological models the constraints are just algebraic equations which one solves explicitly, and then the remaining independent variables can be varied in Equation (5) to obtain the equations of motion for the metric. More particularly, it is useful to choose the independent variables so as to obtain a canonical (Hamiltonian) form. This means that in the action integral $I = \int\omega$ the integrand ω must have the standard form

$$\omega = p_A \, dq^A - H(p,q,t)dt. \tag{6}$$

HAMILTONIAN COSMOLOGY

I want to apply the methods I have just summarized to two metrics, both of the form

$$ds^2 = - dt^2 + \frac{1}{4} R^2 (e^{2\beta})_{ij} \, \sigma_i \, \sigma_j \, . \tag{7}$$

Here R and the traceless diagonal matrix β are functions of t only. The invariant differential forms σ_i are chosen in the first case (Bianchi Type I, flat space-like hypersurfaces) to satisfy

$$\sigma_i = dx_i \quad , \quad d\sigma_i = 0 \tag{8}$$

and

$$\int d^3x = (4\pi)^2 \, . \tag{9}$$

This integral condition corresponds to imposing periodic boundary conditions on the x's in a way that makes all the numerical factors agree with the more interesting second case. In the second case (Bianchi Type IX, closed 3-sphere space-like hypersurfaces) the σ_i satisfy

$$d\sigma_i = \frac{1}{2} \varepsilon_{ijk} \, \sigma_j \wedge \sigma_k \tag{10}$$

and

$$\int \sigma_1 \wedge \sigma_2 \wedge \sigma_3 = (4\pi)^2 \, . \tag{11}$$

The consequence of the trace β=1 condition is that $\sqrt{g} = R^3/8$ so $R(t)$ describes the volume of the universe, and $\beta(t)$ its shape. We carry out the space integration in Equation (5) now to obtain

$$I = (16\pi)^{-1} \int \pi^{ij} \, dg_{ij} \wedge \sigma_1 \wedge \sigma_2 \wedge \sigma_3 \;\; = \pi \int \pi^{ij} \, dg_{ij} . \tag{12}$$

Now use $g_{ij} = \frac{1}{4} R^2 (e^{2\beta})_{ij}$ from Equation (7) to compute

$$dg_{ij} = 2 \, g_{ij} \, d\ln R + 2 \, g_{ik} \, d\beta_{kj} \tag{13}$$

so $I = (2\pi)\int [\pi^k{}_k \, d\ln R + \pi_i{}^k \, d\beta_{ki}]$. This is nearly a canonical form; let us treat $-\ln R$ as a time coordinate, then its coefficient $(2\pi)\pi^k{}_k$ will be a Hamiltonian. Also, since β_{ki} is traceless, only the traceless part of $\pi_i{}^k$ is effective as a coefficient of $d\beta_{ki}$. This motivates the definitions

$$H = (2\pi)\pi^k{}_k \ , \quad R = R_o e^{-\Omega}$$

$$p^i{}_k = (2\pi)\left(\pi^i{}_k - \frac{1}{3}\delta^i{}_k \pi^\ell{}_\ell\right) \tag{14}$$

which give

$$I = \int (p^i{}_k \, d\beta_{ki} - H d\Omega) \tag{15}$$

which is essentially a canonical form. For explicit computations it is often convenient to define two independent components for each of the traceless diagonal matrices $p^i{}_k$ and β_{ik} as follows

$$\beta = \begin{pmatrix} \beta_+ + \beta_-\sqrt{3} & \cdot & \cdot \\ \cdot & \beta_+ - \beta_-\sqrt{3} & \cdot \\ \cdot & \cdot & -2\beta_+ \end{pmatrix}$$

$$6p = \begin{pmatrix} p_+ + p_-\sqrt{3} & \cdot & \cdot \\ \cdot & p_+ - p_-\sqrt{3} & \cdot \\ \cdot & \cdot & -2p_+ \end{pmatrix} \tag{16}$$

Then in the action integral $I = \int \omega$ we have

$$\omega = p_+ \, d\beta_+ + p_- \, d\beta_- - H d\Omega \tag{17}$$

which is the standard canonical form of Equation (6) provided $p_\pm$, $\beta_\pm$, and Ω can be varied independently.

To see which of the $\pi^i{}_k$, g_{ik} variables are independent we must study the constraints (3) and (4). In $\pi^{ik}{}_{|k} = 0$ the stroke "|" indicates a covariant derivative using the space metric g_{ij} . For Type I, the space metric is flat, and the π^{ik} do not depend on the space coordinates so $\pi^{ik}{}_{|k}$ vanishes identically. A more difficult calculation shows that $\pi^{ik}{}_{|k}$ also vanishes identically for the Type IX metric provided $\pi^i{}_k$ and β_{ik} are both diagonal, as we have assumed. The momentum constraints (4) therefore impose no conditions in the present examples, and we turn to the energy constraint (3) which reads

$$g\,{}^3R + \frac{1}{2}(\pi^k{}_k)^2 - \pi^i{}_k \pi^k{}_i = g\,T_{**} \tag{18}$$

if there is matter present. This equation is easily solved to give $H = (2\pi)\pi^k{}_k$ in terms of the other variables. The scalar curvature ${}^3R$ of a hypersurface must be computed for the metrics under consideration and gives $^3R_I = 0$ for the Type I metric and

$$^3R_{IX} = \frac{6}{R^2}(1-V)$$

$$V(\beta) = \frac{1}{3}\text{trace}\,(1-2e^{-2\beta} + e^{4\beta}) \tag{19}$$

for the Type IX metric. We will choose

$$R_o^2 = 2G\hbar/3\pi c^3 \tag{20}$$

in Equation (14) to avoid numerical factors in the Hamiltonian H which results from solving Equation (18), namely (for Type IX),

$$H^2 = p_+^2 + p_-^2 + e^{-4\Omega}(V-1) + \mu e^{-3\Omega} + \Gamma e^{-2\Omega}\,. \tag{21}$$

The last two terms arise from an energy density of the form

$$T_{**} = -T^o{}_o = \frac{9}{16}(\mu e^{-3\Omega} + \Gamma e^{-4\Omega}) \tag{22}$$

corresponding to pressureless matter (μ) and fluid radiation (Γ). For the Type I metric in empty space this reduces to $H = (p_+^2 + p_-^2)^{1/2}$.

The solutions of the Type I Hamiltonian are just $d\beta_\pm/d\Omega = const$ since H is independent of $\beta_\pm$ and Ω. The solution with $\beta_- = const$, $d\beta_+/d\Omega = 1$ where there is expansion only along the z-axis of space, and two equivalent solutions (expanding x or y axes), have the unusual property that there are no particle horizons in the one preferred direction. The absence of particle horizons, so all parts of the universe could have interacted with each other arbitrary early in the expansion, would seem to open significant possibilities for understanding the gross homogeniety of our Universe.[4] We are therefore very interested in many details of the motion of the Type IX universe, some of which we describe below. The aim, ultimately, would be to see that the values of $d\beta_\pm/d\Omega$ change ergodically near the singularity, and that they come sufficiently close for sufficiently long periods to each of the three critical values that horizons will be eliminated in all directions for almost all initial conditions.

Let us then ignore the matter terms and study the approach to the singularity, $\Omega \to \infty$, $R \to 0$. With Ω large, the term $e^{-4\Omega}(V-1)$ can only contribute if $V >> 1$, so we can write

$$H = \sqrt{p_+^2 + p_-^2 + e^{-4\Omega}\,V(\beta_+, \beta_-)} \tag{23}$$

$$ds^2 = -\frac{2}{3\pi}\frac{e^{-6\Omega}}{H^2}\,d\Omega^2 + \frac{1}{6\pi}\,e^{-2\Omega}(e^{2\beta})_{ij}\,\sigma_i\,\sigma_j\,. \tag{24}$$

The metric form (24) above is valid in all cases we have considered; see ref. 2 for the derivation of the $g_{\Omega\Omega}$ component. Any solution of Hamilton's equations corresponding to (23) or (21), when inserted into Equation (24) gives a metric satisfying Einstein's

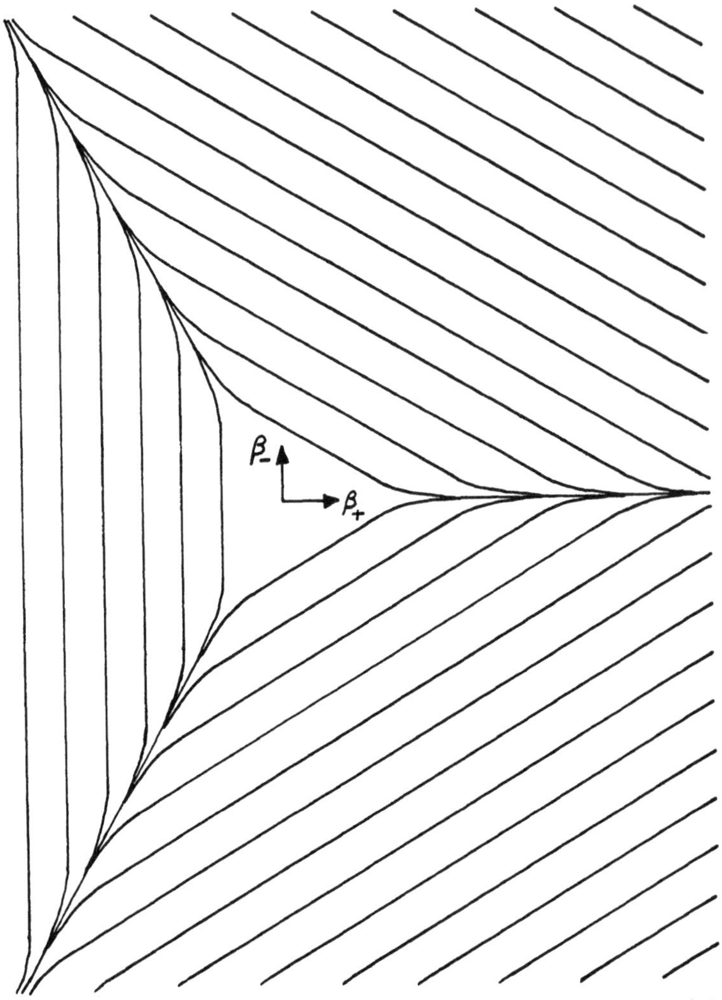

Figure 1. (above) The equipotentials of the function $V(\beta)$ defined in Equation (19) are here sketched from the asymptotic formulae (25) and (26). Equipotentials near the origin of the β-plane are closed curves for $V<1$ and are omitted here. Between the successive equipotentials which are drawn, and which have separations $\Delta\beta=1$, V increases by a *factor* of $e^8 \simeq 3 \times 10^3$.

Figure 2. This diagram from Okerson (ref. 6) shows which directions in the universe of Equation (7) are relatively expanded or compressed for various values of β by sketching, at appropriate positions in the β-plane, the ellipsoid $(e^{2\beta})_{ij}\, y^i\, y^j = 1$. The threefold symmetry of this diagram and of Fig. 1 arises since 120° rotations in the β-plane correspond merely to permuting the principle axes of the β-matrix in Equation (16). The dashed curve represents an interval in a possible history of a Type IX model universe, including eras of near symmetry about the 3-axis, and later about the 2-axis.

equations. The Hamiltonian problem here evidently corresponds to a particle moving in two dimensions in a time dependent potential well. We will build up a solution by considering successively several limiting cases. In the first limit, note that for $\Omega \to \infty$ the $e^{-4\Omega}V$ in Equation (23) should in first approximation be negligible. Then $H = \sqrt{p_+^2 + p_-^2}$ is independent of $\beta_\pm$ and Ω so $p_\pm$ and H are constants of motion. The Hamilton equations $d\beta_\pm/d\Omega = \partial H/\partial p_\pm = p_\pm/H$ show that $\beta'_\pm \equiv d\beta_\pm/d\Omega$ are then constants satisfying $|\beta'|^2 = \beta_+'^2 + \beta_-'^2 = 1$. This approximation will fail if V is sufficiently large, so let us study its asymptotic form. For $\beta_+ << -1$ this is

$$V(\beta) \sim \frac{1}{3} e^{-8\beta_+} \quad , \beta_+ \to -\infty \; . \tag{25}$$

Figure 1 shows this asymptotic form by sketching equipotentials of V. Note from Equations (16)and (24) that $\beta_+ \to -\infty$ corresponds to a stretching of the 3-axis (cigar) relative to the others; the other sides of the triangular equipotentials in the β -plane corresponds to preferential stretching of the other two axes. This is indicated on Okerson's[6] diagram, Figure 2. Another asymptotic

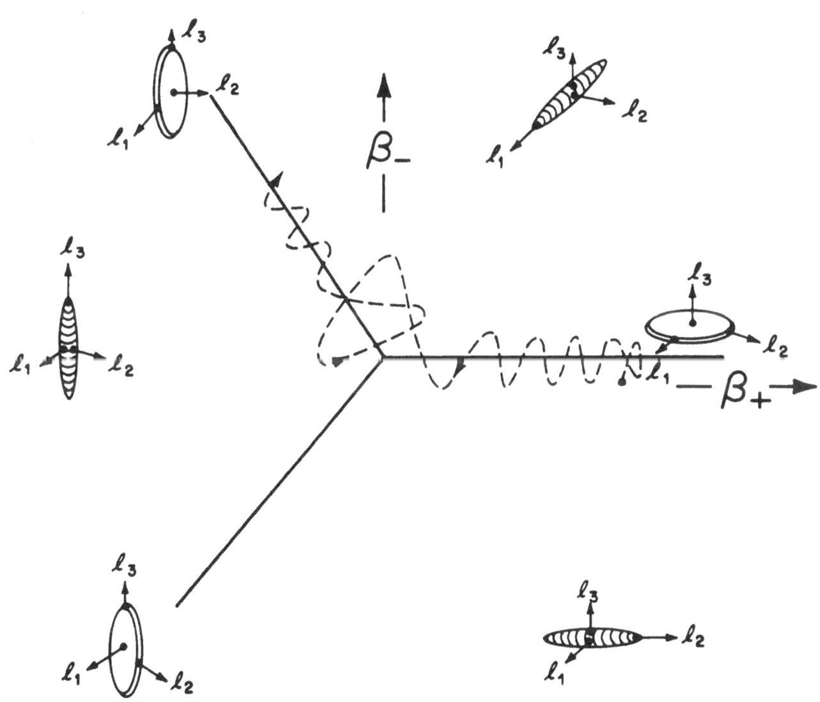

Figure 2

form

$$V(\beta) \sim 1 + 16\,\beta_-^2\, e^{4\beta_+}\,,\quad \beta_+ \to +\infty\,,\quad |\beta_-| \ll 1 \tag{26}$$

corresponds to a relative compression of the 3-axis (pancake), with the other two axis approximately equal, and describes the corners of the triangular equipotentials. The limiting equipotential (the potential "wall") is one which would demand $\beta' = 0$. The Hamilton equations $\beta'_\pm = \partial H/\partial p_\pm = p_\pm/H$ allow Equation (23) to be rewritten as

$$1 = \beta_+'^2 + \beta_-'^2 + H^{-2}\, e^{-4\Omega}\, V \tag{27}$$

so the condition $\beta' = 0$ gives

$$V(\beta_{wall}) = H^2\, e^{4\Omega}. \tag{28}$$

The Ω dependence of H is given by the Hamilton equation $dH/d\Omega = \partial H/\partial\Omega$ which can be rewritten as

$$\frac{d\ln H^2}{d\Omega} = -\,4(1 - \beta'^2). \tag{29}$$

Thus H^2 is nearly constant when $\beta' \simeq 1$ and β is far from the wall of the potential. Under these conditions it is easy to find the motion of the potential walls (and corners) from Equation (25) and (26). The results, sketched in Figure 3, are that the sides of the limiting triangle move outward at velocity $|d\beta_{wall}(d\Omega)| = 1/2$, while the corners correspondingly move at velocity 1.

Since the solution point $\beta(\Omega)$ moves with velocity $\beta' = 1$ when away from the walls, it can catch up with the more slowly moving potential walls and bounce off of them. JMZ[9] found the analysis of this bounce much more transparent using these Hamiltonian methods than with previous techniques.[4,12] Consider the wall at negative β_+, so $e^{-4\Omega}V \sim 1/3\; e^{-4(\Omega+2\beta_+)}$, which suggests the introduction of a new coordinate $b_+ = \beta_+ + (\Omega/2)\Omega$ of position relative to the moving wall. When this b_+ is introduced in the canonical form

$$\begin{aligned}\omega &= p_+\, d\beta_+ + p_-\, d\beta_- - H d\Omega \\ &= p_+\, db_+ + p_-\, d\beta_- - (H + \tfrac{1}{2}\, p_+) d\Omega \end{aligned} \tag{30}$$

we see that

$$K_{1/2} \equiv \frac{1}{2}\, p_+ + H = \frac{1}{2}\, p_+ + \sqrt{p_+^2 + p_-^2 + \frac{1}{3}\, exp(-8b_+)} \tag{31}$$

is a new Hamiltonian with b_+ canonically cojugate to p_+. Since this Hamiltonian $K_{1/2}$ is independent of β_- and Ω there are two corresponding constants of the motion, p_- and $K_{1/2}$. These two constants allow us to determine β' and H after the bounce in terms of their values before. Of course $|\beta'| = 1$ asymptotically both before and after. Ryan, who gave the first derivation of

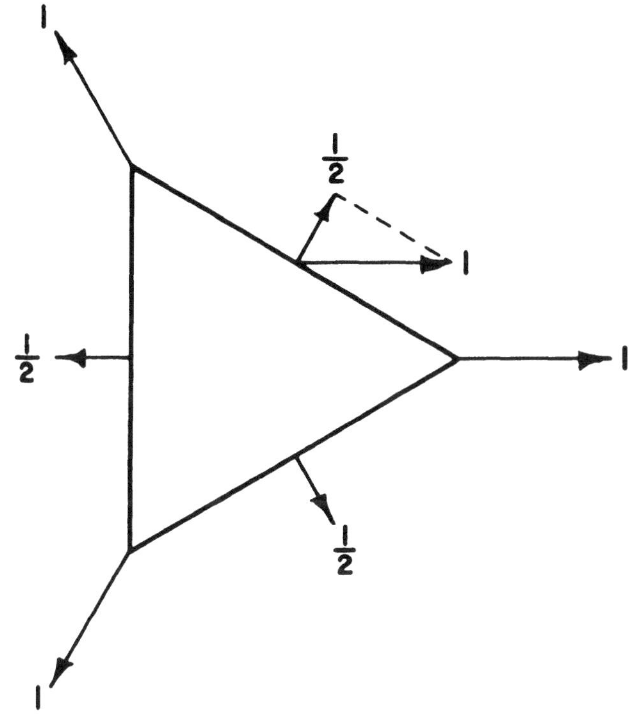

Figure 3. The velocities of the potential walls. From Equation (27) we see that the β-point representing the shape of the universe must (since $\beta'^2 > 0$) always lie inside a certain equipotential which we call the "potential wall" which is defined by $e^{-\Omega}\, V(\beta_{\text{wall}}) = H^2$. The Ω-dependence of this defining equation causes this "potential wall" to move. The simplest and most typical case occurs when the configuration β of the universe is far inside this wall, for then $\beta' \simeq 1$ and Equation (29) shows that $H \simeq$ *const*. For this case the motion of the potential wall is as indicated by the velocity $(d\beta_{\text{wall}}/d\Omega)$ arrows in this diagram. [The shape of the equipotential has been drawn as an equilateral triangle in place of the more accurate form given in Fig. 1] Note that moving a line perpendicular to itself at a velocity 1/2 is equivalent to sliding it at a 30° angle to itself at velocity 1. Thus the corners of the potential wall move outward at velocity 1, while the sides of the triangular potential wall move outward at velocity 1/2.

the bounce law for H, writes the relations as follows (cf. Fig. 4)

$$\sin\theta_f = \frac{3\sin\theta_i}{5-4\cos\theta_i}$$

$$\frac{H_f}{H_i} = \frac{\sin\theta_i}{\sin\theta_f} = \frac{5-4\cos\theta_i}{3} \quad . \tag{32}$$

Lifshitz and Khalatnikov[12] who first computed the bounce law for θ have a way of parameterizing these angles, $\theta = \theta(u)$, for which they find the very convenient relationship $u_f = u_i - 1$.

In case the initial trajectory $\beta(\Omega)$ is aimed at a corner, rather than a side, of the triangular potential well, different considerations are required. Then the appropriate asymptotic form is $e^{-4\beta}V \sim 16\beta_-^2\, e^{4(\beta_+ - \Omega)}$ so one defines $B_+ = \beta_+ - \Omega$ and finds

$$\omega = p_+\, dB_+ + p_-\, d\beta_- - K_1\, d\Omega$$

$$K_1 = -\,p_+ + \sqrt{p_+^2 + p_-^2 + 16\beta_-^2\, exp(4B_+)} \ .$$

Following my analysis[4] of the solution in this case using adiabatic invariants of the β_- oscillations, Chitre[5] found that an analytic solution could be obtained, and the JMZ[9] type canonical transformation

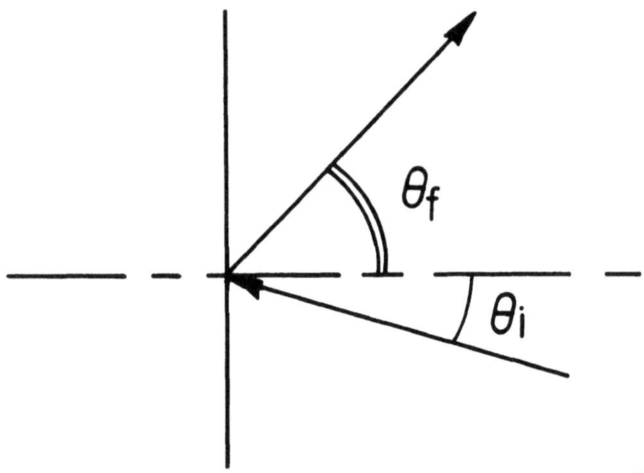

Figure 4. The angles θ_i and θ_f are the angles of incidence and of reflection for a "bounce" from one of the three equivalent walls of the triangular potential $V(\beta)$. Because this wall moves to the left with speed $|\beta'_{wall}| = 1/2$ while the system point $\beta(\Omega)$ moves with speed $|\beta'| \leq 1$, one has a limit $|\theta_i| < 60°$ in order for a bounce against this wall to occur. In other cases the bounce will occur on a different wall (not shown). Note also that $\theta_f > 90°$ is possible because of the motion of the wall.

given above leads to it very quickly. Hamilton's equations from Equation (33) are $B_+' = -1 + (p_+/H) = -K_1/H$, $\beta_-' = p_-/H$, and $p_-' = 16\beta_- e^{4\beta_+}/H$. Using Chitre's trick of introducing B_+ as a new independent variable, and using the fact that K_1 is a constant of motion, then leads easily to the differential equation

$$\frac{d^2\beta_-}{d\beta_+^{\,2}} + \left(\frac{4e^{2\beta_+}}{K_1}\right)^2 \beta_- = 0 \tag{34}$$

which has solutions in terms of Bessel functions

$$\beta_- = Z_o(2e^{2B_+}/K_1) \tag{35}$$

describing the orbit in the β-plane as β drifts slowly out of the corner (B_+' is small and negative). Thus the β_- amplitude grows as B_+ decreases, until when $|\beta_-| > 1$ the motion is controlled again by the flat sides of the potential and the bounce laws (32) apply. Figure 5 shows the results of a numerical integration of the differential equations by Okerson[6] corresponding to this situation. Figure 6 is qualitatively the same situation, except that the numerical approach is more essential here since the approximation $|\beta_-|^2 \ll 1$ needed to obtain Equation (34) would not apply. Figure 7 shows another numerical integration with initial conditions which produce several "bounces" of the type governed by Equations (32).

A simple orbit which can be studied for many successive bounces isaquasi-periodic orbit for which θ_i is the same for every bounce. The condition for this is $\theta_i + \theta_f = 60^o$, which in Equation (32) leads to $\theta_i = 15.5^o$. This simple but unstable quasi-periodic orbit is sketched in Figure 8. Two legs of this orbit are shown in more detail in Figure 9 from which one deduces that the Ω-time required for the runs before (Ω_i) and after (Ω_f) the bounce are related by $\Omega_i \sin\theta_f = \Omega_f \sin\theta_i$. Comparing this to Equation (32) for H_f/H_i we see that

$$H_i\,\Omega_i = H_f\,\Omega_f \quad . \tag{36}$$

This statement implies that $H\Omega$ is an adiabatic invariant for this motion. The behavior of $H\Omega$ for this particular quasi-periodic orbit is sketched in Figure 10, but one expects that $H\Omega$ is also an adiabatic invariant for the general orbit, i.e. a quantity which may jitter a bit but which maintains a fixed average value over the long run.

In the quantum theory one finds that the eigenvalues of the Hamiltonian (23) are, for large Ω,

$$E_{mn} \propto \Omega^{-1}\sqrt{m^2 + mn + n} \tag{37}$$

Therefore the fact that $H\Omega$ is an adiabatic invariant in the classical theory insures that the quantum number m and n are, on the average, time-independent in the quantum theory (at least for large m, n so

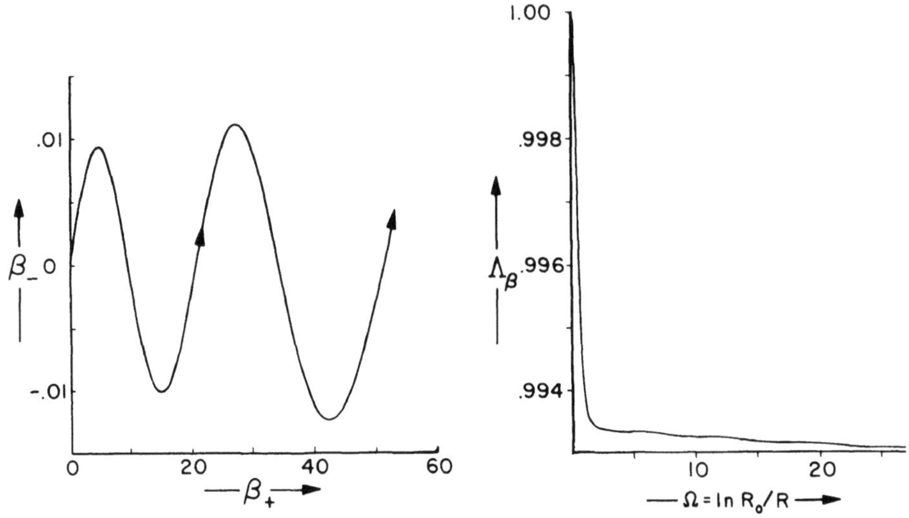

Figure 5

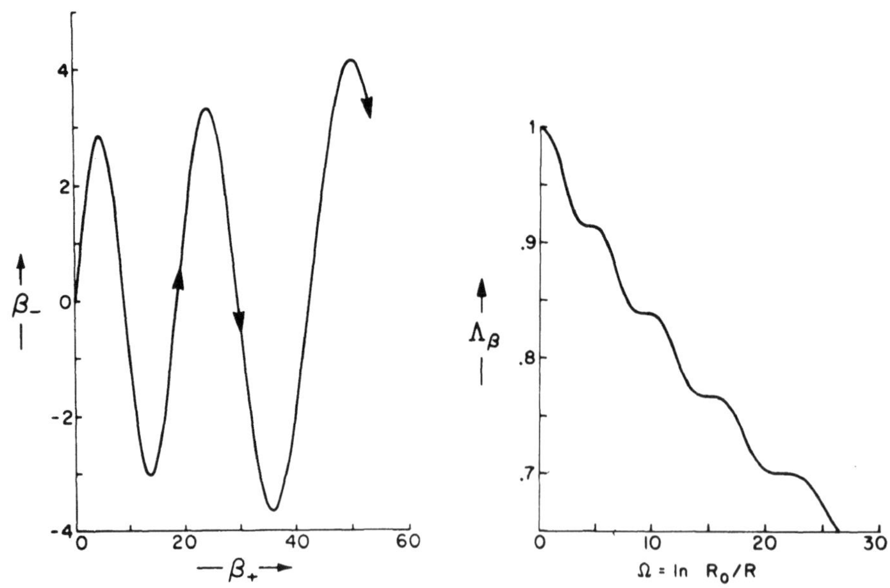

Figure 6

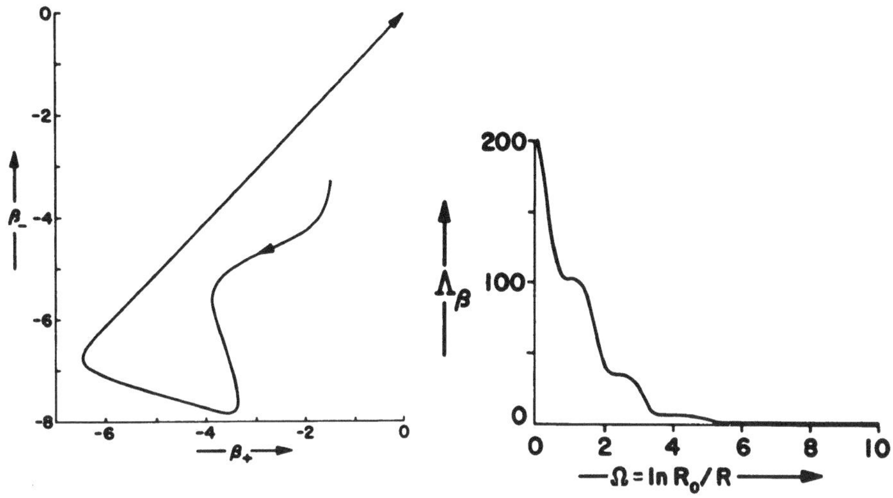

Figure 7 (above). Like Figs. 5 and 6 on the preceeding page, these graphs show solutions obtained by Okerson[6] for $\beta(\Omega)$ and $\Lambda_\beta \equiv H^2(\Omega)$ in a Type IX empty universe. The initial conditions for this solution did not have a high degree of symmetry, so after a few bounces near one corner of the expanding potential wall, the system point moves off a constant velocity toward the opposite side of the potential.

Figure 5 (above, opposite page). This and the next two Figures are taken from Okerson (ref. 6) who used a slightly different notation (from ref. 4 where Λ_β was used for the present H^2, and $\beta_\pm$ for the present $2\beta_\pm$). The solution shown here has initial conditions which make Equations (34) and (35) applicable, and shows a universe in which the 1-axis and the 2-axis are nearly equivalent with their small difference, β_-, oscillating with a slowly increasing amplitude.

Figure 6 (below, opposite page). The initial conditions Okerson has chosen for this solution could represent a much later stage in the oscillation about axial symmetry shown in Fig. 5. The amplitude is now too high for Equations (26) and thus (35) to be applicable, but the qualitative behavior appears similar. The right-hand graph of $\Lambda_\beta \equiv H^2$ shows clearly the decreases of H^2 when the potential becomes important (maxima and minima of β_-), with periods of constant H^2 intervening when the speed $|\beta'|$ acheives its maximum, corresponding to our expectations from Equation (29). The oscillation amplitude will continue to increase until a behavior like that shown in Fig. 7 terminates the era of near axial symmetry.

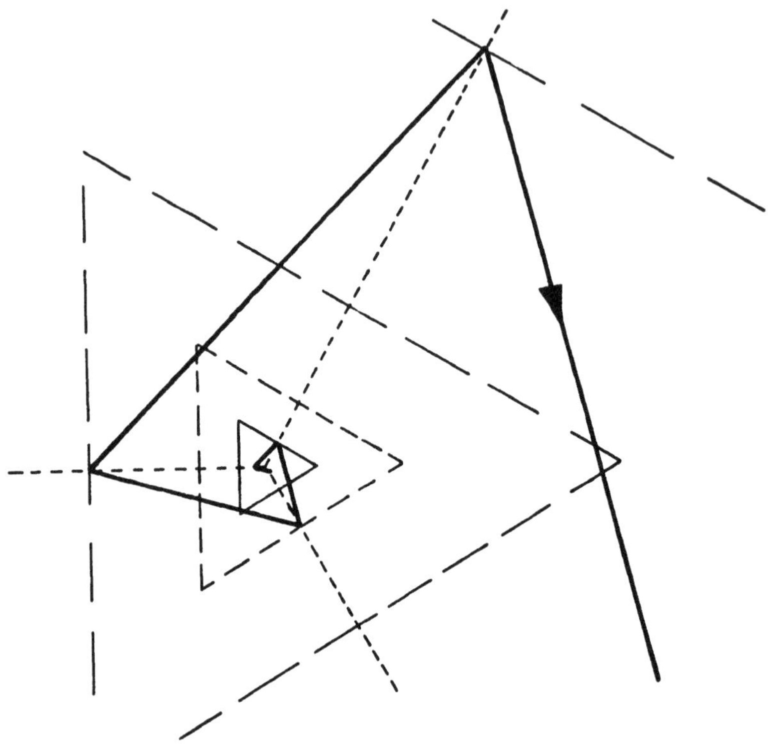

Figure 8. The simplest quasi-periodic orbit for the Hamiltonian of Equation (23) is sketched here in the β-plane in an asymptotic limit, $\Omega \to \infty$, in which the curved portions of the trajectory $\beta(\Omega)$ are on such a small scale compared to the straight portions that they appear as sharp cornered "bounces." The initial conditions are chosen so that $\theta_i = 15.5°$ on one bounce (cf. Fig. 4), and the bounce law of Equation (32) then gives $\theta_i = 15.5°$ for every subsequent bounce. The dashed line triangles indicate equipotentials of $V(\beta)$ at which the bounces occur.

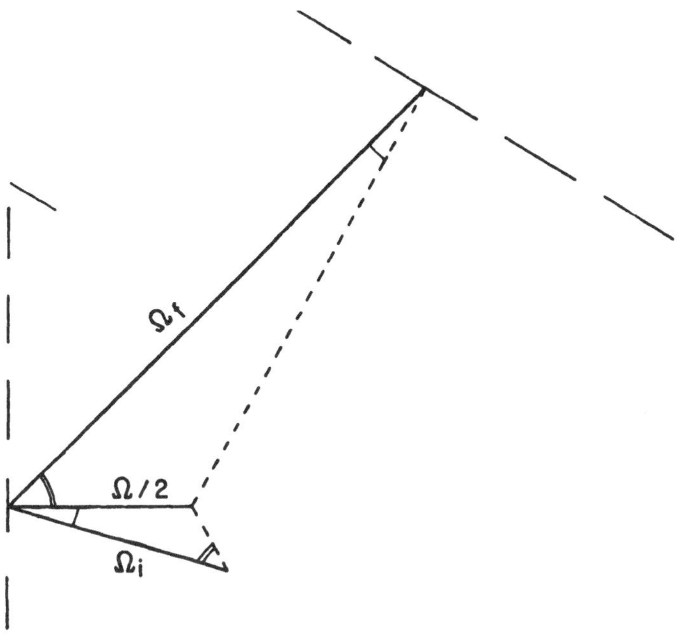

Figure 9. Two legs of the trajectory in Fig. 8 are redrawn here. Note that they form corresponding sides of two similar triangles. Since $|d\beta/d\Omega| = 1$, the lengths $\Delta\beta$ of these legs are just equal to the Ω-time intervals $\Delta\Omega$ spent traversing them, thus $\Delta\Omega = \Omega_i$ before the bounce and $\Delta\Omega = \Omega_f$ after. Because the potential wall against which the bounce occurs moves at velocity $d\beta_{wall}/d\Omega = 1/2$, the common side of these two triangles can be taken to have length $\beta_{wall} = \Omega/2$. The smallest angle in the triangles is $\theta_i = 15.5^o$, the largest is 120°, and the third angle (marked by a double arc) is $\theta_f = 44.5^o$. The law of sines then gives $(\Omega/2)\sin 120^o = \Omega_i \sin\theta_f = \Omega_f \sin\theta_i$ for the two different triangles, which is the relationship we required in deriving Equation (36).

the correspondence principle applies). From this one concludes that if the universe is now in a classical (high quantum number) state, then an extrapolation back to arbitrarily small size near the singularity, $\Omega \to \infty$, leaves it in a classical state. We will return to this question shortly.

Ryan's Rotating Model

The Hamiltonian methods we have just outlined have allowed Michael Ryan[7,8] to give an analytical description of a closed expanding universe containing pressureless fluid matter (dust) which has both shear and rotation. No cosmological constant is used in this model. The metric is again given by Equation (24), but with a non-diagonal β matrix:

$$\beta = \begin{pmatrix} \beta_+ + \sqrt{3}\,\beta_- \cos 2\psi, & \sqrt{3}\,\beta_- \sin 2\psi, & 0 \\ \sqrt{3}\,\beta_- \sin 2\psi, & \beta_+ - \sqrt{3}\,\beta_- \cos 2\psi, & 0 \\ 0, & 0, & -2\beta_+ \end{pmatrix} \tag{38}$$

In this case the momentum constraints $-2\pi^{ik}{}_{|k} = 8\pi T_*{}^i$ are non-trivial and allow the momentum conjugate to ψ to be eliminated in terms of a constant C characterizing the vorticity of the fluid flow. The resulting Hamiltonian is given by

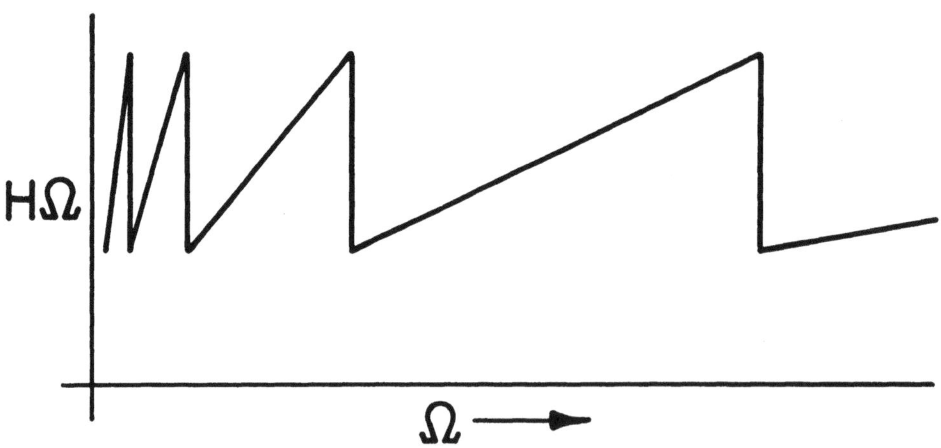

Figure 10. The adiabatic invariant $H\Omega$ is sketched here for the quasi-periodic orbit shown in Figures 8, 9. Although $H\Omega$ is not strictly constant it may be regarded as constant for the purpose of estimating long term secular changes.

$$H^2 = p_+^2 + p_-^2 + \frac{(\mu C)^2}{8\sinh^2(2\sqrt{3}\,\beta_-)} + e^{-4\Omega}(V-1)$$
$$+ \mu e^{-3\Omega}\sqrt{1+(2\ C)^2 e^{2\Omega} e^{4\beta_+}} \qquad (39)$$

The constant μ, as in Equations (21) and (22), characterizes the total amount of matter in the universe. Solutions $\beta(\Omega)$ of Hamilton's equations give $\psi(\Omega)$ from a simple integral

$$\frac{d\psi}{d\Omega} = \frac{\mu C}{8H\sinh^2(2\sqrt{3}\,\beta_-)} \qquad (40)$$

and thus determine the metric $g_{\mu\nu}(\beta,\psi,\Omega)$. The solutions of Hamilton's equations for Equation (39) can be described by the same methods employed in the previous case. The two new terms give rise to two additional potential walls as indicated schematically in Figure 11. The "centrifugal" walls parallel to the β_+ axis arise from the term $(1/6)(\mu C/\sinh 2\sqrt{3}\beta_-)^2$ and are esentially stationary since this term has no explicit time dependence. (They move out slowly as $H \sim \Omega^{-1}$ gradually decreases.) The wall parallel to the β_- axis arises from the square root, kinetic energy of matter, term

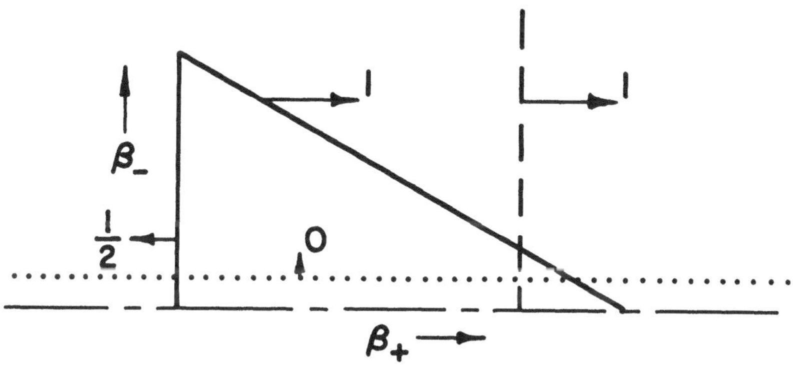

Figure 11. The potentials for Ryan's rotating model are sketched in the same way as was done for the empty model in Figure 3. It is sufficient to consider only the region $\beta_- > 0$ in the present case since a "centrifugal" potential indicated by a horizontal dotted line (wall velocity is zero when H = *const*) prevents the system point $\beta(\Omega)$ from crossing the β_+ axis. The heavy lines are the triangular walls from $V(\beta)$ with the same velocities as before. Finally a term arising from T^{oo}, which includes the kinetic energy of the fluid vorticity, gives rise to a potential wall parallel to the β_- axis, shown by a dashed line, which moves with velocity 1 when H = *const*.

and moves to the right with unit velocity. Significant rotations of the principle axes of the β matrix occur only when the solution point $\beta_{\pm}(\Omega)$ bounces against the horizontal "centrifugal potential".

QUANTUM COSMOLOGY

Because Einstein's equations for these homogeneous cosmological models have been put in Hamiltonian form, it is possible to impose canonical commutation relations and go on to consider the resulting quantum theory. For the cosmological model defined by Equations (23) and (24) one sets

$$p_{\pm} = -i(\partial/\partial\beta_{\pm}) \tag{41}$$

so that H^2 becomes a self-adjoint operator. Since H^2 is positive (at least in the approximation of Equation 23 where $V-1$ is replaced by V), its square root H is also a well defined self-adjoint operator on the space of square-integrable functions of $\beta_{\pm}$. Because H^2 corresponds, in Schroedinger theory, to the Hamiltonian for a particle in a triangular potential well with steep sides, one can guess reliably that its quantum states (except some of the lowest) will correspond closely to the energy levels for a particle is a square box, familiar from the statistical mechanics of an ideal gas. In this analogy Ω is just a parameter fixing the volume (area in two dimensions) of the box. Thus the eigenvalues of H are just $E_n \simeq (2\pi/3^{3/4})|n|\Omega^{-1}$ where $|n|$ is a quantum number which would be $(n^2 + m^2)^{1/2}$ for a square box. The time dependence of $|n|$ for large $|n|$ can be deduced from the correspondence principle as indicated following Equation (37), where we argued that $|n|$ will not change significantly as $\Omega \to \infty$. In this model then, quantum effects at the singularity $\sqrt{g} \to 0$, $\Omega \to \infty$, are not important.

This model theory of quantized relativity deserves more serious study for several reasons: (a) for its implications concerning the nature of the initial singularity; (b) as a pilot plant for discovering and developing new techniques, and for testing and comparing old techniques intended for use in the full theory of quantized gravitational fields; and (c) as a calculable model where concepts and interpretations concerning quantized geometry may be tried out. Not much of this has yet been done, since the Hamiltonian (21) was only discovered in April this year.[13] It is, however, possible to make a few remarks concerning the singularity. Evidently we only learn something about the singularity from this model to the extent the model parallels the full quantum field theory of gravity, yet the model has only two degrees of freedom (β_+, β_-) rather than two per space point as in the full theory. If we regard the model as retaining the two modes of minimum wave number, then most of the omitted modes have very high wave number $K \to \infty$. These short wave length modes have been studied in the classical theory[14] and one finds they act exactly like photons. But a photon gas contributes a term proportional to $(\text{volume})^{-4/3}$ to the energy density, while the homogeneous anisotropy modes $\beta_\pm$ contribute[15] a term proportional to $(\text{volume})^{-2}$, so the contribution from the high modes is negligible near the singularity. This may indicate that we have found one of those fortunate points in the development of physics where one can ignore a host of difficult problems and find further on a manageable problem again. Thus in the study of bulk matter, one would not want to follow a study of crystalline solids with a theory of the quantum structure of liquids as a prerequisite to a study of the still higher energies involved when the same substance boils and becomes a gas. Similarly, the extreme conditions of temperature inside a star actually simplify the theory of stellar structure as compared to the structure of planets where the pressures are much lower. It is appropriate then to try to study the models which present theories of physics allow for the origins of the Universe, to see whether they support in more detail the initial indications that the detailed properties of matter may not play a critical role near the singularity. If this turns out to be the case, it will be significant to study cosmological models even closer to the singularity than the limit $T \le 10^{12}\ {}^\circ K \simeq 100$ Mev imposed by our inability to describe accurately the properties of matter at temperatures above the π-meson rest mass, and our nearly complete ignorance of the properties of matter at energies well above one Gev.

In the hopes, then, of learning something about the origins of the Universe, but with good expectations that we could shed light on some aspects of the classical and quantum theories of gravity, Dr. Jacobs, Professor Zapolsky, and I[9] set out to study some details of the model quantum theory defined by the Hamiltonian of Equation (23). The first step is to think of it as $H = \sqrt{p_+^2 + p_-^2 + V(\beta,\Omega)}$ where the time dependent potential $V(\beta,\Omega)$ is then approximated by a hard-walled potential, $V = 0$ or ∞, with moving walls. For instantaneous eigenfunctions and eigenvalues, $H\phi_n = E_n\phi_n$ is equivalent to $H^2\ \phi_n = E_n^2\phi_n$, which just requires finding eigenfunctions of the two-dimensional Laplacian, $H^2 = p_+^2 + p_-^2 = -(\partial^2/\partial\beta_+^2) - (\partial^2/\partial\beta_-^2)$ with a boundary condition $\phi_n = 0$ on a triangular boundary. Searching the wave-quide literature, we found (i.e. were directed to by Professor J. Weber) the solution of this problem in a book by Schelkunoff.[16] We quote here only the eigenvalues

$$E_{mn} \propto \Omega\ \sqrt{m^2+mn+n^2} \tag{42}$$

where m and m are integers. For the time-dependent problem however, an expansion in these eigenfunctions did not lead to any easy solutions.

Next we considered wave packets bouncing around inside the expanding box, but in the first instance, of course just plane waves.

The condition that a sum of plane waves

$$\psi = \exp[-iK\Omega + i\underset{\sim}{K}\cdot\underset{\sim}{\beta}] - \exp[-i\tilde{K}\Omega + i\underset{\sim}{\tilde{K}}\cdot\underset{\sim}{\beta}] \tag{43}$$

vanish on a moving boundary, $\beta_+ = -(\Omega/2)$ leads to just the classical bounce laws (32), with $K = H$, $K_+ = H\cos\theta$, $K_- = H\sin\theta$. The plane waves in Equation (43) are equally well solutions of the Schroedinger equation $H\psi = i\partial\psi/\partial\Omega$, or the Klein-Gordon equation $H^2\psi = -\partial^2\psi/\partial\Omega^2$. However, for a one dimensional wave packet obtained by doing a Fourier integral of Equation (43) over the single parameter $K = |\underset{\sim}{K}|$ one finds that the reflected pulse does not have the same total probability $\int\psi^*\psi d\beta$ as does the incident pulse. This lack of unitarity can only follow from using a non-hermetian H in the Schroedinger equation, and shows that the boundary condition $\psi = 0$ used to obtain Equation (43) does not adequately reflect some subtleties of the operator theory involved in defining $\sqrt{-\nabla^2}$ as a self-adjoint operator. We abandoned the Schroedinger equation, then, upon finding that the Klein-Gordon equation presents no comparable computational difficulties. No square root appears in the Klein-Gordon equation, of course, and the Klein-Gordon definition of probability density

$$\rho = \frac{i}{2}\left(\psi^*\frac{\partial\psi}{\partial\Omega} - \psi\frac{\partial\psi^*}{\partial\Omega}\right) \tag{44}$$

does give a conserved probability (unitarity) for the wave function of Equation (43).

Evidently many aspects of the expanding potential well problem should first be studied in one dimension, rather than in two. Therefore we studied the one dimensional problem as defined in Figure 12.

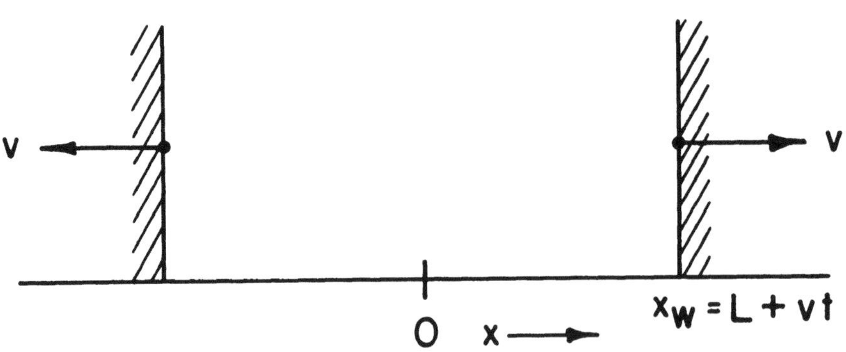

Figure 12. The time-dependent potential of a one-dimensional analogue problem is sketched here. It is a "square-well" potential with $V = 0$ the base of the potential well, and with infinitely high potential walls. The walls move outward with velocity v as indicated.

where the infinite potential walls are located at $x_w = \pm(L+vt)$; we use $H^2 = p^2 = -\partial^2/\partial x$. For the bounce law one finds

$$H_f/H_i = (1-v)/(1+v) \tag{45}$$

which is just what one finds for the energy of a photon reflected from a moving mirror. In fact all our previous calculations of bounces from a moving wall in two dimensions can be repeated for arbitrary wall velocity, and one always finds the relativistic result for a photon. This should not be too surprising since the Hamiltonian for a special relativistic free particle is just $H = \sqrt{p^2+m^2}$, and our Hamiltonians are just the $m = 0$ case. Our problem then is just that of a photon in an expanding box, and we would like to verify the adiabatic result $\lambda \propto L$, or $H \propto 1/L$ where L is the linear scale of the box. Does this conclusion hold for high velocities, $L = vt$, with $v \to 1$? In one dimension the detailed classical particle problem can be solved, and one finds $\langle H\Omega\rangle$ = const as in Figure 10 for all $v < 1$. (If $v > 1$ there are no bounces and $H = const$.) This physical analogy strengthens the case for assuming $\langle H\Omega\rangle = const$ for the triangular expanding well in two dimensions, since all points except the corners expand at less than light velocity, $v < 1$. In one dimension one can go on to find exact "expanding eigenstates". Thus

$$\begin{aligned}\psi_n(x,t) = \exp\{-i\omega_n \ln[(1+v)(\bar{t}-x)]\} \\ -\exp\{-i\omega_n \ln[(1-v)(\bar{t}+x)]\}\end{aligned} \tag{46}$$

satisfies both the Klein-Gordon equation $-(\partial^2\psi/\partial t^2) + (\partial^2\psi/\partial x^2) = 0$, and the boundary conditions $\psi = 0$ at $x = \pm v\bar{t}$, with $\bar{t} = t + (L/v)$, provided

$$\omega_n = n\pi/\ln\left(\frac{1+v}{1-v}\right), \quad n = 1,2,\ldots. \tag{47}$$

In the limit $v \to 0$ at fixed x,t one recovers the usual energy levels and wavefunctions for a static infinite square well in one dimesnion, so one expects this set of states to be complete. The probability-density function from Equation (44) shows more clearly than the wave functions that these states have a stationary shape while expanding at the same rate as the box:

$$\rho \propto \frac{n}{\bar{t}(1-\zeta^2)} \sin^2\left\{\frac{n\pi}{2}\left[1 - \frac{\ln\left(\frac{1-\zeta}{1+\zeta}\right)}{\ln\left(\frac{1-v}{1+v}\right)}\right]\right\} \tag{48}$$

where $\zeta = x/\bar{t}$. The eigenvalue n or ω_n must correspond to a constant of the motion in this problem of the type Komar[17] discussed here in his proposals for quantizing gravity. The existence of these expanding eigenstates corresponds to the validity of the adiabatic law $\langle H\Omega\rangle$ = constant in the classical problem, and we are encouraged to look for similar states in the two dimensional triangular problem.

Quantized Robertson-Walker Universes

The first study of quantum cosmology is due to DeWitt[18] who considered a closed Robertson-Walker (R-W) universe from the point of view of Dirac's Hamiltonian methods for general relativity. From the ADM viewpoint[1] the R-W universes are anomolous since they are determined entirely by the Friedmann equation (the T^{oo} Einstein equation for these models) which is a constraint equation. Thus in the ADM view, an R-W universe presents a Hamiltonian problem with no canonical coordinates or momenta, i.e. no independent dynamical degrees of freedom. Therefore after I became familiar with some anisotropic cosmological models,[15] I concluded that they would be more suitable as models for quantum gravity than the R-W models used by DeWitt, and eventually began to study them with the results described above. With the background provided by these models, however, we can return to the R-W case. Setting $p_{\pm} = 0 = \beta_{\pm}$ in Equation (21) gives the ADM Hamiltonian for R-W models containing pressureless matter and fluid radiation,

$$H = \sqrt{\Gamma e^{-2\Omega} + \mu e^{-3\Omega} - k e^{-4\Omega}} \tag{49}$$

where $k = 0, \pm 1$ gives the space curvature. The only Hamilton equation is $dH/d\Omega = \partial H/\partial\Omega$. We can, nevertheless, write down the quantum theory as the Klein-Gordon equation $-\partial^2\psi/\partial\Omega^2 = H^2\psi$. In the case of a radiation dominated universe ($\mu = 0 = k$) this reduces to

$$-\frac{d^2\psi}{d\Omega^2} = \Gamma e^{-2\Omega}\psi \tag{50}$$

which has a solution in terms of Bessel functions

$$\psi = Z_o(\Gamma^{1/2}\, e^{-\Omega}) \tag{51}$$

The solution representing an expanding universe (positive frequencies only in terms of a time coordinate $\alpha = -\,\Omega$ which increases during the expansion) is a Hankel function

$$\psi = H_o^{(+)}(\Gamma^{1/2}\, e^{-\Omega}) \tag{52}$$

and the Klein-Gordon probability (44) then reduces to the Wronskian for the two independent solutions $H_o^{(+)}$ and $H_o^{(-)} = H_o^{(+)*}$ of Equation (50), which is a constant. Thus

$$\rho(\Omega) = 1, \tag{53}$$

so the probability that this universe attains any given volume $\sqrt{g} = e^{-3\Omega}$ is $\rho(\Omega) = 1$, independent of Ω! The differential equations look more complicated if we choose $k = +1$, so there will be a maximum of the expansion. In this case one expects ψ for $\Omega \to \infty$ to be a linear combination of $H_o^{(+)}(\Gamma^{1/2}e^{-\Omega})$ and $H_o^{(-)}(\Gamma^{1/2}e^{-\Omega})$ with the one term representing the expansion at the beginning of the universe the other the contraction at the end. One could, of course, then impose a boundary condition $\psi \to J_o(\Gamma^{1/2}e^{-\Omega})$ as $\Omega \to \infty$ and thus obtain quantum conditions on the possible values of Γ. This seems to give a closer parallel to DeWitt's results, but there seems no compelling reason to impose this additional condition in the present context. Note that it makes the wave-function finite as $\Omega \to \infty$, but not zero. It gives $\rho(\Omega) \to 0$ for $\Omega \to \infty$, but confusesthe classical limit of this quantum theory by not letting us think of the $H_O^{(+)}$ and $H_O^{(-)}$ components of ψ as the distinct expanding and contracting phases of the evolution of this closed universe. [Mr. L. Fishbone is attempting to find some models of expanding universes which do contain unconstrained dynamical variables and in which a time coordinate can be introduced which can be used at the maximum of expansion. Ω-time is of course singular as a coordinate at the maximum of expansion where by definition $d\Omega = 0$.]

In addition to some differences in viewpoint which the experience of anisotropic models has helped us to take, there are also computational differences in detail between the preliminary results for R-W universes I have given above, and DeWitt's calculations. These, I believe, amount to a different choice of factor orderings in the Hamiltonian constraint equation which is our Klein-Gordon equation. The approach we have taken through the ADM Hamiltonian leads to a natural (or at least obvious) choice of factor orderings distinct from DeWitt's. This difference can also be stated by saying that Jacobs, Zapolsky, and I[9] in writing a Klein-Gordon equation $-(\partial^2\psi/\partial\Omega^2) + (\partial^2\psi/\partial\beta_+^2) + (\partial^2\psi/\partial\beta_-^2) = 0$ for the anisotropic universe have used a metric

$$ds^2 = -d\Omega^2 + d\beta_+^2 + d\beta_-^2 \tag{54}$$

in the "super-space" for this problem. This metric is of course, complete, and therefore differs from the metric for super-space which DeWitt employs in his discussion[18] of canonical quantization.

It is important also to emphasize the similarities here to DeWitt's work. The idea of studying quantum gravity by reducing it to a finite dimensional cosmological analogue is taken from DeWitt, but strengthened in that I have used the model not as an example to illustrate a general theory, but as a manageable first problem in which to discover some of the directions which a full theory should take. Also the "Hamiltonian" constraint (18) plays a different role in the ADM approach than in DeWitt's work following Dirac. In the ADM approach this equation is to be solved and thus eliminated from theory in the process of defining a Hamiltonian H. Therefore it does not appear in the theory which we quantize. Nevertheless, following the path of least resistance, JZM are diverted from Schroedinger quantization, distracted from a Dirac-type square-root process with Pauli matrices, and choose a Klein-Gordon wave equation which is just an operator form of this Dirac Hamiltonian constraint which DeWitt by a more direct path used as the basis for his canonical quantization. In the present approach the use of a Klein-Gordon equation suggests that a Minkowski signature metric in super-space ($\beta_{\pm}\Omega$ space in the example) plays a significant role. DeWitt, in an antiparallel argument, deduces from the form of the Dirac Hamiltonian constraint that a metric is super-space is implied, and concludes as a consequence of the signature of this metric that the operator form of this equation should be treated as a generalized Klein-Gordon equation. I consider the fact, that our rather different route toward a quantum of gravity also leads to a Klein-Gordon type wave equation and to the idea of a metric in super-space, an encouraging confirmation that these two elements of the theory are natural and fundamental.

REFERENCES

1. Arnowitt, R., Deser, S., and Misner, C. W., Chapt. 7 in *Gravitation: an introduction to current research*, edited by L. Witten (Wiley, N.Y. 1962), referred to subsequently as ADM.
2. Misner, C. W., "Quantum Cosmology. I." (submitted to the Physical Review).
3. Misner, C. W., paper in preparation (will be submitted to Proc. Roy. Soc., London).
4. Misner, C. W., Phys. Rev. Letters 22, 1071 (1969).
5. Chitre, D., unpublished Ph.D. Thesis research, University of Maryland.
6. Okerson, D. J., unpublished B.S. Thesis, Princeton University, 1969.
7. Misner, C. W. and Ryan, M., paper in preparation (for submission to Phys. Rev. Letters).
8. Ryan, M., unpublished Ph.D. Thesis research, University of Maryland.
9. Jacobs, K., Misner, C. W. and Zapolsky, H., "Quantum Cosmology. II." paper in preparation, referred to subsequently as JMZ.

10. Misner, C. W., "Gravitational Collapse," in *Astrophysics and General Relativity*, 1968 Brandeis Summer Institute, edited by M. Chretien, S. Deser, and J. Goldstein (Gordon and Breach, New York, 1969), Vol. 1, pp 198-201.
11. Misner, C. W., "The Absolute Zero of Time," submitted to Physical Review.
12. Belinski, V. A. and Khalatnikov, I. M., to be published. I thank Professor K. S. Thorne for making this preprint available to me in an English translation by Dr. A. Pogo.
13. Misner, C. W., "The Absolute Zero of Time," an unsuccessful essay submitted for the 1969 gravity essay competition (Gravity Research Foundation, New Boston, N.H.).
14. Isaacson, R. A., Phys. Rev. 166, 1263, 1272 (1968).
15. Misner, C. W., Ap. J. 151, 431 (1968)
16. Schelkunoff, S. A., *Electromagnetic Waves* (Van Nostrand, N.Y., 1943), pp. 393-4.
17. Komar, A., "The Quantization Program for General Relativity," these proceedings (Relativity Conference in the Midwest, Cincinnati, Ohio, 2-6 June 1969).
18. DeWitt, B. S., Phys. Rev. 160, 1113 (1967).

POST-NEWTONIAN METHODS AND CONSERVATION LAWS

S. Chandrasekhar

University of Chicago

Chicago, Illinois

Abstract

The present paper is concerned with the conservation laws in general relativity as expressed in terms of the Landau-Lifshitz complex and their role in the development of the successive post-Newtonian approximations to the equations of general relativity of a perfect fluid. In § I, the known conservation laws of Newtonian hydrodynamics are formulated in a language that the relativistic laws appear as natural generalizations. In § II, the same laws are considered in the framework of general relativity. In particular the conserved energy is identified as the difference between the (0,0)-component of the Landau-Lifshitz complex and the conserved rest-mass energy ($= c^2\rho u^0\sqrt{-g}$). In § III, the development of the first and the second post-Newtonian approximations to the equations of relativistic hydrodynamics are described and illustrated. And finally in § IV, the manner in which one can obtain the equations of the $2\frac{1}{2}$ - post-Newtonian approximation is described. In this approximation all terms inclusive of $O(c^{-5})$ beyond the Newtonian are retained; it is in this approximation that terms representing the reaction of the fluid to the emission of gravitational radiation by the system first make their appearance. It is shown how the derived radiation-reaction terms of $O(c^{-5})$ contribute to the dissipation of energy and angular momentum in agreement with the predictions of the linearized theory of gravitational radiation.

I. Conservation Laws in the Newtonian Hydrodynamics of a Perfect Fluid

As a preliminary to the discussion in the following sections, we shall first formulate the integrals, which the equations of hydrodynamics in the Newtonian limit admit, in a manner that their relations to the integrals in the various post-Newtonian approximations are manifest.

With the definition of the energy-momentum tensor,

$$T^{00} = \rho c^2, \quad T^{0\alpha} = \rho c \upsilon_\alpha, \quad \text{and } T^{\alpha\beta} = \rho \upsilon_\alpha \upsilon_\beta + p\delta_{\alpha\beta}, \tag{1}$$ [1]

where ρ denotes the density, p the pressure, and υ_α the Cartesian components of the velocity, we can write the Eulerian equations of hydrodynamics in the form

$$T^{ij}{}_{,j} = 0. \tag{2}$$

The existence of the conserved quantities

$$P^i = \int_V T^{0i}\, d\vec{x} = \text{constant}, \tag{3}$$

where the integration is extended over the volume V occupied by the fluid, follows directly from the form of equation (2) and the condition that the pressure vanishes on the boundary of V.

From the symmetry of the energy-momentum tensor, the constancy of the total angular momentum

$$L_\gamma = \varepsilon_{\alpha\beta\gamma} \int_V \rho x_\alpha \upsilon_\beta \, d\vec{x} \tag{4}$$

follows from the space components of equation (2).

---

[1] The convention regarding the indices is the same as in Chandrasekhar (1965): Latin indices take the values 0,1,2, and 3 and Greek indices take only the values 1,2, and 3 referring to the spatial coordinates; and the summation convention will be restricted to their respective ranges. Also, x_0 will be replaced by ct when the notation of ordinary Cartesian tensors is used; and when the notation of Cartesian is used, the Greek indices will always be written as subscripts; and the summation over repeated Greek (Cartesian) indices will also be assumed. The notations of semicolon and comma will be used to indicate covariant and (simple) partial differentiation.

In addition to the integrals included in equations (3) and (4), the equations of motion allow an additional energy-integral if we make the additional assumption that every fluid element, during its motion, preserves its entropy. This assumption, by the first law of thermodynamics, requires that the change in the thermodynamic internal energy Π (per unit mass), which an element of the fluid experiences during its motion, must be traceable directly to the work done by the pressure in expanding (or contracting) its volume; thus

$$\frac{d\Pi}{dt} = -p \frac{d}{dt} \left(\frac{1}{\rho}\right) . \tag{5}$$

On this assumption, we have the integral

$$\mathscr{E} = \int_V \rho\left(\frac{1}{2} \upsilon^2 + \Pi\right) d\vec{x} = \text{constant}. \tag{6}$$

The foregoing equations apply when no external forces are acting on the fluid; indeed, no allowance has been made even for the forces resulting from its own gravitation. If we should now allow for these self-gravitational forces, then the space components of equation (2) must be modified to read

$$T^{\alpha j}{}_{,j} - \rho \frac{\partial U}{\partial x_\alpha} = 0, \tag{7}$$

where U denotes the gravitational potential resulting from the prevailing distribution of ρ:

$$\nabla^2 U = -4\pi G\rho . \tag{8}$$

Equation (7) is not of the form (2); but it can be restored to that form by making use of the identity

$$-\rho \frac{\partial U}{\partial x_\alpha} = \frac{1}{16\pi G} \frac{\partial}{\partial x_\beta} \left[4 \frac{\partial U}{\partial x_\alpha} \frac{\partial U}{\partial x_\beta} - 2\delta_{\alpha\beta} \left(\frac{\partial U}{\partial x_\mu}\right)^2 \right], \tag{9}$$

which obtains by virtue of the "field equation" (8). Thus, by defining the symmetric tensor t^{ij} by

$$t^{00} = 0, \; t^{0\alpha} = 0 \text{ and } t^{\alpha\beta} = \frac{1}{16\pi G} \left[4 \frac{\partial U}{\partial x_\alpha} \frac{\partial U}{\partial x_\beta} - 2\delta_{\alpha\beta} \left(\frac{\partial U}{\partial x_\mu}\right)^2 \right], \tag{10}$$

and letting

$$\Theta^{ij} = T^{ij} + t^{ij}, \tag{11}$$

we can rewrite the equations governing the fluid, once again, in the form

$$\Theta^{ij}{}_{,j} = 0. \tag{12}$$

Therefore, we may call t^{ij} the contribution of the gravitational field to the total energy-momentum complex Θ^{ij}.

With the equations of motion reduced to the form (12), it is manifest that the fluid subject to its own gravitation allows the same linear and angular momentum integrals. Again, if we should supplement equation (12) by the condition (5) that ensures that each fluid element preserves its entropy during its motion, then we should obtain the additional energy integral

$$\mathscr{E} = \int_V \rho\left(\frac{1}{2} \upsilon^2 + \Pi - \frac{1}{2} U\right) d\vec{x} = \text{constant}. \tag{13}$$

II. Conservation Laws in General Relativity

In general relativity, in contrast to the Newtonian theory, the physical character of a system is completely specified by the choice of the expression for the energy-momentum tensor. Thus, for a perfect fluid the expression for the energy-momentum tensor one generally assumes is

$$T^{ij} = \rho(c^2 + \Pi + p/\rho)\, u^i u^j - p g^{ij}, \tag{14}$$

where $u^i (=dx_i/ds)$ is the four velocity, g^{ij} is the contravariant form of the metric tensors, and the remaining symbols have the same meanings as hitherto. While the expression (14) for T^{ij} is a natural "covariant" generalization of the one that one assumes in special relativity, the fact should not be overlooked that it involves the unspecified metric.

When the expression (14) for T^{ij} is inserted in the field equation

$$G^{ij} = R^{ij} - \frac{1}{2} R g^{ij} = - \frac{8\pi G}{c^4} T^{ij}, \tag{15}$$

we no longer have the choices we had in the Newtonian theory. Thus, the equation governing the fluid, namely,

$$T^{ij}{}_{;j} = 0, \tag{16}$$

necessarily includes the effect of the gravitational field on the fluid motions: the terms in the covariant divergence of T^{ij}, besides those included in the ordinary divergence of T^{ij}, are to be attributed, even as in equation (7), to the effect of the gravitational field; indeed, in the Newtonian limit, the additional term is precisely that included in equation (7), namely $-\rho\partial U/\partial x_\alpha$. Similarly, the fact that the expression for T^{ij} includes no dissipative mechanism means that the motions must necessarily be consistent with requirements for isentropic flow: it should not be necessary to supplement equation (16) by any additional statement to ensure that each fluid element, during its motion, preserves its entropy. Thus, writing out equation (16) explicitly, we have

$$[\rho(c^2 + \Pi + p/\rho)u^j]_{;j}\, u^i + \rho(c^2 + \Pi + p/\rho)u^j u^i{}_{;j} - g^{ij}p_{,j} = 0; \tag{17}$$

and contracting this equation with u_i, we obtain

$$[\rho(c^2 + \Pi + p/\rho)\, u^j]_{;j} - u^j p_{,j} = 0. \tag{18}$$

Expanding this last equation and simplifying, we have

$$(\rho u^j)_{;j}(c^2 + \Pi + p/\rho) + \rho u^j[\Pi_{,j} + p(\tfrac{1}{\rho})_{,j}] = 0. \tag{19}$$

From equation (19) it follows that <u>the conservation of mass required by the equation</u>

$$(\rho u^j)_{;j} = 0, \tag{20}$$

<u>is compatible with the equations of motion if, and only if,</u>

$$u^j[\Pi_{,j} + p(\tfrac{1}{\rho})_{,j}] = 0; \tag{21}$$

and this last equation is no more than the requirement that the motion be isentropic (cf. eq. [5]). In other words, <u>the conservation of mass and the conservation of entropy are not independent requirements in the framework of general relativity</u>. And the reason for their <u>independence</u> in the Newtonian limit is that in this limit ("$c^2 = \infty$") equation (19) reduces simply to the equation of continuity.

The question now arises as to what extent the conservation of mass required by equation (20) should be considered as a statement of a fundamental physical law that should supplement the field equations. From the standpoint of physics, a conservation law to which one might accede as "fundamental" is the conservation of the baryon number. In that case, one should rather write for the energy-momentum tensor the expression

$$T^{ij} = (\varepsilon + p)\, u^i u^j - pg^{ij}, \tag{22}$$

where the energy density ε is some function of N (the number of baryons per unit volume), and p (the "equation of state"); thus

$$\varepsilon \equiv \varepsilon(N,p). \tag{23}$$

And we must supplement equation (16) by the requirement

$$(Nu^j)_{;j} = 0. \tag{24}$$

Now treating equation (16), with T^{ij} given by equation (22), exactly as before, we find on using equation (24) that we must have

$$u^j \frac{\partial p}{\partial x_j} - \frac{\gamma p}{\sqrt{-g}} \frac{\partial}{\partial x_j} (u^j \sqrt{-g}) = 0, \tag{25}$$

where

$$\gamma = \frac{1}{p\partial N/\partial p} [N - (\varepsilon+p) \frac{\partial N}{\partial \varepsilon}] \tag{26}$$

is the "ratio of the specific heats" as defined under these circumstances.

It is clear that the conservation of baryon number together with the statement of the first law of thermodynamics in the form of equation (25) and the conservation of mass and the statement of the first law in the form of equation (21) are entirely equivalent if we restrict ourselves to conditions in which the only baryons present are protons and neutrons and we agree to ignore their mass difference. In order not to complicate the analysis by matters that are inconsequential to our main purpose, we shall base our further considerations in this paper on equations (20) and (21).

An equivalent form of equation (20) is

$$(\rho u^j \sqrt{-g})_{,j} = 0; \tag{27}$$

and it follows from this equation that

$$M = \int_V \rho u^0 \sqrt{-g}\, d\vec{x} = \text{constant}, \tag{28}$$

is a conserved quantity.

Since in the framework of Einstein's equations, the conservation of mass implies isentropic motion, and conversely, it follows that the energy integral (13) must be a derivable consequence of the field equations in an appropriate limit. And if this is the case, what is the generalization of this energy integral in the framework of the exact theory? To answer this question, we must turn to the exact conservation laws that obtain in the general theory.

The question of the conservation laws in general relativity is related to well known questions concerning the "pseudo tensors" and the ambiguities associated with their definitions. While there is a vast literature on the subject (and the writer makes no claim that he understands it all), it appears that the essential formal content of the theory is very simple.

A mathematical identity that is directly and easily verifiable is (Synge 1960)

$$\frac{1}{2}(gU^{imkl})_{,l,m} = gG^{ik} - \frac{8\pi G}{c^4} t^{ik}, \tag{29}$$

where

$$U^{imkl} = g^{il}g^{km} - g^{ik}g^{lm}, \tag{30}$$

$$t^{ik} = \frac{c^4}{16\pi G}(U^{imkl}D_{ml} + U^{ilmn}E^k{}_{lmn} + U^{klmn}E^i{}_{lmn} - U^{lmnp}\Gamma^i_{lp}\Gamma^k_{mn}), \tag{31}$$

$$D_{ml} = 2\Gamma^m{}_{lm}y_n - y_m y_l - \Gamma^n{}_{lp}\Gamma^p{}_{mn},$$

$$E^i{}_{lmn} = \Gamma^i{}_{lm}y_n + \Gamma^p{}_{lm}\Gamma^i{}_{pn}, \text{ and } y_n = \frac{\partial \log\sqrt{-g}}{\partial x_n}; \tag{32}$$

(the factor $c^4/16\pi G$ has been introduced in the definition of t^{ik} for later convenience.) If we now replace the Einstein tensor G^{ik} by $-8\pi G T^{ik}/c^4$, in accordance with the field equation, we obtain

$$\frac{1}{2}(gU^{imkl})_{,l,m} = \frac{8\pi G}{c^4}(-g)(T^{ik} + t^{ik}). \tag{33}$$

The essential features of equation (33) that are to be noted are (1) it is not a tensor equation, (2) U^{imkl} has the symmetries of the Riemann-Christoffel tensor: it is antisymmetric in (i,m) and (k,l) and symmetric for the simultaneous interchanges of i and k and l and m, (3) t^{ik} as defined in equation (31) is manifestly symmetric in i and k but it is not a tensor: it is in fact the "pseudo tensor" of Landau and Lifshitz.

Defining the energy - momentum complex

$$\Theta^{ik} = (-g)(T^{ik} + t^{ik}), \tag{34}$$

we can rewrite equation (33) in the form

$$\frac{1}{2}(gU^{imkl})_{,l,m} = \frac{8\pi G}{c^4}\Theta^{ik}. \tag{35}$$

From the antisymmetry of U^{imkl} in (i,m) and (k,l) it now follows that

$$\Theta^{ik}{}_{,k} = \Theta^{ik}{}_{,i} = 0. \tag{36}$$

Thus the energy-momentum complex satisfies the same equation as in the Newtonian theory; and as in the Newtonian theory its derivation has required the explicit use of the field equation. And we may now, as then, call t^{ik} the contribution of the gravitational field to the total energy-momentum complex Θ^{ik}.

From the form of equations (36) we can infer the existence of the conserved quantities

$$P^i = \int \Theta^{0i} d\vec{x} = \text{constant} \tag{37}$$

and

$$L_\gamma = \varepsilon_{\alpha\beta\gamma} \int \Theta^{0\alpha} x_\beta \, d\vec{x} = \text{constant}, \tag{38}$$

provided, of course, the integrals defining these quantities converge when integrated over the whole of space. However, an important difference between these conserved quantities and those that occur in the Newtonian theory should be noted: the integrations in equations (37) and (38) have to be effected over the whole of the three-dimensional space and not only over the volume V occupied by the fluid: Θ^{ik}, unlike T^{ik}, need not necessarily vanish outside V.

We can now answer the question as to what the conserved quantity is that in the exact theory is the analogue of the energy integral (13) in the Newtonian theory. It is given by

$$\mathscr{E} = \int (\Theta^{00} - c^2 \rho u^0 \sqrt{-g}) \, d\vec{x} = \text{constant}. \tag{39}$$

An observation with respect to the energy integral (39) is here pertinent: it is, in the context of the perfect fluid, the content of the first law of thermodynamics implicit in the theory; and as such it gives to the Landau-Lifshitz complex a physical significance that cannot be evaded.

III. The First and the Second Post-Newtonian Approximations to the Equations of Relativistic Hydrodynamics

The motivations underlying the attempts that have been made to develop a sequence of post-Newtonian approximations are the following.

In the Newtonian framework when we are given a well-defined physical system, such as n-mass points under their mutual gravitational attractions or a mass of perfect fluid subject to internal stresses and its own gravitation, we can write down a set of equations of motion which governs all possible motions that can occur in the system. We ask: Can we write down a similar set of equations in the framework of general relativity? It appears that, in general, we cannot - at least not in as explicit a manner as in the Newtonian framework. A more modest inquiry under the circumstances would be: Can we write down an explicit set of equations of motion which will govern, in a well-defined scheme of successive post-Newtonian approximations, the departures from the Newtonian motions resulting from the effects of general relativity?

A related question concerns the integral properties of a system - such as its mass, its energy, its linear momentum, and its angular momentum - which are conserved and are constants of the motion. We ask: Can we specify quantities, which are generalizations of the corresponding Newtonian quantities and which are constants of the extended post-Newtonian equations of motion?

A further question of some basic interest concerns radiation reaction. It is generally believed that gravitating systems are capable of radiating energy in the form of gravitational waves. And if they do, What are the radiative reactions on the motions of the system?; and, Can we write the equations of motion in a high enough order that terms representing the reaction of the system to the emission of gravitational radiation occur explicitly in them and can be unmistakably recognized as such? In particular, does the system lose energy and angular momentum as the result of these terms and are the rates of loss of energy and angular momentum in agreement with the predictions of the linearized theory of gravitational radiation?

The pioneering investigation on the development of a scheme of approximation that could answer the kind of questions that we have raised is, of course, that of Einstein, Infeld, and Hoffmann on the n-body problem. They were, in fact, able to write down the first post-Newtonian Lagrangian which differs from the Newtonian Lagrangian by quantities of the order of v^2/c^2, U/c^2, and comparable ones. The equations of motion derived from their first post-Newtonian Lagrangian suffice, for example, to derive the precession of the Keplerian orbit of two finite mass points about one another. The extension of the original theory of Einstein, Infeld, and Hoffmann to higher orders has never been completely or satisfactorily accomplished. The theory, moreover, suffers formal difficulties connected with the fact that mass points are not concepts that are strictly consistent with the spirit of general relativity. And as Bondi has stated in another connection, "General relativity is a peculiarly complete theory and may not give sensible solutions for situations too far removed from what is physically reasonable." The concept of a perfect fluid defined in terms of an isotropic pressure and a conserved rest-mass density does not suffer from similar limitations; and we do not also have to contend with δ- or $\hat{\delta}$- functions! In any event the rest of the present discussion will be devoted to the relativistic hydrodynamics of a perfect fluid.

We start, then, with a system specified by the same energy momentum tensor,

$$T^{ij} = \rho(c^2 + \Pi + p/\rho)u^i u^j - pg^{ij}, \tag{40}$$

considered in § II; and as in § II we shall supplement the assumptions implicit in the choice of this form for T^{ij} by law of the conservation of the rest-mass energy:

$$c^2(\rho u^i \sqrt{-g})_{,i} = 0. \tag{41}$$

Equations (40) and (41) together with the field equation,

$$G^{ij} = -\frac{8\pi G}{c^4} T^{ij} , \tag{42}$$

complete the set of equations at our disposal. The basic question concerns how the field equation (42) is to be solved for the metric coefficients in order that the equation governing the fluid, namely,

$$T^{ij}_{\ \ ;j} = 0, \tag{43}$$

when written out explicitly, will provide the required equations of motion as a power series in a suitable expansion-parameter.

The basis for an expansion procedure of the type envisaged is provided by two considerations. The first is that under "normal conditions," i.e. under conditions of common occurrence in the universe around us, the energy ρc^2 associated with the rest mass dominates by far all the other forms of energy that may be present. Thus in the case of a perfect fluid we may consider the kinetic energy of mass motions $(1/2)\rho v^2$, the potential energy $-(1/2)\rho U$, the internal energy $\rho\Pi$, and the energy of molecular motions as measured by p/ρ, all of them, as small compared to ρc^2, but comparable among themselves. This fact clearly provides the basis for an expansion procedure. The "smallness" parameter is then any of the following:

$$v^2/c^2 , \ U/c^2 , \ \Pi/c^2 , \text{ and } p/\rho c^2. \tag{44}$$

Formally we may specify the expansion parameter as c^{-1} though it has to be understood that the power to which c^{-1} occurs is merely a label for the order of the approximation to which the quantity in question has been carried.

A parenthetical remark is pertinent here. Since v/c is in particular considered as an expansion parameter, the approximation based on the smallness of the quantities listed in (44) is called a <u>slow-motion</u> approximation.

The second consideration relates to what we can infer about the geometry of space-time from the most elementary requirements of physics. It is an immediate consequence of the principle of equivalence (in its "weakest form") that rates of clocks must depend on the gravitational potential of their locations; and this inference translated in terms of proper times implies that

$$g_{00} = 1 - 2U/c^2 + O(c^{-4}). \tag{45}$$

In other words, the metric one must associate with the Newtonian theory of gravitation is

$$g_{00} = 1 - 2U/c^2 \ , \ g_{0\alpha} = 0 \ , \text{ and } g_{\alpha\beta} = -\delta_{\alpha\beta}. \tag{46}$$

We thus have a basis for an expansion procedure provided by the dominance of the rest energy over all other forms; and we also have the first departure of the metric from the Minkowskian determined by the principle of equivalence. And as we shall explain these two considerations suffice to develop an entirely consistent scheme of successive post-Newtonian approximations.

Before we proceed further it is necessary to specify how we shall distinguish the different post-Newtonian approximations. We shall distinguish them by the largest power of c^{-1} that is consistently retained beyond the Newtonian terms in the equations of motion. Thus, the first post-Newtonian equations will contain all terms of $O(c^{-2})$ beyond the Newtonian, while the second post-Newtonian equation will contain terms of $O(c^{-4})$ as well. (As we shall presently see, to obtain the equations of motion to a given order, the different metric coefficients must be known to different orders; and the orders to which the metric coefficients must be known should not be confused with the order of the equations of motion which provides the only basic criterion.)

Returning to the development of the different post-Newtonian equations of motion, we first observe that if we ignore all contributions to the (0,0)-component of the energy momentum tensors in comparison with ρc^2, then

$$T_{00} = T = \rho c^2 + O(1); \tag{47}$$

and the (0,0)-component of the field equation,

$$R_{ij} = -\frac{8\pi G}{c^4}\left(T_{ij} - \frac{1}{2} T g_{ij}\right), \tag{48}$$

combined with the expression (45) for g_{00}, reduces to Poisson's equation

$$\nabla^2 U = -4\pi G\rho \tag{49}$$

governing U, and confirms that the Newtonian equations are indeed the "zero-order" solutions of Einstein's field equations. The argument can to some extent be reversed. If we had written the field equation in the form

$$G^{ij} = \kappa T^{ij}, \tag{50}$$

where κ is some unspecified "coupling constant," then the requirement, that in the limit (47) Einstein's equations together with equation (45) reduce to those of Newton, determines the coupling constant κ to be $-8\pi G/c^4$.

We can now proceed further. With the same underlying assumption concerning the dominance of ρc^2 over all other forms of energy, the (α,β)-component of equation (48) gives

$$\nabla^2 g_{\alpha\beta} - \frac{\partial}{\partial x_\alpha}\left(\frac{\partial g_{\beta\mu}}{\partial x_\mu} - \frac{1}{2}\frac{\partial g^\sigma{}_\sigma}{\partial x_\beta}\right) - \frac{\partial}{\partial x_\beta}\left(\frac{\partial g_{\alpha\mu}}{\partial x_\mu} - \frac{1}{2}\frac{\partial g^\sigma{}_\sigma}{\partial x_\alpha}\right) = \frac{8\pi G}{c^2}\rho\delta_{\alpha\beta}. \tag{51}$$

With the coordinate condition

$$\frac{\partial g_{\alpha\mu}}{\partial x_\mu} - \frac{1}{2}\frac{\partial g^\sigma{}_\sigma}{\partial x_\alpha} = 0, \tag{52}$$

the required solution for $g_{\alpha\beta}$ to $O(c^{-2})$ is

$$g_{\alpha\beta} = -\delta_{\alpha\beta}\left(1 + \frac{2U}{c^2}\right). \tag{53}$$

(We shall presently return to the meaning of imposing coordinate conditions, such as eq. (52), on the solutions.)

As is well known the curvature of space implied by the solution (53) is precisely what is at the base of the deflection of light by a gravitational field predicted by general relativity; and it is a non-trivial consequence of the field equations.

Now, we have a little more knowledge of the metric coefficients than we started with; and we need no longer be satisfied with the bare statement that ρc^2 is the dominant term in T^{ij}. With the improved knowledge of the metric coefficients (namely that provided by eq. (53)) we find that consistent with the definition (40) of T^{ij}, we can now write

$$T^{00} = \rho c^2\left[1 + \frac{1}{c^2}(v^2 + 2U + \Pi)\right] + O(c^{-2}),$$

$$T^{0\alpha} = \rho c v_\alpha\left[1 + \frac{1}{c^2}\left(v^2 + 2U + \Pi + \frac{p}{\rho}\right)\right] + O(c^{-3}), \tag{54}$$

and

$$T^{\alpha\beta} = \rho v_\alpha v_\beta + p\delta_{\alpha\beta} + \frac{1}{c^2}\left[\rho\left(v^2 + 2U + \Pi + \frac{p}{\rho}\right) v_\alpha v_\beta - 2pU\delta_{\alpha\beta}\right] + O(c^{-4}).$$

We now find that these expressions for the components of T^{ij}, when inserted on the right-hand side of equation (48), enable us to determine the terms of $O(c^{-4})$ in g_{00} and of $O(c^{-3})$ in $g_{0\alpha}$. We find

$$g_{00} = 1 - \frac{2U}{c^2} + \frac{2}{c^4}(U^2 - 2\Phi) \text{ and } g_{0\alpha} = \frac{P_\alpha}{c^3}, \tag{55}$$

where

$$\nabla^2\Phi = -4\pi G\rho(\upsilon^2 + U + \frac{1}{2}\Pi + \frac{3}{2}\frac{p}{\rho}) = -4\pi G\rho\phi$$

and

$$\nabla^2 P_\alpha = -16\pi G\rho\upsilon_\alpha + \frac{\partial^2 U}{\partial t \partial x_\alpha}. \tag{56}$$

We find next that the present improved knowledge of the metric coefficients enables us to evaluate the Christoffel symbols with an accuracy just sufficient to write out equation (43) to include consistently all terms of $O(c^{-2})$ beyond the Newtonian. Thus the α-component of equation (43), in the first post-Newtonian approximation, is found to be (Chandrasekhar 1965)

$$(\frac{d}{dt} + \text{div}\,\vec{\upsilon})\Pi_\alpha - \rho\frac{\partial U}{\partial x_\alpha} + \frac{\partial}{\partial x_\alpha}[(1 + \frac{2U}{c^2})p]$$

$$- \frac{2}{c^2}\rho(\phi\frac{\partial U}{\partial x_\alpha} + \frac{\partial \Phi}{\partial x_\alpha}) + \frac{1}{c^2}\rho\upsilon_\beta\frac{\partial P_\beta}{\partial x_\alpha} = 0, \tag{57}$$

where

$$\Pi_\alpha = \rho\upsilon_\alpha + \frac{1}{c^2}\rho[\upsilon_\alpha(\upsilon^2 + 6U + \Pi + \frac{p}{\rho}) - P_\alpha]. \tag{58}$$

The 0-component of equation (43) gives no more information than that the integral of the density $\rho u^0\sqrt{-g}$ is conserved. With our present knowledge of the metric coefficients, we find

$$\rho u^0\sqrt{-g} = \rho + \frac{1}{c^2}\rho(\frac{1}{2}\upsilon^2 + 3U); \tag{59}$$

and this density satisfies the "equation of continuity"

$$(\frac{d}{dt} + \text{div}\,\vec{\upsilon})\,\rho[1 + \frac{1}{c^2}(\frac{1}{2}\upsilon^2 + 3U)] = 0. \tag{60}$$

And we need not stop here. With the improved knowledge of the metric coefficients provided by equation (55), we can evaluate the terms in the components of the energy-momentum tensor beyond those given in equations (54). This improved knowledge of the energy-momentum tensor will in turn enable us to solve the field equations for the metric coefficients to higher orders; precisely, we can determine the terms of $O(c^{-6})$ in g_{00}, of $O(c^{-5})$ in $g_{0\alpha}$, and of $O(c^{-4})$ in $g_{\alpha\beta}$. This improved knowledge of the metric coefficients will enable us to obtain equations governing the fluid in the second post-Newtonian approximations, i.e. inclusive of terms of $O(c^{-4})$ beyond the Newtonian. (For the details of this development, see Chandrasekhar and Nutku 1969.)

And we need not stop here either! We might continue and obtain the equations in the third post-Newtonian approximation. But what of the radiative corrections which must appear at the $2\frac{1}{2}$ post-Newtonian approximation if one believes in the linearized theory of gravitational radiation? Before we turn to this question, we shall clarify two aspects of the first and the second post-Newtonian approximations completed so far.

a) The Gauge Conditions

Since the field equations of general relativity are invariant to arbitrary coordinate transformations, any solution of the field equations must involve four arbitrary functions. And this fact entitles us to impose "coordinate conditions" on the solutions - such as the one we imposed in obtaining the solution (53). Quite generally we find that, when solving the field equations appropriately for the successive post-Newtonian approximations, the solutions involve four arbitrary functions in a way that if the solutions have been determined with particular coordinate or gauge conditions, then the transformations needed to go to a different gauge can be explicitly given. Thus if $Q_{\alpha\beta} c^{-n}$ is the term of $O(c^{-n})$ in $g_{\alpha\beta}$, then $Q_{\alpha\beta}$ satisfies an equation of the form (cf. eq. (51))

$$\nabla^2 Q_{\alpha\beta} - \frac{\partial}{\partial x_\alpha}\left(\frac{\partial Q_{\beta\mu}}{\partial x_\mu} - \frac{1}{2}\frac{\partial Q_{\sigma\sigma}}{\partial x_\beta}\right) - \frac{\partial}{\partial x_\beta}\left(\frac{\partial Q_{\alpha\mu}}{\partial x_\mu} - \frac{1}{2}\frac{\partial Q_{\sigma\sigma}}{\partial x_\alpha}\right) = S_{\alpha\beta}, \tag{61}$$

where $S_{\alpha\beta}$ is a function determined by the solutions obtained in the earlier approximations. It can be readily verified that there is an integrability condition for the solvability of equation (61): equation (61) allows non-trivial solutions if and only if

$$\frac{\partial}{\partial x_\alpha}\left(S_{\alpha\beta} - \frac{1}{2}\delta_{\alpha\beta} S_{\sigma\sigma}\right) = 0. \tag{62}$$

And it is found that this integrability condition is satisfied by virtue of the equations of motion in the earlier approximations that have determined $S_{\alpha\beta}$.

With the integrability condition (62) satisfied, it follows that if $Q_{\alpha\beta}$ is a solution of equation (61) then so is

$$Q_{\alpha\beta} + \frac{\partial W_\alpha}{\partial x_\beta} + \frac{\partial W_\beta}{\partial x_\alpha}, \tag{63}$$

where W_α is an arbitrary vector function in space-time. (It is often necessary to require that W_α is of the nature of the support functions in the theory of distributions.)

Similarly, when seeking for solutions for the term $Q_{0\alpha}/c^{n+1}$ in $g_{0\alpha}$, we find that if $Q_{0\alpha}$ is a solution of the appropriate equation, then so is

$$Q_{0\alpha} + \frac{\partial W}{\partial x_\alpha} + \frac{\partial W_\alpha}{\partial t} , \tag{64}$$

where W is a further arbitrary function. The explicit way in which the four arbitrary functions W_α and W occur in the solutions for the metric coefficients appropriate for a particle post-Newtonian approximation makes it a relatively simple matter to determine the effect of the choice of gauge on the conclusions one may draw from the solutions.

b) The Conserved Quantities

In the first post-Newtonian approximation the equation of motion (57) is sufficiently simple that one can, by "simple inspection," determine the quantities that are conserved. For example, it is not difficult to show that if one integrates equation (57) over the volume V occupied by the fluid then all the terms except the first vanish identically so that

$$\int_V (\frac{d}{dt} + \text{div } \vec{v})\Pi_\alpha \, d\vec{x} = 0; \tag{65}$$

or alternatively,

$$\frac{d}{dt} \int_V \Pi_\alpha \, d\vec{x} = 0. \tag{66}$$

This last equation expresses the conservation of the linear momentum Π_α integrated over the volume occupied by the fluid. Similarly, it can be verified that by multiplying equation (57) by x_β, integrating over the volume occupied by the fluid, and antisymmetrizing the result with respect to α and β, we obtain the result

$$\frac{d}{dt}\int_V (\Pi_\alpha x_\beta - \Pi_\beta x_\alpha)\, d\vec{x} = 0. \qquad (67)$$

In other words, the total angular momentum defined in terms of Π_α is also conserved.

However, the method of inspection appears impractical when one has to deal with equations as complicated as those one obtains in the second post-Newtonian approximation. But then the evaluation of the Landau-Lifshitz complex Θ^{ij} provides a straightforward algorism for determining the conserved quantities. For further details on this manner of determining the conserved quantities the reader must be referred to the original papers (Chandrasekhar 1969b and Chandrasekhar and Nutku 1969). However, one remark concerning the conserved energy, $\mathscr{E} = \Theta^{00} - c^2\rho u^0\sqrt{-g}$, requires to be made. To obtain the conserved energy in the first post-Newtonian approximation, for example, it is necessary to know Θ^{00} and $c^2\rho u^0\sqrt{-g}$ to $O(c^{-2})$ (i.e. in the second post-Newtonian approximation) in order that their difference may be known to $O(c^{-2})$ appropriate for the first post-Newtonian approximation. From these remarks it follows that by evaluating Θ^{00} and $c^2\rho u^0\sqrt{-g}$ in the first post-Newtonian approximation we shall only be determining the energy conserved in the Newtonian approximation; and these results confirm the remarks made in § II.

And finally a fact of some interest is that both in the first and in the second post-Newtonian approximations it is found that the equations,

$$\Theta^{ij}{}_{,j} = 0 \quad \text{and} \quad T^{ij}{}_{;j} = 0, \qquad (68)$$

are identically the same.

IV. THE EQUATIONS OF THE $2\frac{1}{2}$-POST-NEWTONIAN APPROXIMATION AND THE RADIATION-REACTION TERMS

First it is useful to clarify why the post-Newtonian scheme as outlined in § III automatically generates for the equations of motion an even series in c^{-1}.

As we have stated, the starting point of the post-Newtonian scheme is provided by the initial values specified in equation (46). And as we have also seen, the first iteration of Einstein's field equations with the initial values (46) leads to an improvement of the metric coefficients by determining further terms of $O(c^{-4})$ in g_{00}, of $O(c^{-3})$ in $g_{0\alpha}$, and of $O(c^{-2})$ in $g_{\alpha\beta}$. Had we supposed, in developing the expansion for the metric coefficients, that terms of $O(c^{-3})$ in g_{00}, of $O(c^{-2})$ in $g_{0\alpha}$, and of $O(c^{-1})$ in

$g_{\alpha\beta}$ occur, then we should have found that these terms satisfy homogeneous equations (unlike the terms of one higher order which satisfy inhomogeneous equations); they can consequently be set equal to zero by a suitable choice of gauge. The same phenomenon will be repeated when we proceed to the second post-Newtonian approximation: the terms of $O(c^{-5})$ in g_{00}, of $O(c^{-4})$ in $g_{0\alpha}$, and of $O(c^{-3})$ in $g_{\alpha\beta}$ will again satisfy homogeneous equations; and again they can be set equal to zero by a suitable choice of gauge. By induction it follows that we shall continue to skip the "odd steps" indefinitely if we continue the scheme of iterations without any modifications. The question arises: How are we to break this chain by providing a non-zero source that will provide "starting values" for a first non-trivial step (even as the principle of equivalence originally provided the "source" $-2U/c^2$ in g_{00} for starting the even series)?

Clearly the reason why the post-Newtonian scheme as outlined in § III fails to provide a source for a non-trivial odd step is that nowhere in the scheme do we impose on the solutions the Sommerfeld radiation-condition. This condition cannot however be applied in any straightforward manner to the solutions obtained in a "slow-motion" approximation as the post-Newtonian approximations are: for these approximations based as they are on the assumption that $v/c < 1$, require that the operation of $\partial/\partial x_0 (= \partial/c\partial t)$ on any quantity lowers its order by one. This last fact implies that the solutions obtained on the basic assumptions of the post-Newtonian scheme can be valid only in the near zone where $r < ct$. What is required, then, is a "matching" (in the sense of Thorne 1969 and Burke 1969) of the solutions appropriate to the near zone with those appropriate for the far zone (where the Sommerfeld condition is to be imposed). A method by which this "matching" could be accomplished was described by Trautman in two early investigations (1958a, b) in the context of the original theory of Einstein, Infeld, and Hoffmann on the n-body problem. However, by an oversight, Trautman failed to get agreement with the predictions of the linearized theory of gravitation. When this oversight is corrected, Trautman's procedure (as applied and extended in a recent paper by Chandrasekhar and Esposito 1970 in the framework of hydrodynamics) becomes consistent with the predictions of the linearized theory. Here we shall give a brief account of this investigation. But one result established by Trautman is certainly correct and relevant: the laws of the conservation of mass and of linear momentum, in their Newtonian forms, enable one to infer that the lowest order non-vanishing terms in the metric coefficients, that result from the imposition of the Sommerfeld radiation-condition at infinity, are $O(c^{-5})$ in g_{00}, $O(c^{-6})$ in $g_{0\alpha}$, and $O(c^{-5})$ in $g_{\alpha\beta}$. We shall denote these lowest order terms by

$$\underset{5}{g_{00}} = \frac{1}{c^5} \overset{(5)}{Q_{00}} , \quad \underset{6}{g_{0\alpha}} = \frac{1}{c^6} \overset{(6)}{Q_{0\alpha}} , \quad \text{and} \quad \underset{5}{g_{\alpha\beta}} = \frac{1}{c^5} \overset{(5)}{Q_{\alpha\beta}} . \tag{69}$$

(By the numerals such as 5 and 6 below g_{00} and $g_{0\alpha}$ we indicate that we are referring to the terms of orders c^{-5} and c^{-6} in the series expansion of these quantities in inverse powers of c. This notation for referring to the terms of the different orders in the expansion of a quantity will be used in the rest of this paper.) The problem now is to obtain unique expressions for these coefficients.

Let

$$\mathfrak{g}^{ik} = g^{ik}\sqrt{-g} = \mathfrak{g}^{ik}_{(2)} - \gamma^{ik}, \tag{70}$$

where $\mathfrak{g}^{ik}_{(2)}$ represents the solution for metric coefficients in the second post-Newtonian approximation and γ^{ik} is the lowest order term in $\mathfrak{g}^{ik}$ that derives from the imposition of the boundary condition at infinity. In accordance with the orders specified in equation (69), we expect the orders of the γ^{ik}'s to be

$$\gamma^{00} = O(c^{-5}) , \quad \gamma^{0\alpha} = O(c^{-6}) , \quad \text{and} \quad \gamma^{\alpha\beta} = O(c^{-5}). \tag{71}$$

We now make the substitution (70) in the field equation written in terms of the Landau-Lifshitz complex in the form (cf. eq. [29])

$$(\mathfrak{g}^{ik}\mathfrak{g}^{lm} - \mathfrak{g}^{il}\mathfrak{g}^{km})_{,l,m} = \frac{16\pi G}{c^4}\,\Theta^{ik}, \tag{72}$$

and linearize with respect to γ^{ik}: the terms that will thus be ignored will be of orders higher than any that are retained. We shall then obtain

$$\begin{aligned}(\mathfrak{g}^{ik}_{(2)}\mathfrak{g}^{lm}_{(2)} - \mathfrak{g}^{il}_{(2)}\mathfrak{g}^{km}_{(2)} - \mathfrak{g}^{ik}_{(2)}\gamma^{lm} - \mathfrak{g}^{lm}_{(2)}\gamma^{ik} + \mathfrak{g}^{il}_{(2)}\gamma^{km} + \\ \mathfrak{g}^{km}_{(2)}\gamma^{il})_{,l,m} = \frac{16\pi G}{c^4}\,\Theta^{ik}.\end{aligned} \tag{73}$$

So long as the $\gamma^{\cdot\cdot}$'s are of the orders specified in (71) - and we shall presently verify that they are - we may consistently replace the $\mathfrak{g}^{\cdot\cdot}_{(2)}$'s, that occur as factors of the $\gamma^{\cdot\cdot}$'s, by the corresponding Minkowskian coefficients. Also, we may impose on the $\gamma^{\cdot\cdot}$'s a gauge independently of the one chosen in the solution for the $\mathfrak{g}^{\cdot\cdot}_{(2)}$'s: there will be no conflict since the $\mathfrak{g}^{\cdot\cdot}_{(2)}$'s do not include any terms of the orders of the $\gamma^{\cdot\cdot}$'s. And we shall find it convenient to impose on the $\gamma^{\cdot\cdot}$'s the de Donder condition

$$\gamma^{ik}{}_{,k} = 0. \tag{74}$$

On these assumptions and restrictions, equation (73) becomes

$$(\mathfrak{g}^{ik}_{(2)}\mathfrak{g}^{lm}_{(2)} - \mathfrak{g}^{il}_{(2)}\mathfrak{g}^{km}_{(2)})_{,l,m} + (\nabla^2 - \frac{1}{c^2}\frac{\partial^2}{\partial t^2})\,\gamma^{ik} = \frac{16\pi G}{c^4}\,\Theta^{ik}. \tag{75}$$

The terms in the $\mathfrak{g}^{\cdot\cdot}_{(2)}$'s on the left-hand side of equation (75) do not include any of the orders specified in (71); and, moreover, these terms (derived from the $\mathfrak{g}^{\cdot\cdot}_{(2)}$'s), when consistently expanded, will just make up $16\pi G\Theta^{ik}_{(1)}/c^4$, where $\Theta^{ik}_{(1)}$ is the Landau-Lifshitz complex in the <u>first</u> post-Newtonian approximation. Accordingly, we may expect γ^{ik} to be determined by the equation

$$(\nabla^2 - \frac{1}{c^2}\frac{\partial^2}{\partial t^2})\,\gamma^{ik} = \frac{16\pi G}{c^4}\,\Theta^{ik}_{(1)}\;. \tag{76}$$

Notice that by virtue of the property,

$$\Theta^{ik}{}_{,k} = \Theta^{ik}{}_{,i} = 0, \tag{77}$$

of the Landau-Lifshitz complex, equation (75) is consistent with the de Donder condition imposed on the $\gamma^{\cdot\cdot}$'s (to the order required).

The solution of equation (76) which satisfies the outgoing-radiation condition is

$$\gamma^{ik}(\vec{x},t) = -\frac{4G}{c^4}\int\frac{d\vec{x}'}{|\vec{x}-\vec{x}'|}\;\Theta^{ik}_{(1)}\;(\vec{x}',t-|\vec{x}-\vec{x}'|/c). \tag{78}$$

By this choice of the particular solution of equation (76) expressed in terms of the "retarded potentials," we have automatically excluded the possibility of any incoming radiation.

We now expand $\Theta^{ik}_{(1)}(\vec{x}',t-|\vec{x}-\vec{x}'|/c)$ as a series in inverse powers of c in a manner that is appropriate for the near zone. Thus,

$$\gamma^{ik}(\vec{x},t) = -\frac{4G}{c^4}\int \frac{\Theta^{ik}_{(1)}(\vec{x}',t)}{|\vec{x}-\vec{x}'|}\,d\vec{x}'$$
$$+\frac{4G}{c^4}\sum_{n=1}^{\infty}\frac{(-1)^{n+1}}{n!\,c^n}\frac{\partial^n}{\partial t^n}\int \Theta^{ik}_{(1)}(\vec{x}',t)\,|\vec{x}-\vec{x}'|^{n-1}d\vec{x}'. \tag{79}$$

Considering first the (0,0)-component of equation (79), we have

$$\gamma^{00} = \frac{4G}{c^5}\frac{\partial}{\partial t}\int \Theta^{00}_{(1)}(\vec{x}',t)\,d\vec{x}' + \frac{2G}{3c^7}\frac{\partial^3}{\partial t^3}\int \Theta^{00}_{(1)}(\vec{x}',t)\,|\vec{x}-\vec{x}'|^2 d\vec{x}', \tag{80}$$

where we have not written the terms of even order. (It may be recalled here that the dominant term in Θ^{00} is ρc^2.)

For $\Theta^{00}_{(1)}$ determined in the first post-Newtonian approximation

$$\frac{\partial}{\partial t}\int \Theta^{00}_{(1)}(\vec{x}',t)\,d\vec{x}' = O(c^{-2}). \tag{81}$$

The first term on the right-hand side of equation (80) is, therefore, of $O(c^{-7})$ in contrast to the second term which is of $O(c^{-5})$. We thus obtain without any ambiguity that

$$\underset{5}{\gamma^{00}} = \frac{2G}{3c^5}\frac{\partial^3}{\partial t^3}\int_V \rho(\vec{x}',t)\,|\vec{x}-\vec{x}'|^2\,d\vec{x}'. \tag{82}$$

Considering next the $(0,\alpha)$-component of equation (79), we similarly obtain

$$\gamma^{0\alpha} = \frac{4G}{c^5}\frac{\partial}{\partial t}\int \Theta^{0\alpha}_{(1)}(\vec{x}'t)d\vec{x}' + \frac{2G}{3c^7}\frac{\partial^3}{\partial t^3}\int \Theta^{0\alpha}_{(1)}(\vec{x}'t)\,|\vec{x}-\vec{x}'|^2\,d\vec{x}', \tag{83}$$

where we have not written the terms of even order. (It may be recalled here that the dominant term in $\Theta^{0\alpha}$ is $\rho c v_\alpha$.)

For $\Theta^{0\alpha}_{(1)}$ determined in the first post-Newtonian approximation,

$$\frac{\partial}{\partial t}\int \Theta^{0\alpha}_{(1)}(\vec{x}',t)\,d\vec{x}' = O(c^{-3}). \tag{84}$$

The first term on the right-hand side of equation (83) is, therefore, of $O(c^{-8})$ in contrast to the second which is of $O(c^{-6})$.

We thus obtain, again, without any ambiguity that

$$\underset{6}{\gamma}^{0\alpha} = \frac{2G}{3c^6}\frac{\partial^3}{\partial t^3}\int \rho(\vec{x}',t)\upsilon_\alpha(\vec{x}',t)|\vec{x}-\vec{x}'|^2\, d\vec{x}'. \qquad (85)$$

Finally, considering the (α,β)-component of equation (79), we obtain

$$\underset{5}{\gamma}^{\alpha\beta} = \frac{4G}{c^5}\frac{\partial}{\partial t}\int \Theta^{\alpha\beta}(\vec{x}',t)\, d\vec{x}', \qquad (86)$$

where it will suffice for our present purpose to substitute for $\Theta^{\alpha\beta}$ its Newtonian expression (cf. eq. [9] and [10])

$$\Theta^{\alpha\beta} = \Theta_{\alpha\beta} = \rho\upsilon_\alpha\upsilon_\beta + p\delta_{\alpha\beta} + t_{\alpha\beta},$$

where (87)

$$t_{\alpha\beta} = \frac{1}{16\pi G}\left[4\frac{\partial U}{\partial x_\alpha}\frac{\partial U}{\partial x_\beta} - 2\delta_{\alpha\beta}\left(\frac{\partial U}{\partial x_\mu}\right)^2\right].$$

The "oversight" in Trautman's treatment to which we have referred earlier occurs precisely here: he has $T^{\alpha\beta}$ where he should have had $\Theta^{\alpha\beta}$.

The expressions (82), (85), and (86) we have found for the $\gamma^{\cdot\cdot}$'s can be transformed into expressions for the corresponding $g^{\cdot\cdot}$'s with the aid of the formulae

$$\underset{n}{g}_{00} = \frac{1}{2}(\underset{n}{\gamma}^{00} + \underset{n}{\gamma}^{\alpha\alpha}), \quad \underset{n}{g}_{0\alpha} = -\underset{n}{\gamma}^{0\alpha},$$

and (88)

$$\underset{n}{g}_{\alpha\beta} = \underset{n}{\gamma}^{\alpha\beta} + \frac{1}{2}\delta_{\alpha\beta}(\underset{n}{\gamma}^{00} - \underset{n}{\gamma}^{\mu\mu}),$$

which follow from the general relation (Trautman 1958b, eq. (27))

$$\underset{n}{g}_{ij} = \eta_{ik}\eta_{jl}(\underset{n}{\gamma}^{kl} - \frac{1}{2}\eta^{kl}\eta_{rs}\underset{n}{\gamma}^{rs}). \qquad (89)$$

We thus obtain

$$Q_{00}^{(5)} = \frac{1}{3}G\frac{\partial^3}{\partial t^3}\int \rho(\vec{x}',t)\,|\vec{x}-\vec{x}'|^2 d\vec{x}' + 2G\frac{\partial}{\partial t}\int \Theta_{\mu\mu} d\vec{x}, \tag{90}$$

$$Q_{0\alpha}^{(6)} = -\frac{2}{3}G\frac{\partial^3}{\partial t^3}\int \rho(\vec{x}',t)\upsilon_\alpha(\vec{x}',t)\,|\vec{x}-\vec{x}'|^2 d\vec{x}', \tag{91}$$

and

$$Q_{\alpha\beta}^{(5)} = 4G\frac{\partial}{\partial t}\int \Theta_{\alpha\beta} d\vec{x} + \delta_{\alpha\beta}\Big[\frac{1}{3}G\frac{\partial^3}{\partial t^3}\int \rho(\vec{x}',t)\,|\vec{x}-\vec{x}'|^2 d\vec{x}'$$

$$-2G\frac{\partial}{\partial t}\int \Theta_{\mu\mu} d\vec{x}\Big]. \tag{92}$$

It is satisfactory in many ways that the expressions which we have found for the lowest order terms in the metric coefficients in the near zone, which reflect the outgoing-radiation condition at infinity, are of purely Newtonian origin in the sense that they do not involve explicitly any quantity defined in the higher approximations.

By using the relations familiar in the theory of the tensor virial theorem in Newtonian hydrodynamics (for a brief account of this theory, see Chandrasekhar 1969a), equations (90) - (92) can be brought to the forms

$$Q_{00}^{(5)} = \frac{4}{3}G\frac{d^3 I_{\mu\mu}}{dt^3}, \tag{93}$$

$$Q_{0\alpha}^{(6)} = \frac{2}{3}Gx_\mu\frac{d^4 I_{\mu\alpha}}{dt^4} - \frac{2}{3}G\frac{d^3}{dt^3}\int_V \rho\upsilon_\alpha|\vec{x}|^2 d\vec{x}, \tag{94}$$

and

$$Q_{\alpha\beta}^{(5)} = 2G\frac{d^3 I_{\alpha\beta}}{dt^3} - \frac{2}{3}G\,\delta_{\alpha\beta}\frac{d^3 I_{\mu\mu}}{dt^3}, \tag{95}$$

where

$$I_{\alpha\beta} = \int_V \rho x_\alpha x_\beta d\vec{x}, \tag{96}$$

denotes the moment of inertia tensor.

An important consequence of the formula (95) for $Q_{\alpha\beta}^{(5)}$ is

$$Q_{\mu\mu}^{(5)} = 0, \tag{97}$$

i.e. $Q_{\alpha\beta}^{(5)}$ is traceless. It appears that this traceless character of $Q_{\alpha\beta}^{(5)}$ is essential for the self-consistency of the whole development.

Further consequences of the foregoing formulae are

$$\frac{dQ_{00}^{(5)}}{dt} = 2\,\frac{\partial Q_{0\alpha}^{(6)}}{\partial x_\alpha} \quad \text{and} \quad \frac{\partial Q_{0\alpha}^{(6)}}{\partial x_\beta} = \frac{\partial Q_{0\beta}^{(6)}}{\partial x_\alpha}. \tag{98}$$

The first of these relations is equivalent to the de Donder condition that was imposed on the γ^{ik}'s in obtaining the solutions; and the second is an expression of the conservation of the Newtonian angular momentum.

We have now seen how the metric coefficients $g_{0\alpha}$ and $g_{\alpha\beta}$, to orders required for the $2\frac{1}{2}$- post-Newtonian approximation, can be deduced by supplementing the solutions obtained in the second post-Newtonian approximation by terms of $O(c^{-6})$ in $g_{0\alpha}$ and of $O(c^{-5})$ in $g_{\alpha\beta}$-terms which are required by the outgoing-radiation condition at infinity. But the equations of motion in the $2\frac{1}{2}$-post-Newtonian approximation cannot be written down without a knowledge of the term of $O(c^{-7})$ in g_{00}, i.e. we need to know "$Q_{00}^{(7)}$": the knowledge of $Q_{00}^{(5)}$ will not suffice.

To determine $Q_{00}^{(7)}$ we must appeal, once again, to the field equation. And it appears that our present knowledge of $Q_{00}^{(5)}$, $Q_{0\alpha}^{(6)}$, and $Q_{\alpha\beta}^{(5)}$ is just sufficient to determine $Q_{00}^{(7)}$ with the aid of the field equation. We find (for details, see Chandrasekhar and Esposito 1970)

$$Q_{00}^{(7)} = -2Q_{00}^{(5)}U - Q_{\mu\nu}^{(5)}\mathscr{V}_{\mu\nu} + \frac{1}{60}G\frac{\partial^5}{\partial t^5}\int \rho(\vec{x}',t)\,|\vec{x}-\vec{x}'|^4\,d\vec{x}' \tag{99}$$

$$+\frac{1}{3}G\frac{\partial^3}{\partial t^3}\int \Theta_{\mu\mu}(\vec{x}',t)\,|\vec{x}-\vec{x}'|^2 d\vec{x}',$$

where

$$\mathscr{V}_{\mu\nu}(\vec{x}) = G\int \rho(\vec{x}')\,\frac{(x_\mu - x'_\mu)(x_\nu - x'_\nu)}{|\vec{x}-\vec{x}'|^3}\,d\vec{x}' \tag{100}$$

is the tensor potential defined in the theory of the tensor virial equations.

With the knowledge of the requisite metric coefficients we now have, we can obtain the equations of motion in the $2\frac{1}{2}$ -post-Newtonian approximation.

Considering the (0,0)-component of equation (43), we find that there are no additional terms beyond those included in the second post-Newtonian approximation. The reason for this circumstance is that the terms of $O(c^{-5})$ in the conserved density $\rho u^0\sqrt{-g}$ vanishes:

$$\underset{5}{\rho u^0\sqrt{-g}} = 0. \tag{101}$$

In other words, no alteration in the baryon number, conserved in the second post-Newtonian approximation, is introduced in this higher approximation. One can convince oneself that this fact is necessary for the self-consistency of the theory.

Considering next the α-component of equation (43), we find that terms of $O(c^{-5})$, that must follow the terms of $O(c^{-4})$ in the same equation in the second post-Newtonian approximation (given explicitly in Chandrasekhar and Nutku 1969, eq. [54]), are

$$\begin{aligned}\frac{1}{c}\underset{4}{T}{}^{\alpha j}{}_{;j} = \frac{1}{c^5}\Big[&-\rho Q^{(5)}_{00}\frac{d\upsilon_\alpha}{dt} - \frac{1}{2}\rho\upsilon_\alpha\frac{dQ^{(5)}_{00}}{dt} - \rho\frac{d}{dt}(\upsilon_\mu Q^{(5)}_{\mu\alpha}) \\ &- \frac{1}{2}\rho Q^{(5)}_{\mu\nu}\frac{\partial\mathfrak{V}_{\mu\nu}}{\partial x_\alpha} + \frac{1}{5}\rho x_\alpha G\frac{d^5 I_{\mu\mu}}{dt^5} - \frac{3}{5}\rho x_\mu G\frac{d^5 I_{\mu\alpha}}{dt^5} \\ &- \frac{1}{30}\rho G\frac{d^5}{dt^5}\int\rho|\vec{x}|^2 x_\alpha d\vec{x} + \frac{2}{3}\rho G\frac{d^4}{dt^4}\int\rho|\vec{x}|^2\upsilon_\alpha d\vec{x} - \frac{1}{3}\rho G\frac{d^3}{dt^3}\int\Theta_{\mu\mu}x_\alpha d\vec{x}\Big].\end{aligned} \tag{102}$$

Two further facts concerning the equations of the $2\frac{1}{2}$ -post-Newtonian approximation should be noted. The first is that

$$\underset{3}{\Theta}{}^{00} = 0 \quad\text{and}\quad \underset{4}{\Theta}{}^{0\alpha} = 0; \tag{103}$$

in other words, the (0,0)- and the $(0,\alpha)$-components of the Landau-Lifshitz complex retain in the $2\frac{1}{2}$ -post-Newtonian approximation the same values as in the second post-Newtonian approximation. And second that

$$\frac{1}{c}\underset{4}{\Theta}{}^{\alpha j}{}_{,j} = \underset{5}{\Theta}{}^{\alpha\beta}{}_{,\beta} = \frac{1}{c}\underset{4}{T}{}^{\alpha j}{}_{;j} = 0; \tag{104}$$

in other words the identity of the equations, $\Theta^{\alpha j}{}_{,j} = 0$, and $T^{\alpha j}{}_{;j} = 0$, verified in the first and the second post-Newtonian approximations continues to be maintained in the $2\frac{1}{2}$ - post-Newtonian approximation.

Finally, we turn to the question of the physical consequences to the system that result from the presence of the terms (102) in the equations of motion. It can be shown that by contracting the terms (102) with υ_α and integrating them over the volume occupied by the fluid, we obtain the rate of change of the energy $\mathscr{E}_{(2)}$ that is conserved in the second post-Newtonian approximation. Thus

$$\frac{d}{dt}\int \mathscr{E}_{(2)} d\vec{x} + \frac{1}{c}\int_V \upsilon_\alpha T^{\alpha j}{}_{;j}\, d\vec{x} = 0. \tag{105}$$

The explicit form of equation (105), in a frame of reference in which the center of mass is at rest, is found to be

$$\frac{d}{dt}\int \mathscr{E}_{(2)} dx - \frac{1}{c^5}\frac{d}{dt}\left(\overset{(5)}{Q_{00}}\mathscr{T}_{\mu\mu} + 2\overset{(5)}{Q_{\alpha\mu}}\mathscr{T}_{\alpha\mu}\right)$$
$$+ \frac{G}{30c^5}\frac{d}{dt}\left(\frac{d^3D_{\alpha\beta}}{dt^3}\frac{d^2D_{\alpha\beta}}{dt^2} - \frac{d^4D_{\alpha\beta}}{dt^4}\frac{dD_{\alpha\beta}}{dt}\right) = -\frac{G}{45c^5}\frac{d^3D_{\alpha\beta}}{dt^3}\frac{d^3D_{\alpha\beta}}{dt^3}, \tag{106}$$

where

$$\mathscr{T}_{\alpha\beta} = \frac{1}{2}\int_V \rho\upsilon_\alpha\upsilon_\beta d\vec{x} \tag{107}$$

denotes the kinetic-energy tensor and

$$D_{\alpha\beta} = 3I_{\alpha\beta} - \delta_{\alpha\beta}I_{\sigma\sigma} \tag{108}$$

defines the quadrupole moment of the system.

It appears that we must associate with the $2\frac{1}{2}$ -post-Newtonian approximation the energy

$$\int \mathscr{E}_5 dx = \frac{1}{c^5}\left[-\left(\overset{(5)}{Q_{00}}\mathscr{T}_{\mu\mu} + 2\overset{(5)}{Q_{\alpha\mu}}\mathscr{T}_{\alpha\mu}\right) - \frac{1}{3}G\frac{d^4I_{\alpha\beta}}{dt^4}\frac{dI_{\alpha\beta}}{dt}\right.$$
$$\left. + \frac{1}{18}G\frac{d^3D_{\alpha\beta}}{dt^3}\frac{d^2D_{\alpha\beta}}{dt^2} - \frac{1}{90}G\frac{d^4D_{\alpha\beta}}{dt^4}\frac{dD_{\alpha\beta}}{dt}\right]. \tag{109}$$

Including this energy together with $\mathscr{E}_{(2)}$, we can rewrite equation (106) in the form

$$\frac{d}{dt}\int \mathscr{E}_{(2.5)}dx = -\frac{G}{45c^5}\frac{d^3D_{\alpha\beta}}{dt^3}\frac{d^3D_{\alpha\beta}}{dt^3} + \frac{G}{45c^5}\frac{d^2}{dt^2}\left(\frac{d^3D_{\alpha\beta}}{dt^3}\frac{dD_{\alpha\beta}}{dt}\right)$$

$$-\frac{G}{6c^5}\frac{d^2}{dt^2}\left(2\frac{d^3I_{\alpha\beta}}{dt^3}\frac{dI_{\alpha\beta}}{dt} - \frac{d^2I_{\alpha\beta}}{dt^2}\frac{d^2I_{\alpha\beta}}{dt^2}\right). \tag{110}$$

We observe that the first term on the right-hand side of equation (110) is <u>negative definite</u>. It therefore represents <u>a secular decrease of the integrated energy of the system</u>. In contrast to the first term, the two remaining terms (being total time derivatives) may be expected to vanish when averaged over a long enough interval of time. We may therefore write

$$\left\langle \frac{d}{dt}\int \mathscr{E}_{(2\cdot5)}dx\right\rangle = -\frac{G}{45c^5}\left\langle\left|\frac{d^3D_{\alpha\beta}}{dt^3}\right|^2\right\rangle. \tag{111}$$

<u>This result is in exact agreement with the rate of emission of the gravitational radiation energy predicted by the linearized theory of gravitational radiation.</u>

In a similar way it can be shown that the angular momentum $L_{\gamma(2)}$ of the system, that is conserved in the second post-Newtonian approximation, is secularly affected in the center of mass frame in accordance with the equation

$$\left\langle \frac{d}{dt}\int L_{\gamma(2)}dx\right\rangle = -\frac{2G}{5c^5}\left\langle \frac{d^2I_{\alpha\mu}}{dt^2}\frac{d^3I_{\beta\mu}}{dt^3} - \frac{d^2I_{\beta\mu}}{dt^2}\frac{d^3I_{\alpha\mu}}{dt^3}\right\rangle \tag{112}$$

($\alpha \neq \beta \neq \gamma$, and α,β,γ in cyclical order).

This result is again in agreement with the predictions of the linearized theory of gravitational radiation.

Finally in Table 1 we summarize the extent to which the post-Newtonian scheme for relativistic hydrodynamics has now been carried out.

TABLE 1

INFORMATION ON THE METRIC COEFFICIENTS THAT IS

NEEDED IN THE VARIOUS APPROXIMATIONS

Equations of motion	Orders of the metric coefficients needed		
	$g_{\alpha\beta}$	$g_{0\alpha}$	g_{00}
Newtonian	$-\delta_{\alpha\beta}$	0	$1-2U/c^2$ *
$\frac{1}{2}$ -post-Newtonian	0 †	0 †	0 †
1 -post-Newtonian	$-2U\delta_{\alpha\beta}c^{-2}$	$P_\alpha c^{-3}$	$2(U^2-2\Phi)c^{-4}$
$1\frac{1}{2}$ -post-Newtonian	0 †	0 †	$Q_{00}^{(5)}c^{-5}$ ††
2 -post-Newtonian	$(\dots.)c^{-4}$	$(\dots)c^{-5}$	$(\dots.)c^{-6}$
$2\frac{1}{2}$ -post-Newtonian**	$Q_{\alpha\beta}^{(5)}$ ††	$Q_{0\alpha}^{(6)}$ ††	$Q_{00}^{(7)}c^{-7}$
3 -post-Newtonian	$(\dots.)c^{-6}$ §		
$3\frac{1}{2}$ -post-Newtonian	$Q_{\alpha\beta}^{(7)}c^{-7}$ §		

* The term $-2U/c^2$ in g_{00} is demanded by the principle of equivalence.

† These terms vanish by virtue of the laws of the conservation of mass and of linear momentum.

†† These are the lowest order terms in the metric coefficients that derive from the imposition of the Sommerfeld radiation-condition at infinity. (All the other terms in the table are obtained by solving the field equations.)

§ These terms are needed to determine the energy in the one lower approximation.

** This is the approximation in which radiative-reaction terms first appear.

The research reported in this paper has in part been supported by the Office of Naval Research under contract Nonr-2121(24) with the University of Chicago.

REFERENCES

Burke, W. 1969, "The Coupling of Gravitational Radiation to Non-relativistic Sources," Ph.D. Thesis, California Institute of Technology preprint.

Chandrasekhar, S. 1965, Ap. J., 142, 1488.

Chandrasekhar, S. 1969a, Ellipsoidal Figures of Equilibrium (New Haven: Yale University Press).

Chandrasekhar, S. 1969b, Ap. J. 158, 45.

Chandrasekhar, S., and Esposito, F. Paul 1970, Ap. J. 159.

Chandrasekhar, S., and Nutku, Y. 1969, Ap. J., 158, 55.

Synge, J. L. 1960, Relativity: The General Theory (Amsterdam: North-Holland Publishing Company); see pp. 252-255.

Thorne, K. S. 1969, Ap. J. 158, 1.

Trautman, A. 1958a, Bull. Acad. Polon. Sci., 6, 627.

Trautman, A. 1958b, Lectures on General Relativity; mimeographed notes (London: King's College).

RELATIVISTIC BOLTZMANN THEORY AND THE GRAD METHOD OF MOMENTS

James L. Anderson

Stevens Institute of Technology

Hoboken, New Jersey

INTRODUCTION

It has been the custom, when writing on relativistic Boltzmann theory, to justify such studies by recounting the various physical systems to which they can apply. By now one is sufficiently acquainted with relativistic plasmas, massive stellar systems and the like to make such justifications unnecessary. While applicability is of course the final justification for any physical theory, there is one other that I would like to mention briefly. It is the aesthetic appeal, so often emphasized by Dirac. The relativistic Boltzmann equation is both simple and elegant. From it one can obtain many beautiful results, such as those of Ehlers, Gerun and Sachs. It brings into play virtually the whole of relativity theory in one way or another and is amenable to analysis by such modern mathematical techniques as fiber bundle theory. It also affords a unifying view that is lacking in the classical theory. One need only compare the classical and relativistic treatments of radiative transfer theory, which is a special application of the Boltzmann equation to zero rest-mass particles to appreciate this fact. Finally I would mention that, even in the more mundane matter of finding approximate solutions, there are decided advantages to a relativistic treatment over the corresponding classical treatment.

Having established the desirability of studying the relativistic Boltzmann equation I can now address myself to the main subject of this paper, namely the relativistic version of the Grad[1] method of moments for constructing approximate solutions of this equation. The Grad method has much to recommend it over the older method of Chapman-Enskog-Hilbert (referred to hereafter as the CEH method.) It is at the same time simpler and more informative. The CEH method is restricted to normal solutions, namely those distribution functions that can be expressed as functionals of its first five moments. The Grad method, on the other hand, allows one to construct solutions that depend on a much larger set of moments. Furthermore, the Grad method yields closed form expressions for the various transport coefficients that arise in the case of normal solutions for arbitrary cross-sections while the CEH method gives them as a power series and then only for Maxwellian cross-sections. Finally it is not even clear that these series converge, at least in the relativistic case. The most extensive discussion to date of the relativistic CEH method has been given by Israel[2)] who showed that a relativistic perfect gas possessed a bulk viscosity. However, the method was so cumbersome that Israel was unable to obtain (or at least did not give in his paper) an expression for the shear viscosity. Furthermore, his results were restricted, as in the classical case, to Maxwellian cross-sections. Somewhat later Chernikov[3)] published a relativistic version of the Grad method. However, his development contained a number of inconsistencies and, among other things, he did not predict a bulk viscosity for the perfect gas. Recently Marle[4)] and John Stewart and myself have developed similar versions of the Grad method that are free of the inconsistencies in the Chernikov work. Marle so far has applied his version to a relaxation-time approximation of the relativistic Boltzmann equation while Stewart and I have applied it to the full equation, that is, to the Boltzmann equation with a collision term that takes account of binary collisions. It is this latter work that I shall report on here.

THE RELATIVISTIC BOLTZMANN EQUATION

Before one can write down a Boltzmann equation of any kind it is first necessary to introduce a one-particle distribution function. In classical theory this distribution function is taken to be a function of the particle position $\underset{\sim}{x}$, particle velocity $\underset{\sim}{v}$ and the time t. Then $f(\underset{\sim}{x},\underset{\sim}{v},t)d^3\underset{\sim}{x}d^3\underset{\sim}{v}$ is defined to be the number of particles in $d^3\underset{\sim}{x}$, $d^3\underset{\sim}{v}$ at $\underset{\sim}{x}$, $\underset{\sim}{v}$ and t. In the relativistic case the distribution function is taken to be a function of the four-coordinates of the particle x^μ and the four-momentum p^μ with the proper-time as the path parameter so that $p^\mu \equiv m\dot{x}^\mu$ where m is the rest-mass of a particle and a dot over a quantity indicates differentiation with respect to proper time. Then

$$dN \equiv f(x,p)p^\mu \, dS_\mu \, dP \tag{1}$$

is the number of world-lines transversing an element dS_μ on a space-like hypersurface with four-momentum in dP. The element dP in momentum space is given by

$$dP = \delta^{+}(p^2 - m^2)d^4p \tag{2}$$

where the positive-frequency delta-function $\delta^{+}(z^2 - a^2)$ is defined by

$$\delta^{+}(z^2 - a^2) \equiv \frac{1}{|a|}\ \delta(z-|a|). \tag{3}$$

It is included in the expression for dP to take account of the fact that the particles of the gas (all assumed to have equal mass) are constrained to move on the mass-shell $p^2-m^2=0$ in momentum space. Since dN, $p^{\mu}dS_{\mu}$ and dP are all obviously scalars under a mapping of the space-time manifold into itself it follows that f is also a scalar under such a mapping. If an observer is locally at rest with respect to the element of hypersurface dS_{μ} then $p^{\mu}dS_{\mu}=p^{0}d^3x$. If we then make use of the expression (2) for dP together with that for $\delta^{+}(z^2-a^2)$ given by Eq. (3) we see that the expression (1) for dN reduces to the usual non-relativistic form.

In the absence of interactions the one-particle distribution function must satisfy the relativistic Liouville equation first given by Walker:[5)]

$$m\frac{D}{D\tau}f \equiv p^{\mu}\frac{\partial f}{\partial x^{\mu}} + mF^{\mu}\frac{\partial f}{\partial p^{\mu}} = 0 \tag{4}$$

where F^{μ} is the external force acting on the particles of the gas. For particles moving in a combined gravitational, electromagnetic field mF^{μ} is given by

$$mF^{\mu} = eF^{\mu}{}_{\nu}p^{\nu} + \{{}^{\mu}_{\rho\sigma}\}p^{\rho}p^{\sigma} \tag{5}$$

where $F^{\mu}{}_{\nu}$ is the electromagnetic field tensor and $\{{}^{\mu}_{\rho\sigma}\}$ are the Christoffel symbols associated with the gravitational field.

Let us now try to take account of the interactions between particles in the gas. One way of doing this is by means of the so-called "self-consistent field" approximation. In this approximation one assumes that each particle "sees" only an average force produced by all of the other particles. In the case of an electromagnetic interaction the field $F^{\mu}{}_{\nu}$ is determined by the Maxwell equations with the gas as a whole acting as its source. Thus we have, as the equation determining this field

$$F^{\mu\nu}{}_{;\nu} = e\int p^{\mu}f\,dP \tag{6}$$

where the semicolon denotes covariant differentiation. Eqs. (4) and (6) together constitute the relativistic generalization of the classical Vlasov equations. If there is also a graviataional interaction then the gravitational field $g_{\mu\nu}$ is determined by the

Einstein equations

$$R^{\mu\nu} - \frac{1}{2}g^{\mu\nu}R = \kappa\int p^\mu p^\nu f dP \tag{7}$$

where $\int p^\mu p^\nu f dP$ is the stress-energy tensor for the gas. The justification for using the particular source terms in Eqs. (6) and (7) follows from the fact that they are both covariantly conserved as a consequence of Eq. (4).

While the Vlasov-type equations take account partially of the long-range electromagnetic and gravitational interactions (one would have to include Fokker-Planck type terms in a more accurate description) they do not properly take account of short-range interactions, e.g., scattering processes. One can try to include these effects in an approximate way by means of a relaxation-time approximation as Marle has done. However, the appropriate relativistic generalization of the classical relaxation-time approximation is not completely straightforward since among other things, as we shall see, there are three relaxation times associated with a relativistic gas. A more accurate description is obtained with the help of the collision matrix $W(p,p';p'',p''')$, where p, p' are the momenta of the incident particles and p'', p''' are the momenta of the scattered particles. The quantity $W(p,p';p'',p''')\ f(x,p)f(x,p')dPdP'dP''dP'''d\tau$ gives the number of collisions in the four-volume $d\tau$ at x^μ between particles with initial momenta p, p' in the ranges dP, dP' and final momenta p'', p''' in the ranges dP'', dP'''. If the scattering matrix is invariant under both time reversal and space reflection it satisfies the relation

$$W(p,p';p'',p''') = W(p'',p''';p,p'). \tag{8}$$

With the help of the collision matrix and the assumption of molecular chaos one can determine the change in the distribution function due to collisions in a way that is completely analogous to the classical derivation.[7] The result is the full Boltzmann equation

$$p^\mu \frac{\partial f}{\partial x^\mu} + mF^\mu \frac{\partial f}{\partial p^\mu} = \iiint W(p,p';p'',p''')$$

$$(f''f''' - ff')dP'dP''dP''' \tag{9}$$

where f', etc., indicates that f is a function of the momentum p', etc. In this equation F can either be an external or a self-consistent force as in the Vlasov approximation. The "full" Boltzmann equation of course still affords only an approximate description of the gas since it only takes into account binary collisions and neglects particle correlations. It would be desirable to obtain the relativistic Boltzmann equation from a more fundamental set of equations such as the BBGKY hierarchy as is done in the classical theory in order to assess properly its range of validity. However, to date no one has succeeded in constructing such a hierarchy for a relativistic system.

Before we examine some of the immediate consequences of the relativistic Boltzmann equation it will be desirable to set down the relation between the collision matrix W and the differential cross-section σ since there seems to be some question concerning this point in the literature. To do so we define first the following quantities:

$$g^\mu \equiv p^\mu - p'^\mu, \quad g'^\mu \equiv p''^\mu - p'''^\mu \tag{10}$$

$$g \equiv (-g^\mu g_\mu)^{1/2} = (-g'^\mu g'_\mu)^{1/2} \equiv g' \tag{11}$$

$$q^\mu \equiv Q^{-1}(p^\mu - p'^\mu) = Q'^{-1}(p''^\mu - p'''^\mu) \equiv q'^\mu \tag{12}$$

where $Q = (4 + g^2/m^2)^{1/2}$, and

$$\cos\Theta \equiv g^\mu g'_\mu/g^2. \tag{13}$$

The equalities in Eqs. (11) and (12) follow from the conservation laws that hold during a collision

$$p^\mu + p'^\mu = p''^\mu + p'''^\mu. \tag{14}$$

By making use of the definition of the differential cross-section $\sigma(g,\Theta)$ for an axial symmetric scattering and the identity

$$dPdP' = \frac{1}{2m} Qg^2 dg d\Omega d\Pi, \tag{15}$$

where $d\Omega$ is an element of solid angle about g'^μ and $d\Pi = \delta^+(q^2 - m^2)$, one can show that

$$W(p,p';p'',p''') = \frac{m^2}{g}\delta_Q(q^\mu - q'^\mu)\delta(g-g')\sigma(g,\Theta). \tag{16}$$

In this last equation $\delta_Q(q^\mu - q'^\mu)$ is the delta-function on the mass-shell with the property that

$$\int \Xi(q)\delta_Q(q^\mu - q'^\mu)d\Pi = \Xi(q'). \tag{17}$$

ELEMENTARY PROPERTIES

One can derive a number of importantconservation laws from the Boltzmann equation. The particle number flux vector is defined to be

$$N^\mu = \int p^\mu f dP. \tag{18}$$

If we make use of Eqs. (2) and (3) we find that its conponents are of the form

$$N^{\mu} = (\int fd^3p,\ \int \underset{\sim}{v} fd^3p) \tag{19}$$

where the three-velocity $\underset{\sim}{v} = \underset{\sim}{p}/p^o$. If now we integrate Eq. (9) over momentum space and make use of the symmetry relation (8) we find that

$$N^{\mu}{}_{,\mu} = 0, \tag{20}$$

which expresses the conservation of particle number.

Likewise we can define a stress-energy tensor

$$T^{\mu\nu} = \int p^{\mu}p^{\nu}fdP. \tag{21}$$

Again using Eqs. (2) and (3) we find, for the components of $T^{\mu\nu}$

$$T^{\mu\nu} = \begin{matrix} \int p^o fd^3p & \int \underset{\sim}{p} fd^3p \\ \int \underset{\sim}{p} fd^3p & \int \underset{\sim}{p}\underset{\sim}{v} fd^3p \end{matrix} \tag{22}$$

If we multiply Eq. (9) by p^{ν} and integrate over momentum space we find, using the symmetry relation (8) and the conservation laws (14), that

$$T^{\mu\nu}{}_{;\nu} = 0 \tag{23}$$

giving the conservation laws for energy and momentum.

One can also define an entropy flux vector

$$s^{\mu} \equiv \int p^{\mu} f \ln f\ dP \tag{24}$$

with components

$$s^{\mu} = (\int f \ln f d^3p,\ \int v f \ln f d^3p). \tag{25}$$

Although s^{μ} is not in general covariantly conserved it can be shown, as a consequence of the Boltzmann equation that

$$s^{\mu}{}_{;\mu} \geq 0. \tag{26}$$

If s^{μ} is covariantly conserved, corresponding to zero entropy production or equilibrium, then $\ln f$ must be of the form

$$\ln f = \alpha(x) - \beta^{\mu}(x)p_{\mu} \tag{27}$$

where $\alpha(x)$ is an arbitrary space-time function and $\beta^{\mu}(x)$ is an arbitrary time-like vector. The requirement that β^{μ} be time-like is necessary in order that f be bounded for arbitrarily large momenta. An f of this form will cause the right side of the Boltzmann equation to vanish for arbitrary α and β^{μ}. However, the left side will only vanish in the absence on an electromagnetic force if

$$\alpha_{,\mu}=0 \text{ and } \beta_{\mu;\nu} + \beta_{\nu;\mu} = 0. \tag{28}$$

Thus β^{μ} must be a time-like Killing vector of whatever gravitational field, external or self-consistent, is present. If the gas consists of zero rest-mass particles β^{μ} need only be a conformal time-like Killing vector, i.e., it need only satisfy the condition

$$\beta_{\mu;\nu} + \beta_{\nu;\mu} = \frac{1}{4} g^{\rho\sigma}\beta_{\rho;\sigma}g_{\mu\nu}. \tag{29}$$

The fact that β^{μ} must be a time-like Killing vector poses certain problems for cosmological theory as was first pointed out by Ehlers. Since no evolutionary cosmological models possess a time-like Killing vector there are no equilibrium solutions to the Boltzmann equations for these models. The only exception is the case of a radiation filled (or ultra-relativistic gas filled) universe, since some cosmological models, e.g., the Friedmann models, do admit conformal time-like Killing vectors. The reason for this exception is fairly clear. In the Friedmann models the momentum of a freely falling body scales inversely as the radius R. Hence, for zero rest-mass particles their energy, which is equal to the magnitude of their momentum, also scales as R^{-1}. Now for an equilibrium distribution, $\ln f \sim -E/kT$. Therefore if the temperature also scales as R^{-1}, $\ln f$ will remain constant and there will be no new entropy production. However, for finite rest-mass particles $E = \sqrt{(p^2 + m^2)}$ and there is no scaling of T that will keep $\ln f$ constant. Only in the extreme relativistic (and nonrelativistic) limits will such scalings exist. Therefore if there are epochs during the evolution of the universe when matter at relativistic temperatures dominates, or at least plays a significant role, in the evolution the matter will not be in equilibrium and there will be a corresponding entropy production. One such epoch might be that just preceding the period of helium production when electron pairs are in abundance. Since the amount of helium produced depends critically on how long the universe spends in this epoch and dissipation tends to slow down the expansion it may be necessary to take account of this slowing down in calculating the helium abundance in various models.

One might ask what mechanism is responsible for the generation of entropy in a gas filled universe. Classically, the two mechanisms, at least for an ideal gas, are heat conduction and shear viscosity. But in the Friedmann models at least, neither of these two mechanisms can play a role since these models are both isotropic and shear free. However, in a relativistic gas there is a third dissipative mechanism, namely, bulk viscosity. Shücking has suggested that this is just the mechanism that is responsible for the absence of equilibrium in the evolutionary models. As we shall see, the coefficient of bulk viscosity for a relativistic gas has just the right temperature dependence to support this suggestion since it goes to zero in both the low- and high- temperature limits.

EQUILIBRIUM DISTRIBUTION

Although we shall be mainly concerned with non-equilibrium distributions some of the properties of the equilibrium distribution will be needed in our discussion. We shall outline these properties in this section.

If the one-particle distribution function is an equilibrium distribution it must have the form given by Eq. (27), that is, it must be of the form

$$f_{eq} = e^{\alpha-\beta u^{\mu}p_{\mu}} \tag{30}$$

where

$$\beta^2 = \beta^{\mu}\beta_{\mu} \qquad \text{and} \quad u^{\mu} = \beta^{\mu}/\beta. \tag{31}$$

The particle number flux vector, defined by Eq. (18) has, in this case, the form

$$N^{\mu}{}_{eq} = \rho u^{\mu} \tag{32}$$

where

$$\rho = 4\pi m^2 e^{\alpha} K_2(m\beta)/\beta \tag{33}$$

and K_2 is a modified Bessel function of order two. We see that u^{μ} is just the mean velocity of the gas and ρ is the rest-particle number density. Likewise, the stress-energy tensor given by Eq. (21) has the form

$$T^{\mu\nu}{}_{eq} = (\mu + P)u^{\mu}u^{\nu} - Pg^{\mu\nu} \tag{34}$$

where

$$\mu + P = 4\pi e^{\alpha}m^3 K_3(m\beta)/\beta \tag{35}$$

and

$$P = 4\pi e^{\alpha}m^2 K_2(m\beta)/\beta^2. \tag{36}$$

This is just the pehnomenological form for the stress-energy tensor of a perfect fluid with μ the rest-energy density and P the pressure. We note, for future reference that, in the local rest-frame, i.e., in the frame in which $u^\mu = (1,\underset{\sim}{0})$, there is neither a net particle flux nor a net momentum flux.

THE GRAD METHOD

To obtain information about non-equilibrium solutions of the Boltzmann equation it is in general necessary to resort to approximation methods. In this section I shall outline the relativistic version of the Grad method as developed by Stewart and myself.[8] In this method one starts with a zero-order distribution function f_o which may, but need not, be a local equilibrium distribution. In applying the method to a stellar system one might use for f_o a truncated equilibrium distribution, that is, a distribution function that is equal to an equilibrium distribution for momenta less than the escape momentum from the system and zero otherwise. With f_o as a weight function one then proceeds to construct an orthogonal set of polynomials in the momentum by a Schmidt orthogonalization process. We will designate a polynomial of this set by

$$\overset{m}{H}{}^{\alpha}(x,p)$$

where m is the order of the polynomial and α is a generic index appropriate to that order. The orthogonality of these polynomials is expressed by the requirement that

$$(\overset{m}{H}{}^{\alpha},\ \overset{n}{H}{}^{\beta}) \equiv \int f_o \overset{m}{H}{}^{\alpha}\overset{n}{H}{}^{\beta} dP = \overset{n}{M}{}^{\alpha\beta}\delta^{mn}. \tag{37}$$

One begins the process by taking $\overset{o}{H} = 1$. The first order polynomials are taken to be of the form

$$\overset{1}{H}{}^{\mu} = p^\mu - a^\mu \overset{o}{H}. \tag{38}$$

The requirement that $(\overset{o}{H},\ \overset{1}{H}{}^{\mu}) = 0$ determines a^μ to be

$$a^\mu = N^\mu_{\ o}/A_o \tag{39}$$

where naught subscript refers to moments of f_o, e.g.,

$$A_o = \int f_o dP. \tag{40}$$

One then proceeds to construct $\overset{2}{H}{}^{\mu\nu}$ by taking it to be of the form

$$\overset{2}{H}{}^{\mu\nu} = p^\mu p^\nu - a^{\mu\nu}{}_\rho \overset{1}{H}{}^\rho - b^{\mu\nu}\overset{o}{H} \tag{41}$$

and requiring that $(\overset{o}{H}, \overset{2}{H}{}^{\mu\nu}) = (\overset{1}{H}{}^\rho, \overset{2}{H}{}^{\mu\nu}) = 0$. These latter conditions again determine the coefficients $a^{\mu\nu}{}_\rho$ and $b^{\mu\nu}$ in terms of moments of f_o. In a like manner one can continue the process to construct all of the higher order polynomials. However, for most applications only the first three will usually be needed.

Having constructed the $\overset{m}{H}{}^\alpha$ as above, one now expands the actual distribution function f as

$$f = f_o\left(\sum_{m=0}^{\infty} \overset{m}{a}_\alpha(x)\overset{m}{H}{}^\alpha(x,p)\right). \tag{42}$$

By making use of the orthogonality relations (37) one can determine the coefficients in the expansion in terms of moments of f and vice versa. In general one finds that the nth order coefficients depend on the nth and lower order moments of f. By multiplying the Boltzmann equation by the various order polynomials and integrating over momentum space one obtains a hierarchy of equations for the coefficients or, equivalently, for the moments of f. Multiplying by $\overset{o}{H}$ yields the conservation law

$$N^\mu{}_{,\mu} = 0 \tag{43}$$

while multiplying by $\overset{1}{H}{}^\mu$ gives

$$T^{\mu\nu}{}_{;\nu} = 0. \tag{44}$$

Multiplication by $\overset{2}{H}{}^{\mu\nu}$ leads to the first equation that involved the collision matrix,

$$S^{\mu\nu\rho}{}_{;\rho} = \sum_{r,s} \overset{2rs}{B}{}^{\mu\nu\alpha\beta} \overset{r}{a}_\alpha \overset{s}{a}_\beta \tag{45}$$

where $S^{\mu\nu\rho} = \int p^\mu p^\nu p^\rho f dP$ is the third moment of f and

$$\overset{nrs}{B}{}^{\alpha\beta\gamma} = \frac{1}{2}\iiiint W(p,p';p'',p''')(\overset{n}{H}{}''^{\alpha} - \overset{n}{H}{}^{\alpha}) f_o f'_o \times \overset{r}{H}{}^\beta \overset{s}{H}{}^\gamma dP dP' dP'' dP'''. \tag{46}$$

Multiplication by higher order polynomials yields equations similar to Eq. (45). On the left side of the nth order equation appear nth and lower order coefficients (or moments) while on the right side coefficients (or moments) of all orders appear quadratically.

As they stand, the totality of these equations are equivalent to the original Boltzmann equation but are no easier to solve. The Grad approximation consists in setting all of the coefficients past a certain order equal to zero thereby reducing the infinite set of equations to a finite number. This approximation also allows one to express the higher moments of f in terms of a finite set of lower moments. It is in no way equivalent, therefore, to setting all of the moments of f equal to zero past a certain order as is sometimes done.

Since one only reproduces the perfect fluid theory by taking $\overset{n}{a}_\alpha = 0$ for $n \geq 2$, the first deviations from this theory arises when one takes $\overset{n}{a}_\alpha = 0$ for $n \geq 3$. It is in this approximation that one first gets non-zero contributions to those parts of the stress-energy tensor that correspond to heat conduction and viscosity. It is perhaps worthwhile to point out that in the corresponding classical case it is necessary to retain a subset of the third order coefficients in order to take account of heat conduction. The reason for this difference lies in the fact that in the classical case one works with the three-momentum while in the relativistic case one works with the four-momentum. Thus classically, heat conduction will involve third moments of the distribution function, i.e., $\int p^2 \underset{\sim}{p} f d^3p$ while in the relativistic case it will involve only second moments of the form $\int p^o p^r f dP$. For this reason the whole formalism and many of the calculations of the Grad method are simpler and more symmetric in the relativistic case. Vivre relativity.

The question of the validity of the Grad approximation is not easy to answer in general. The best that one can do at present is to increase the order of the approximation and calculate the effect on whatever quantities one is interested in computing. This, at least in the Grad method is doable although the effort involved is rather enormous. Alternately, one can try to compare the predictions of the approximation with experiment. Al least in the classical case these predictions agree remarkably well with experiment. By using an appropriate zero-order distribution function, Mintzer[9)] has obtained good agreement with experiments on the noble gases all the way down to the Knudsen limit.

When one takes $\overset{n}{a}_\alpha = 0$ for $n \geq 3$ one is left with 14 unknowns, one more than in the classical case because of the bulk viscosity. The unknowns can be taken to be either the coefficients $\overset{0}{a}$, $\overset{1}{a}_\mu$, $\overset{2}{a}_{\mu\nu}$ with $g^{\mu\nu}\overset{2}{a}_{\mu\nu} = 0$ or the moments A (= $\int fdP$), N^μ, $T^{\mu\nu}$ with $g_{\mu\nu}T^{\mu\nu} = m^2 A$. The one set can be expressed as linear functions of the other as explained above. For some calculations, such as those for relaxation processes, the coefficients are easier to work with while for others, such as those for transport processes, the moments are to be preferred since they have a more direct physical interpretation.

The equations for these quantities are given by Eqs. (43), (44) and (45) where, in Eq. (45), the third moment $S^{\mu\nu\rho}$ is expressed in terms of lower moments by making use of the fact that $\overset{3}{a}_{\mu\nu\rho} = 0$. One finds thereby that

$$S^{\mu\nu\rho} = S_o^{\ \mu\nu\rho} A/A_o + b^{\mu\nu\rho}{}_{\sigma}(N^{\sigma} - N_o^{\ \sigma}A/A_o) +$$

$$a^{\mu\nu\rho}{}_{\sigma\tau}\{T^{\sigma\tau} - a^{\sigma\tau}{}_{\lambda}(N^{\lambda} - N_o^{\ \lambda}A/A_o) -$$

$$T_o^{\ \sigma\tau}A/A_o\} \tag{47}$$

where the naught subscripts again refer to moments of f_o and where $b^{\mu\nu\rho}{}_{\sigma}$ and $a^{\mu\nu\rho}{}_{\sigma\tau}$ are the coefficients of $\overset{1}{H}{}^{\sigma}$ and $\overset{2}{H}{}^{\sigma\tau}$ in the polynomial $\overset{3}{H}{}^{\mu\nu\rho}$.

Depending on the choice of f_o, the 14 equations (43), (44) and (45) plus boundry and/or initial conditions may or may not be sufficient to determine the unknowns. This is because these equations determine the unknowns in terms of moments of f_o which might itself contain unknown functions. Thus, if we take f_o to be a local Maxwellian distribution of the form given by Eq. (30), the 5 quantities α, β and u^{μ} must also be determined as functions of position and time. We can fix these quantities by requiring that certain moments of f be equal to the corresponding moments of f_o. To a certain extent which moments one chooses for the matching are arbitrary. The natural requirement that the particle-number density and energy density calculated using f be the same functions of α and β as they are when calculated using f_o leads to the two matching conditions

$$u_{\mu}N^{\mu} = u_{\mu}N_o^{\ \mu} \tag{48}$$

and

$$u_{\mu}u_{\nu}T^{\nu\mu} = u_{\mu}u_{\nu}T_o^{\ \mu\nu}. \tag{49}$$

However, three more conditions are required to completely specify the problem. These latter conditions must define the local rest frame, i.e., must fix u_{μ}. For this prupose one can follow Eckart[10]) and require, in addition to the condition (49), that

$$N^{\mu} = N_o^{\ \mu}. \tag{50}$$

In this case there will be no net particle flux in the local rest frame although there will be, in general, a net momentum flux. Or one can follow the procedure of Landau and Lifshitz[11)] and require condition (48) together with the condition

$$u_\mu T^{\mu\nu} = u_\mu T_o^{\mu\nu}, \tag{51}$$

in which case there will be no net momentum flux and a finite particle flux in the local rest frame. As we shall see, the two alternate matching conditions lead to identical results for one special set of solutions.

NORMAL SOLUTIONS AND TRANSPORT EQUATIONS

The existence of normal solutions presupposes that the actual distribution f differs from the local equilibrium distribution by a small amount. Accordingly we can take the unknown moments of f to differ from the corresponding moments of f_{eq} by small amounts. Thus we take

$$A/A_{eq} = 1 - \varepsilon\gamma \tag{52}$$

$$N^\mu = N_{eq}^{\ \mu} + \varepsilon(\xi u^\mu + \nu^\mu) \tag{53}$$

and

$$T^{\mu\nu} = T_{eq}^{\ \mu\nu} + \varepsilon\{(\delta+\tau)u^\mu u^\nu - \tau g^{\mu\nu} + q^\mu u^\nu + q^\nu u^\mu + \pi^{\mu\nu}\} \tag{54}$$

where

$$u_\mu \nu^\mu = u_\mu q^\mu = u_\mu \pi^{\mu\nu} = g_{\mu\nu}\pi^{\mu\nu} = 0 \text{ and } 1 \gg \varepsilon. \tag{55}$$

If we impose the matching conditions (48), (49) we must take

$$\xi = \delta = 0. \tag{56}$$

The Eckart condition (50) further requires us to take

$$\nu^\mu = 0 \tag{57}$$

while the Landau-Lifshitz condition (51) demands that

$$q^\mu = 0. \tag{58}$$

The remaining moments can now be determined as functions of α, β, u^μ and their gradients by means of the moment equation (46).

With Eckart matching one finds

$$\varepsilon\tau = -\xi\theta, \quad \theta = u^{\mu}{}_{;\mu} \tag{59}$$

$$\varepsilon\pi^{\mu\nu} = -\eta\sigma^{\mu\nu}, \quad \sigma^{\mu\nu} = (u^{\mu;\nu} + u^{\nu;\mu} - \dot{u}^{\mu}u^{\nu} - \dot{u}^{\nu}u^{\mu}) \tag{60}$$

where $\dot{u}^{\mu} = u^{\mu}{}_{;\nu}u^{\nu}$ and

$$\varepsilon q^{\mu} = -\kappa(u^{\mu}u^{\nu} - g^{\mu\nu})\{\dot{u}_{\nu}/m\beta + (1/m\beta)_{,\nu}\}. \tag{61}$$

From their form and their appearance in $T^{\mu\nu}$ we see that $\varepsilon\tau$ is the bulk stress, $\varepsilon\pi^{\mu\nu}$ the shear stress and εq^{μ} the energy (heat) flux. The corresponding transport coefficients ξ, η and κ are functions of α, β and the $\overset{202}{B}{}^{\mu\nu\rho\sigma}$. The latter quantities can be shown to depend linearly on three collision integrals involving the differential cross-section σ. This is in contrast to the classical case where the analogous quantities depend on only one collision integral. Since these quantities play the role of relaxation times as well in the moment equations we see that there are three relaxation times in the relativistic case as opposed to only one in the classical case.

If one uses the Landau-Lifshitz matching conditions one recovers Eqs. (59) and (60) while Eq. (61) gets replaced by the equation

$$\varepsilon\nu^{\mu} = \lambda(u^{\mu}u^{\nu} - g^{\mu\nu})\{u_{\nu}/m\beta + (1/m\beta)_{,\nu}\} \tag{62}$$

for the particle number flux vector $\varepsilon\nu^{\mu}$. Although $q^{\mu} = 0$ with the Landau-Lifshitz conditions one can still define an energy flux relative to the particle number flux, i.e., one takes the quantity

$$\varepsilon\bar{q}^{\mu} = (N^{\mu}N^{\nu} - g^{\mu\nu})N^{\rho}T_{\nu\rho}, \quad N^{\mu} \equiv N^{\mu}/|N^{\rho}N_{\rho} \tag{63}$$

to represent this flux. To the extent that one neglects terms of order ε^2 in computing this quantity it yields the same result as that given for εq^{μ} by Eq. (61). Since we have consistently neglected terms of this order in obtaining Eqs. (59-62) we see that the Eckart and Landau-Lifshitz matchings lead to equivalent results. Ever since Eckart and Landau and Lifshitz first proposed their phenomenological theories there has been a question as to which one is the more correct since in general the two descriptions are not equivalent. Our analysis leads to the conclusion that the two descriptions are equivalent when they are applicable and inequivalent when they are not.

ASYMPTOTIC FORMS

We conclude this brief description of the Grad method by

giving asymptotic forms for the transport coefficients ξ, η and κ. For $mc^2 >> kT$ one finds

$$\eta = \frac{5}{8}(\pi mkT)^{1/2} J^{-1}\{1 - (J^{-1}K_{1/4} - \frac{121}{16})(kT/mc^2 + \ldots)\} \tag{64}$$

$$\kappa = \frac{75}{32} (\pi kT/m)^{1/2} kJ^{-1}\{1 - (J^{-1}K_{-1/4} - \frac{77}{16})(kT/mc^2 + \ldots)\} \tag{65}$$

$$\xi = \frac{375}{384} mc^{-6}(kT/m)^{7/2} J^{-1}\{1-(J^{-1}K_{-3/4} + \frac{142}{16})(kT/mc^2 + \ldots)\} \tag{66}$$

where the dots indicate higher order terms in kT/mc^2,

$$J = \int_0^\infty e^{-s^2} s^7 \bar{\sigma}(2\sqrt{kT/m}s)ds \tag{67}$$

and

$$K_a = \int_0^\infty e^{-s^2}(as^9 + \frac{1}{4} s^{11})\bar{\sigma}(s\sqrt{kT/m}s)ds \tag{68}$$

with

$$\bar{\sigma}(z) \equiv 2\pi \int \sigma(z,\Theta)\sin^3\Theta d\Theta . \tag{69}$$

In the classical limit $c \to \infty$ we find that the coefficients of shear viscosity η and heat conduction κ reduce to their classical values as calculated from the classical Grad theory while the coefficient of bulk viscosity goes to zero as it should.

In the extreme relativistic limit $kT >> mc^2$ one finds the asymptotic forms

$$\eta \sim T \tag{70}$$

$$\kappa \sim \text{const.} \tag{71}$$

and

$$\xi \sim T^{-2} \tag{72}$$

so that in this limit ξ also goes to zero. Thus the bulk viscosity has just the right asymptotic limits to account for the dissipation in evolving models of the universe.

With the presentation of the above asymptotic forms our story has run its course for the moment. In many ways it must be considered as only a prologue to future work since much work remains to be done. At relativistic temperatures one must take account of elementary processes such as pair production and bremsstrahlung. Likewise one would have to study normal solutions of the Boltzmann equation in a cosmological field if one wants to take account of the effect of transport processes in the evolution of the universe. I believe that the relativistic Grad approximation will serve as a valuable tool in these investigations and it is for this reason that I have troubled the reader with this short introduction.

REFERENCES

1. For a discussion of this and other approximation methods together with further references see H. Grad, in Hb. d. Phys. XII, S. Flügge, ed. (Springer, Berlin, 1958).

2. W. Israel, J. Math. Phys., 4, 1163 (1963).

3. N. A. Chernikov, Acta Phys. Polonica, 27, 465 (1964).

4. C. Marle, Ann. Inst. Henri Poincaré, X, 67 (1969).

5. A. G. Walker, Proc. Edinburgh Math. Soc., 4, 238 (1936).

6. S.T. Beliaev and G. I. Budker; Soviet Physics, Doklady (translation) 1, 218 (1956).

7. see, for example, W. Israel, loc. cit.

8. to be published

9. D. Mintzer, Phys. of Fluids, 8, 1076 (1965).

10. C. Eckart, Phys. Rev. 58, 919 (1940)

11. L. Landau and E. M. Lifshitz, Fluid Mechanics (Addison-Wesley, Cambridge, Mass., 1959)

A LEMMA ON THE EINSTEIN-LIOUVILLE EQUATIONS

R. Berezdivin and R. K. Sachs*

University of California at Berkeley

1). Introduction

General relativistic kinetic theory provides an elegant approach to treating matter and gravity simultaneously; it is slightly less phenomenological than hydrodynamics. Since Anderson gave such a nice summary of the physics involved[1] we shall concentrate on some essentially geometric questions. The main question is to what extent symmetries of the matter are related to symmetries (isometries) of the space time. We start with some geometric preliminaries, then define matter symmetries, then state a lemma.

2). The Tangent Bundle[2]

Let M be a space-time, taken oriented and time oriented. The tangent bundle, or phase space, above M, call it T(M), is an 8

*Research supported in part by NSF grant GP-12996

[1] For surveys see J. Anderson's article in this volume, or J. Ehlers, and R. K. Sachs, article in the 1968 Brandeis summer school lectures.

[2] The geometry of the tangent bundle is discussed, for example, by P. Dombrowski, Journal fur die reine und angewandte Mathematik, Vol. 210, Book 1/2, (1962), p. 73. Physicists who find this article too heavily mathematical can rederive the essential results starting from the appendix to the paper by R. Lindquist, Annals of Physics 37, (1966), 487.

dimensional space one of whose points determines both a space-time point x and a vector at x. Thus if x^a (a = 1...4) are local coordinates for space-time then $y^A = [x^a, p^a]$ (A = 1...8) are acceptable local coordinates for phase space, where p^a are the contravariant components of a vector at x^a. We shall call p^a the momentum.

The tangent bundle has many special properties that a general 8-space lacks; one way to see these special properties is to note that the y^A defined above are a preferred kind of local coordinates with the special transformation law $\bar{y}^A = [\bar{x}^a(x^b), \frac{\partial \bar{x}^a}{\partial x^b} p^b]$. Thus there is a well defined "projection operator", π which takes each point P of T(M) into a point πP of M; in the special coordinates $(x^a, p^a) \overset{\pi}{\to} x^a$ (Fig. 1). Similarly there is a "drop" D which takes P into a vector of M at πP, namely $(x^a, p^a) \overset{D}{\to} p^a$. Moreover, consider the "fibre" or momentum space $\pi^{-1}\pi P$ above some point $x = \pi P$ of M; it consists of all vectors at x; therefore the fibre has linear structure - it makes sense to add two momenta at the same point and zero momentum is well defined - and the fibre is also a flat Minkowski space with metric

$$\phi = g_{ab}(x^c)dp^a dp^b \tag{2.1}$$

where $g_{ab}dx^a dx^b$ is the space-time metric.

The quantity

$$p^2 = g_{ab}(x^c)p^a p^b \tag{2.2}$$

is a scalar function on T(M).

We shall also need some properties of vector fields on T(M). Recall that if we have any contravariant vector field in any manifold, say $Z^A(u^B)$ (A, B = 1...N) then the differential operator $Z^A \frac{\partial}{\partial u^A} = Z$ acts on scalar functions $\psi(u^A)$ to take them into scalar functions $Z^A \psi_{,A}$: the operator has the properties

$$Z(\phi\psi) = (Z\phi)\psi + \phi(Z\psi) \tag{2.3}$$

$$Z(a\phi + b\psi) = aZ\phi + bZ\psi$$

for any scalar functions ϕ, ψ and numbers a,b. Conversely, giving

Figure 1 Phase Space T(M)

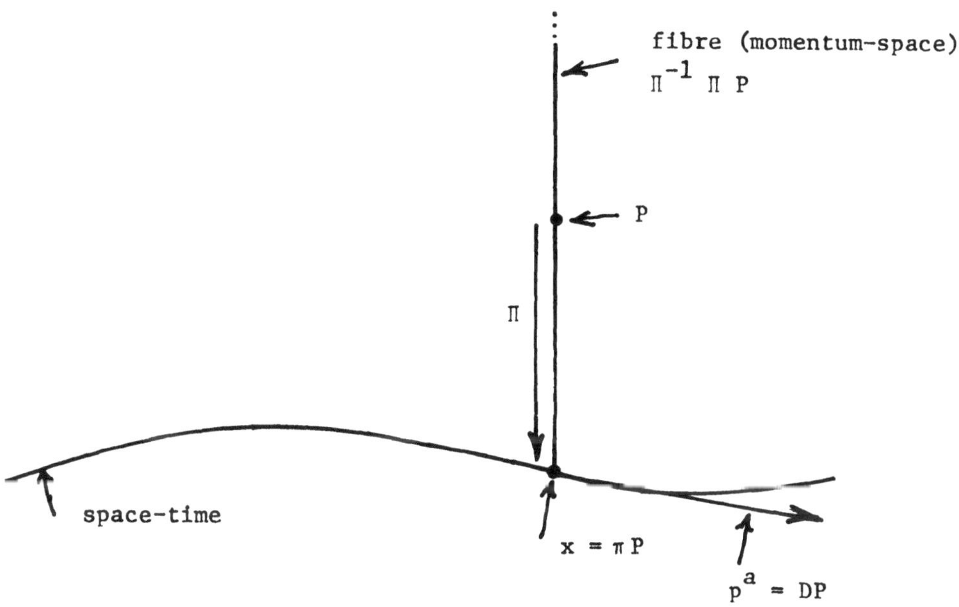

such a differential operator determines Z^A. In the tangent bundle we know what it means to have a "vertical" vector field - only the momenta vary, so that $Z = Z^a(y^B)(\partial/\partial p^a)$ characterizes a vertical vector field. Similarly, a "horizontal" vector field is one along which the momenta are parallel displaced, characterized by the equation $Z = Z^a(y^B)\nabla_a \equiv Z^a[(\partial/\partial x^a) - \Gamma_{ac}{}^{b} p^c(\partial/\partial p^b)]$, where $\Gamma_{bc}{}^{a}$ is the affine connection of space-time. Thus any vector field can be decomposed into its horizontal and vertical parts $Z = Z^A(\partial/\partial y^A) = H^a\nabla_a + V^a(\partial/\partial p^a)$ and then $Z^A = [H^a, V^a - \Gamma_{bc}{}^{c} p^b V^c]$. The differential operator notation is much easier to use than the component notation because of the following lemma: suppose $\omega_a(x^b)$ is any covariant vector field on M and we know $Z\psi$ for all those scalar functions $\psi(y^B)$ which have the special form $\psi = \omega_a p^a$. Then Z is completely determined. For example, the reader may check that

$$Z(\omega_a p^a) = \omega_{a;b} H^b p^a + \omega_a V^a \tag{2.4}$$

where $H^a\nabla_a$ and $V^a(\partial/\partial p^a)$ are the horizontal and vertical parts of Z and the semicolon is the usual covariant derivative for M. Now to calculate $Z[\omega_{ab}(x^c)p^a p^b]$ we can use (2.3) and (2.4) since a covariant tensor ω_{ab} of M can always be written as a sum of terms of the form $\omega_a\gamma_b$. The result is

$$Z(\omega_{ab}p^a p^b) = \omega_{ab;c}p^a p^b H^c + \omega_{ab}(p^a V^b + V^a p^b). \tag{2.5}$$

A special horizontal vector field on T(M) is the "geodesic spray" or Liouville operator

$$\mathcal{L} = p^a\nabla_a \tag{2.6}$$

Please work out the following characterization of $\mathcal{L}$:

(a) The projection πc, of a curve c to which $\mathcal{L}$ is everywhere tangent, is a geodesic of M; (b) The drop of a point P on c is the tangent to πc at πP. Note that $\mathcal{L}p^2 = 0$. Another kind of useful vector field can be obtained from a vector field $\xi^a(x^b)(\partial/\partial x^a)$ on M. The vector field $W = \xi^a\nabla_a + \xi^a{}_{;b}p^b(\partial/\partial p^a)$ is called the natural lift of $\xi^a(\partial/\partial x^a)$ from M to T(M). Note that if ξ^a is Killing, e.g.

$\xi_{a;b} = -\xi_{b;a}$ then displacements along W carry fibres into fibres linearly and isometrically. By a short calculation one deduces that the most general vector field W with these properties must have the form

$$W = H^a(x^b)\nabla_a + A^a_{\ .b}(x^d)p^b(\partial/\partial p^a) \tag{2.7}$$

where H^a is a vector field and $A_{ab} \equiv g_{ac}H^c_{\ .b}$ an antisymmetric tensor field on M.

Finally, recall that the Lie Bracket of any two vector fields, say W and $\mathscr{L}$, is defined by $[W,\mathscr{L}]\psi = W(\mathscr{L}\psi) - \mathscr{L}(W\psi)$ for all scalers ψ. From (2.6) and (2.7) we have, using (2.3),

$$[W,\mathscr{L}] = (H^a_{\ ;b} - A^a_{\ .b})\, p^b\nabla_a + (R^a_{\ .bcd}H^c - A^a_{\ .b;c})p^bp^c\nabla_a \tag{2.8}$$

where $R^a_{\ .bcd}$ is the curvature tensor of M.

3). Kinetic Theory and Matter Symmetries

In kinetic theory the matter is described by a distribution function f which is a scalar function on T(M). f is different from zero only if

$$p^ap^bg_{ab} \geq 0,\ p^4 > 0 \tag{3.1}$$

In general, the matter equation (Boltzmann equation) is $[\mathscr{L} - eF^a_{\ .b}p^b(\partial/\partial p^a)]\, f = (\delta f/\delta v)_{coll.}$ We shall here neglect collisions and assume no macroscopic electromagnetic field F_{ab} so our equation is the Liouville equation:

$$\mathscr{L}f = 0 \tag{3.2}$$

The self-consistent gravitational field generated by f is obtained from the Einstein field equations

$$G^{ab}(x) = -\int_{\pi^{-1}(x)} p^ap^bf\,dp \qquad (3.3$$

where G^{ab} is the Einstein tensor and $dp = \sqrt{-g}\, d^4p$ is the natural volume element for the fibre $\pi^{-1}(x)$ with metric (2.1). We shall now define what is meant by a "matter symmetry." Suppose an observer at a point x in space-time chooses a local Lorentz frame L and measures f on the fibre $\pi^{-1}(x)$. Suppose now that there is a point x' (which may, but need not, be the same as x) and a local Lorentz frame L' at x' such that f measured in L' on $\pi^{-1}(x')$ is identical with f measured in L on $\pi^{-1}(x)$. Then there is a matter symmetry from (x,L) to (x',L').

Please check the following statement: f permits a one parameter group of matter symmetries if and only if there is a vector field W on T(M) of the form (2.7) with

$$Wf = 0 \tag{3.4}$$

Note that if W obeys (2.7) and (3.4) so does $\psi(x)W$, where ψ is any scalar function on M.

4). A Lemma

We can now ask how closely the Liouville and Einstein equations (3.2) and (3.3) inter-relate matter symmetries with actual symmetries of space-time. More specifically, given (2.7), (3.2), (3.3), and (3.4) as we shall assume henceforth can we deduce $H^a_{;b} = A^a_{.b}$ so that W is the natural lift of $H^a(\partial/\partial x^a)$ and, from the antisymmetry of A_{ab}, H^a is Killing? That such inter-relations sometimes exist is shown by the following theorem:[3] If f permits a three parameter group of vertical matter symmetries, with the structure and action of SO_3 and $f \neq 0$ for some points in T(M) off the light cone ($p^2>0$), and M is not stationary then M is Robertson-Walker. Thus in this case $H^a = 0$ for each of three generators W and given P in T(M) there are always three Killing vectors ξ^a of M such that $\xi^a]_{\pi P} = H^a]_{\pi P} = 0$, $\xi^a_{;b}]_{\pi P} = A^a_{.b}]_{\pi P}$ for each of the three W.

We have proved an analogous result for another special case. Lemma: If f permits a one parameter group of horizontal matter symmetries ($A^a_{.b} = 0$) and W and $\mathcal{L}$ are surface forming then $(\psi(x)H^a)_{;b} = 0$ for suitable choice of ψ. (Thus also in this case $H^a_{;b} = A^a_{.b}$ when W is scaled appropriately).

---

[3] J. Ehlers, P. Geren, and R. K. Sachs, Jour. Math. Phys. 9, #9, Sept. 1968, 1344.

We sketch the proof. From the surface forming condition

$$[\mathcal{L},W] = \mathcal{L} + \beta W \tag{4.1}$$

Using $A^a_{.b} = 0$, (2.8), and (4.1) we get, after considerable algebra,

$$H_{a;b} = \omega_b(x)H_a + \beta(x)g_{ab} \tag{4.2a}$$

$$H^a R_{abcd} = 0 \tag{4.2b}$$

Thus $H = H_a dx^a$ is hypersurface orthogonal ($H \wedge dH = 0$) and therefore by rescaling, $H'^a = \psi H^a$, and dropping primes we can demand

$$H_{a;b} = \omega(x)\, H_a H_b + \beta g_{ab}. \tag{4.3}$$

Consider now a four dimensional region throughout which $H^a H_a$ is everywhere positive, negative, or zero. It turns out that the first possibility, H^a timelike is excluded by the energy inequalities (3.1) together with (3.3) and (4.2a). If H^a is lightlike or space-like we can find appropriate coordinate systems, adapted to H^a, in which the Einstein-Liouville equations simplify drastically. The analysis is different for the two cases and quite tedious in both so we merely quote the results. If H^a is lightlike (and scaled appropriately) then $H^a_{;b} = 0$ and $f \propto \delta(p^2)$ where δ is the Dirac delta function. More specifically, the solutions are plane wave metrics, $x^a = (x,y,u,v)$ and $ds^2 = -dx^2 - dy^2 + dudv + A(u,x,y)\, du^2$ with $H_a\, dx^a = du$, $f = f(u,p^3)\, \delta(p^2)\, \delta(p_1)\, \delta(p_2)$ and $(\partial^2/\partial x^2 + \partial^2/\partial y^2)A \propto \int_0^\infty f p^3\, dp^3$.

If H_a is space-like then also $H_{a;b} = 0$, $f \propto \delta(p^2)$. The most general solution has the form, $x^a = (x,x^\alpha)$ ($\alpha = 2,3,4$), $ds^2 = -dx^2 + g_{\alpha\beta}(x^\gamma)\, dx^\alpha dx^\beta$ with $f = \delta(p_1)\delta(p^2)g(x^\alpha,p^\beta)$ and $g\delta(p^2)$, $g_{\alpha\beta}$ are solutions of the Einstein-Liouville equations in 3 dimensions. Specific examples can be constructed; for example certain spatially homogeneous metrics of Bianchi type I. In all cases discussed here, dropping either the Liouville equation or the Einstein field equations breaks the interrelation between $A^a_{\bullet b}$ and $H^a_{;b}$; thus the Einstein Field equations here act as integrability conditions that the matter symmetry be the natural lift of an isometry.

GRAVITATIONAL RADIATION EXPERIMENTS * †

J. Weber

University of Maryland

College Park, Maryland

ABSTRACT

A description is given of the gravitational radiation experiments involving detectors at opposite ends of a 1000 kilometer baseline, at Argonne National Laboratory and the University of Maryland. Sudden increases in detector output are observed roughly once in several days, coincident within the resolution time of 0.25 seconds. The statistics rule out an accidental origin and experiments rule out seismic and electromagnetic effects. It is reasonable to conclude that gravitational radiation is being observed.

INTRODUCTION

Some years ago apparatus was proposed[1] to measure the Fourier transform of the Riemann Tensor and search for gravitational radiation. It employs an electronically instrumented solid with normal modes driven by the curvature tensor. A rigorous analysis has been given, employing the well known methods[2] for deducing equations of motion from Einstein's field equations. A detector was developed for operation in the vicinity of 1662 Hertz because a convenient size mass could be instrumented and because frequencies in this range are swept through in a supernova collapse. A high frequency source was developed for dynamic gravitational fields and the detector tested by doing a communications experiment[3] with high frequency Coulomb fields. Analysis has also been made of the expected response of such detectors to collapse of a supernova or a double neutron star. Present designs are optimum for this kind of radiation.

Recent studies of expected gravitational radiation from the Pulsars indicates that they are very promising sources. Present detector designs should be able to observe gravitational radiation from the Pulsars[5] if an extension in detector size is made of roughly one order. The most promising objects are the Crab Nebula Pulsar and CP 1919. We have been working actively on designs. For an assumed radiated power of 10^{38} ergs per second for the Crab Pulsar the Riemann Tensor amplitude on earth at 60 Hertz is given by R_{1010} with

$$R_{1010} = 4 \times 10^{-41} \text{ cm}^{-2}$$

An elastic solid detector will have strain ε induced by the radiation with amplitude

$$\varepsilon = 3 \times 10^{-20}$$

for a $Q \simeq 10^5$. For a detector cooled to 4°K a mass of 20 tons is required and an integration time of one year. "Lock in" detection would be employed with a doubled frequency radioastronomy signal as reference. Computer programs would be written for the detection including corrections for the motion of the antenna. For a radiated power of 10^{35} ergs per second 20,000 tons are required. These numbers for the mass and integration time become much smaller if Fairbank's suggestions are followed, for going to much lower temperatures.

At frequencies between one cycle per hour and one cycle per minute we plan to use as antenna the earth[1], the moon[6], and the planets.

EXPERIMENTS AT 1662 HERTZ

These detectors have mass of the order of 10^6 grams. It is a formidable task to instrument them so that their sensitivity is at the thermal fluctuation[4] limits--implying detection of relative displacements of 10^{-14} cms over a two meter end to end distance. The normal output is Gaussian noise. There has never been any extended period when a well instrumented detector had above normal output, and careful measurements show absence of any diurnal effects for long period averages of the output.

Earlier observations indicated that on rare occasions the detector output was higher than expected for short periods--seldom longer than the detector relaxation time. The noise output can in fact achieve arbitrarily large values, over a sufficiently long time, purely as a result of the thermal fluctuations. For this reason and to rule out environmental effects two detectors were[7,8] employed at separated locations a few kilometers apart. One of these was large, the other was small. One quite large coincidence was observed (signal to noise power of 18) over a years observations, and coincidences were observed at irregular intervals of about six weeks, as a sudden apparently simultaneous increase in detector outputs. Additional detectors were developed, with larger mass. The extremely difficult problems of instrumentation to observe onset of oscillations of a 10^6 gram mass to a precision of roughly 0.2 seconds with superb noise performance, with automatic tuning to permit unattended operation, have been solved. Most of the technology is completely new and does not employ cryogenics. I hope to publish an account of these developments during the coming year.

My definition of an event is that a detector output voltage crosses some threshold in the positive direction. A coincidence of two or more detectors is defined as their crossing a given threshold in the positive direction within some specified time interval.

The present experiments employed four detectors, all with length 153 centimeters. One has diameter 96 centimeters and employs cryogenic electronics. Two have diameter 66 centimeters and one has diameter 61 centimeters. Coincidence resolving time between 0.44 and 0.20 seconds were employed at various times. One of the 66 centimeter detectors was installed at the Argonne National Laboratory, near Chicago. All other detectors are at College Park, Maryland. A one way telephone line connects the two locations. It is gratifying that the increased sensitivity and extended baseline--to 1000 kilometers--have yielded a higher coincidence rate than the earlier, smaller baseline and smaller sensitivity experiment. Enclosed Figure 1 is a list of some of the coincidences during the early months of 1969. Figure 2 shows recorder traces for one of the coincidences. Observation of coincidences alone proves nothing. Detailed analysis of the intensities must show that the relative frequency of simultaneous crossings with the observed amplitude is so small that an accidental coincidence with given amplitudes is overwhelmingly unlikely. In addition we must have means for ruling out seismic, electromagnetic, and cosmic ray effects.

An initial series of experiments during the first three months of 1969 was done with a coincidence resolving time of 0.44 seconds. This is much shorter than the propagation time for seismic signals over the 1000 kilometer baseline. A seismic array[9] at College Park continuously observes earth acceleration along three axes over a wide band of frequencies. It is known that only the most violent local earth motion excites the detectors. There is no evidence that any of the usual earthquakes or underground nuclear explosions excites them.

Figure 1 Table of Recorded Coincidences During the First Three Months of 1969.

DATE MONTH/DAY/YEAR	UNIVERSAL TIME	NUMBER OF TIMES PER DAY COINCIDENCE AMPLITUDE IS EXCEEDED		PERIOD PER ACCIDENTAL COINCIDENCE
		MARYLAND 66 CM DETECTOR	ARGONNE 66 CM DETECTOR	
12/30/68*	1033	25	15	18 YEARS
1/1/69**	0052	6	88	8×10^4 YEARS
1/6/69**	0025	110	4	230 YEARS
1/28/69	1546	24	5	720 DAYS
1/30/69	1656	1	5	48 YEARS
2/5/69*	2221	30	30	7 YEARS
2/6/69	0447	150	4	144 DAYS
2/16/69	0130	20	72	3×10^4 YEARS (for both 2/16/69 entries)
2/16/69	0130.5	200	200	
2/16/69	0159	1	24	10 YEARS
2/21/69	0634	26	12	280 DAYS
2/23/69*	1218	40	12	15 YEARS
3/4/69	0913	30	15	190 DAYS
3/15/69	0341	75	6	190 DAYS
3/20/69*	$1741\frac{1}{2}$	140	96	7×10^7 YEARS (for both 3/20/69 entries)
3/20/69*	1744	60	125	
3/21/69**	0311	48	2	4×10^4 YEARS

* TRIPLE COINCIDENCE ** QUADRUPLE COINCIDENCE

—NOTE—

TIME RESOLUTION FOR TWO 66 CM DETECTORS IS 0.4 SECONDS. THE 61 CM AND 96 CM DETECTORS ARE NOT COUPLED TO A COINCIDENCE COUNTER. THEIR THRESHOLD CROSSING TIME IS NOT ACCURATELY KNOWN AND THIS IS TAKEN INTO ACCOUNT IN COMPUTING FREQUENCY OF ACCIDENTAL 3 AND 4 DETECTOR COINCIDENCES.

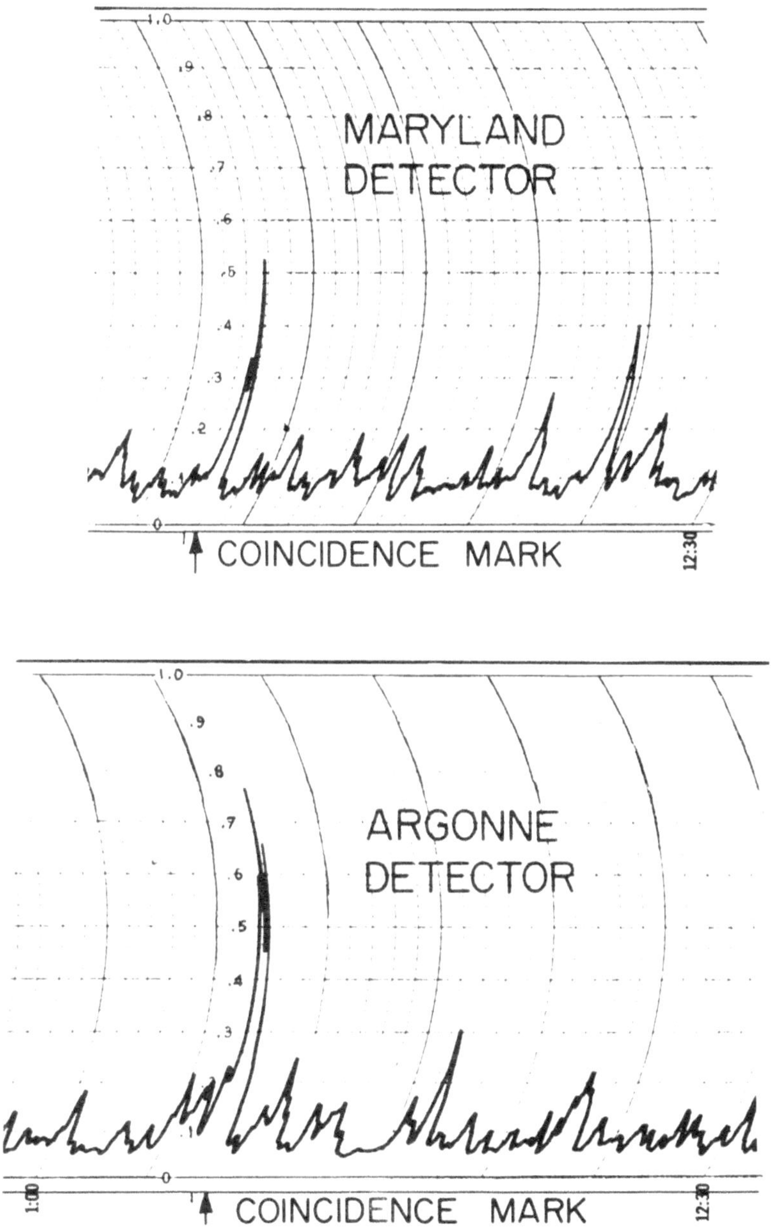

Figure 2 Argonne and Maryland Detector Coincidence. Time Runs from Right to Left, Full Scale is About 25 Minutes, Output Voltage is the Abscissa.

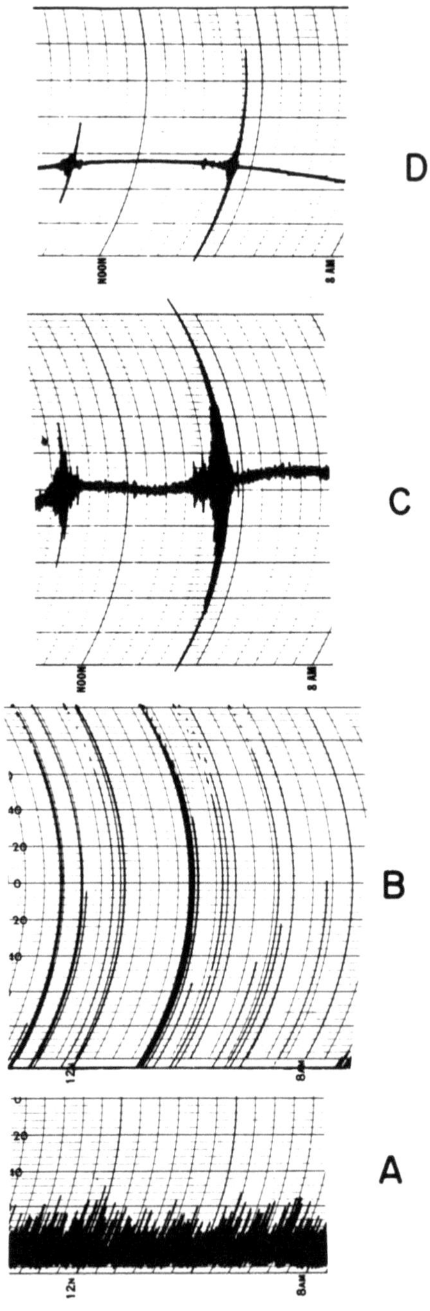

Figure 3 Seismic and Gravitational Radiation Detector Response for the Underground Nuclear Explosion of April 28, 1969.

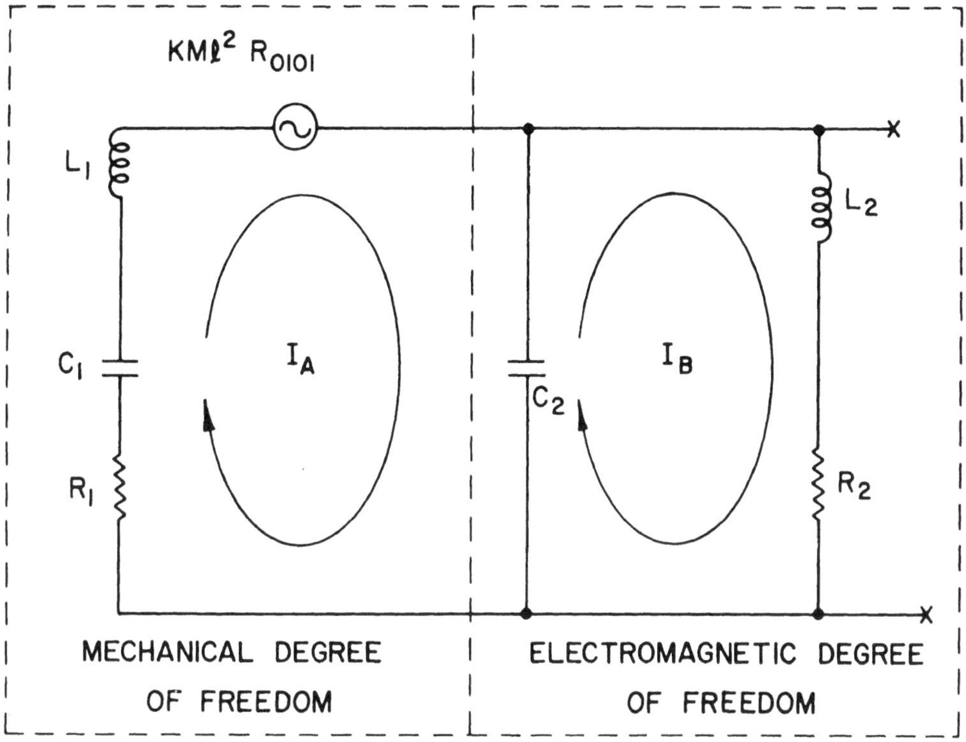

Figure 4 Gravitational Radiation Detector Driven by the Riemann Tensor, with Piezoelectric Coupling and External Inductance.

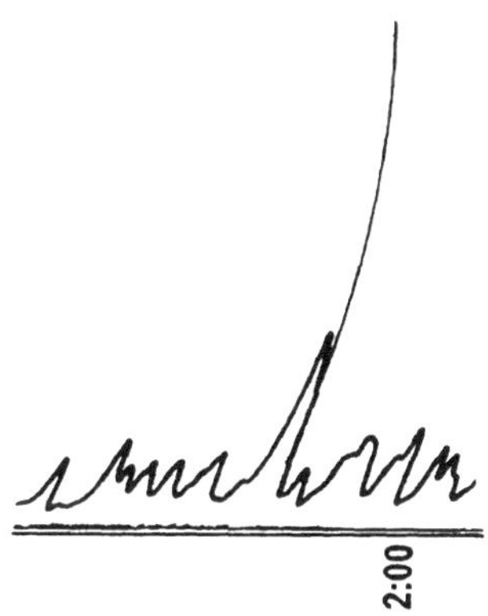

Figure 5 Long Vertical Coincidence Marker Followed by Detector Peak Eleven Seconds Later. Time Runs from Right to Left.

Figure 3A is the response of a vertical axis long period seismograph to the underground nuclear explosion of April 26, 1968. The curved baseline respresents the change in vertical acceleration due to the tides. The large vertical response is the seismic signal propagation via shortest path and the smaller response is the seismic signal arriving via a longer propagation path. Figure 3B shows the response of a horizontal seismograph. Figure 3C shows the response of a high frequency seismograph tuned to the detector frequency. Figure 3D is the lack of response of the gravitational radiation detector during the same period. For a seismic explanation to be acceptable we must conclude that some kind of seismic signal is eluding the seismic array and that the signals originate in some zone of the earth roughly halfway between Maryland and Argonne. Calculations show that such a zone of the earth has roughly 10^{-3} of the total earth volume. Similar effects from the remainder of the earth would be expected to give a high noise level due to non coincident events at each detector. This is not observed. Noise measurements indicate that the input of each detector is a heat reservoir close to room temperature. These noise measurements pose special problems not ordinarily encountered in fields such as radioastronomy and a long series of noise measurements was made at each location including effect of the telephone line. Far more difficult are the possible electromagnetic excitations and geomagnetic effects. Considerable effort has gone into this part of the investigation. The following evidence leads me to rule out an electromagnetic origin.

One detector employs cryogenically cooled electronics. It has a mechanical degree of freedom which is coupled to a long relaxation time (superconducting) electromagnetic degree of freedom. (Figure 4) Mathematical analysis indicates that if a delta function Riemann Tensor excites the cylinder, the electromagnetic output will require eleven seconds to build up. However a delta function electromagnetic excitation does not have the eleven second delay. The other detectors do not employ cryogenically cooled electronics, and are essentially one degree of freedom devices. Their electromagnetic output builds up within 0.1 seconds after excitation of their mechanical systems. Thus if the coincidences are due to electromagnetic effects we should expect the output of the cryogenically cooled detector to lag the others by 11 seconds. This lag is observed for about 1/4 of the coincidences, and is shown in Figure 5. Considerations of noise indicate that the delay will not always be observed. For example initial conditions of noise might result in one or more detectors excited to a minimum while others are in coincidence. Also some coincidences are due to two or more detectors having their mechanical systems excited above the mean but below threshold. Small electromagnetic input circuit noise pulses can then excite both detectors giving a recorded coincidence. Both kinds of coincidence were observed in this and earlier experiments.

Several improvements in noise performance have been made. In each instance the total input noise remained constant since it is associated with the heat reservoir temperature. Noise improvement increased the fraction of the input noise coming from the cylinder. An early electromagnetic shielding experiment reduced the coincidence rate. Subsequent study showed that currents flowing in the shields changed the receiver couplings and degraded the noise performance. When this coupling was reduced by additional shielding the noise performance and earlier coincidence rate were restored. A succession of shielding experiments and rearrangement of grounds substantially reduced the susceptability to electromagnetic effects, but did not affect the coincidence rate.

COSMIC RAY RESPONSE

Let us imagine that an energetic particle comes to rest in one of the detectors, what will the output be ? Let the displacement of a mass point within the detector be ξ then

$$\xi = \sum a_{k\ell m} f_{k\ell m}(\bar{r})\sin(\omega_{k\ell m}t + \phi_{k\ell m}) \qquad (1)$$

In (1) the $a_{k\ell m}$ are amplitude coefficients, $f_{k\ell m}$ are the normal mode solutions and the sum is over all modes. The fundamental mode f_{100} is observed. The velocity is $\dot{\xi}$ with

$$\dot{\xi} = \sum a_{k\ell m} f_{k\ell m}(\bar{r})\omega_{k\ell m} \cos(\omega_{k\ell m}t + \phi_{k\ell m}) \qquad (2)$$

Multiplying by f_{100}, integrating over volume and making use of the orthogonality of $f_{k\ell m}$ gives for the amplitude

$$a_{100} = \int \frac{\dot{\xi} f_{100} d^3x}{\omega_{100}} \qquad (3)$$

We take our origin at one end. Then $f_{100} = \cos kz$ with the cylinder axis in the z direction. For the extreme case where the particle comes to rest within the detector we have

$$a_{100} \simeq \frac{\langle\dot{\xi}\rangle}{\omega_{100}} \qquad (4)$$

In (4) $\langle\dot{\xi}\rangle$ is the volume average of the velocity imparted to elements of the cylinder by the particle. From momentum conservation we have

$$\langle\dot{\xi}\rangle = \frac{\textit{Momentum of Particle}}{\textit{Mass of Cylinder}} \qquad (5)$$

Equation (5) implies that to obtain a coincidence, a 10^{20} electron volt particle must come to rest within each detector during the resolution time. This is extraordinarily unlikely.

The experiment has been improved steadily. Thus the shielding was incomplete during the early months. The statistics could have been affected by variations in gain of the telephone line. However each coincidence was verified by inspection of a recorder at Argonne National Laboratory.

At present the gain and phase modulation anomalies of the long telephone line are automatically compensated. For these reasons it is possible that some of the early coincidences shown in Table One might have been spurious or might have been mistaken accidental coincidences in consequence of instrumental problems. The persistence of the coincidences throughout the period of improvement makes it overwhelmingly unlikely that all are accidental or of electromagnetic or seismic origin. It has in any case been clear for some time that the overall program has succeeded. As I remarked earlier the advent of the pulsars has brought possible sources for detectors similar to those employed in this experiment, but of larger mass and extension. Cryogenic cooling of two of our present detectors is a logical next step along with search at other frequencies, and using modes responsive only to the scalar components of a scalar tensor theory[10,11].

Not so long ago reputable physicists were saying that an expenditure comparable to our gross national product would be needed for major progress in this field. This is in marked contrast to the present austere level of effort. There are no student or junior faculty assistants. The test equipment which I carry on frequent trips to Argonne National Laboratory is borrowed. I do not know whether to feel shocked or flattered by a long telephone call from a physicist last week. He seemed greatly disappointed that there was no accurate information on the location of the sources of the coincidences, their power spectrum and polarization!

REFERENCES

* Work supported in part by the National Science Foundation.

† Part of these results have appeared in Phys. Rev. Lett. 22, 1320, 1969.

1. J. Weber, Phys. Rev. 117, 306 (1960). See also, J. Weber, General Relativity and Gravitational Waves (Interscience Publishers, Inc., New York, 1961), Chap. 8.
2. J. Weber, Relativity Groups and Topology (Gordon and Breach Publishers, Inc., New York, 1964), p. 875.
3. J. Sinsky and J. Weber, Phys. Rev. Letters 18, 795 (1967); J. Sinsky, Phys. Rev. 167, 1145 (1968).
4. J. Weber, Phys. Rev. Letters 17, 1228 (1966).
5. J. Weber, Phys. Rev. Letters 21, 395 (1968).
6. J. Weber, in Physics of the Moon, edited by S. F. Singer (American Astronautical Society, Hawthorne, Calif., 1967), p. 199.

7. J. Weber, Phys. Rev. Letters 18, 498 (1967).
8. J. Weber, Phys. Rev. Letters 20, 1307 (1968).
9. J. Weber and J. V. Larson, J. Geophys. Res. 71, 6005 (1966).
10. R. H. Dicke, Ref. 2, p. 165.
11. D. C. Robinson and J. Winicour, Phys. Rev. Letters 22, 198 (1969).

GENERAL RELATIVITY EXPERIMENTS USING LOW TEMPERATURE TECHNIQUES*

C.W.F. Everitt, William M. Fairbank, William O. Hamilton

Department of Physics, Stanford University

Stanford, California

In this paper we will describe an experiment to test Einstein's theory of general relativity in a satellite in space by means of a nearly perfect gyroscope.[1] This experiment is uniquely made possible by complete use of a low temperature environment and the properties of superconductors including the use of zero magnetic fields and ultrasensitive magnetometry. In the last part of the talk we will mention other relativity experiments which make use of low temperature physics and ultra-sensitive magnetometers.

L. I. Schiff has proposed[2] an experiment to check the equations of motion in Einstein's general theory of relativity by means of a gyroscope which is forced to go around the earth either in a stationary laboratory fixed to the earth or a satellite. Schiff has calculated from Einstein's theory of general relativity that a perfect gyroscope subject to no external torques will experience an anomalous precession with respect to the fixed stars as it travels around the earth. A second anomalous precession arises due to the rotation of the earth. The relativity effect on a gyroscope in a 500 mile polar orbit is illustrated in Figure 1. The spin axis of

---

*Supported in part by NASA grants NGR-05-020-015, NGR-05-020-019 and Air Force contract AF33(615)67C-1245.

[1]William M. Fairbank, W. O. Hamilton, C.W.F. Everitt, "Quantized Flux to the Free Precession Nuclear Gyro", Proc. O.A.R. Research Appl. Conf. (Washington D.C. 1966) p. 153.

[2]L. I. Schiff, Proc. Nat'l Academy 46, 871 (1960).

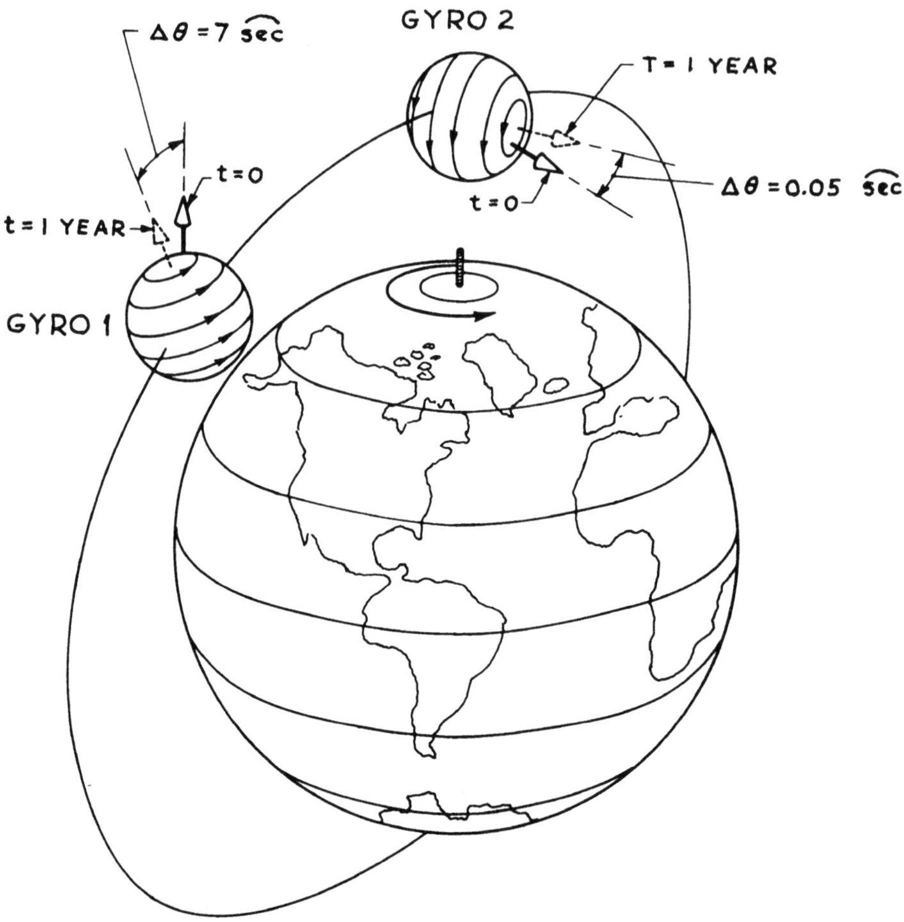

Figure 1

PRINCIPLE OF LONDON-MOMENT READOUT

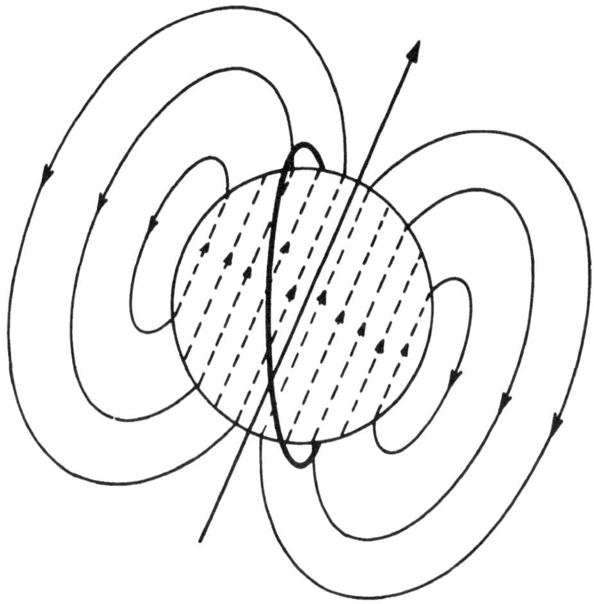

LONDON-MOMENT FIELD $H = 10^{-7}\omega$ GAUSS

Figure 2

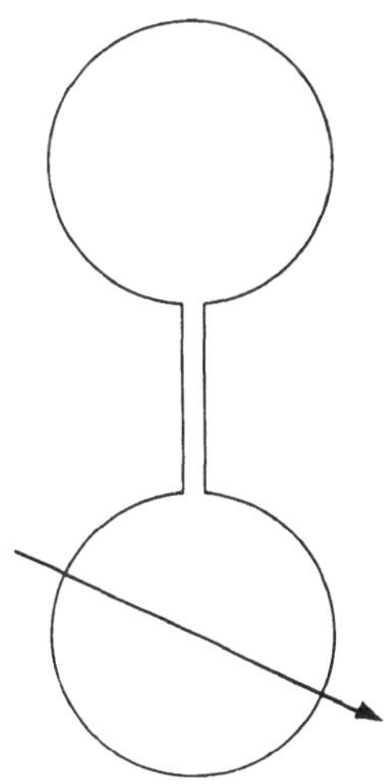

Figure 3

gyro 1 is parallel to the axis of the earth and at right angles to the axis of the orbit. The predicted precession of such a gyroscope is 7 arc/sec per year in the direction shown. This is the effect due to the rotation 15 times a day around the earth. The axis of gyro 2 is oriented parallel to the satellite orbit and at right angles to the axis of the earth. The predicted precession of the axis of this gyro due to the rotation of the earth is 0.05 arc/sec per year in the direction shown in the diagram. Under the conditions shown the two effects do not interfere.

The Einstein theory of general relativity has three important parts: the special theory of relativity which has been very accurately verified by experiment, the equivalence of gravitational and inertial mass which has been verified to a few parts in 10^{11} by the experiments of Dicke[3] and the third part which involves the equations of motion of matter through a gravitational gradient. The experiment proposed by Schiff checks the third part of the theory of general relativity and has become of particular interest since this part of the theory has been checked accurately only by the precession of the perihelion of the planet Mercury around the sun. The orbit of Mercury is an ellipse and due to the effects of the other planets this ellipse gradually precesses with time. The observed precession is different from the precession calculated by Newtonian mechanics by 43 seconds of arc per century. This correction is exactly predicted to 1% accuracy by Einstein's theory of general relativity. Recently Brans and Dicke[4] have proposed that there may be a scalar part of general relativity and that this should cause a correction to the precession of the perihelion of Mercury. Dicke has suggested that perhaps the sun has a bulge at the equator which gives a Newtonian correction to the precession of the perihelion of Mercury which accidently exactly makes up for any change that would be observed due to a scalar term. Dicke and Goldenberg[5] have observed the shape of the sun and have obtained evidence for a bulge sufficient to give a correction of about 8% to the theory of general relativity assuming the bulge corresponds to a mass quadrupole moment. Measurements on the shape of the sun are very difficult due to the complications at the surface of the sun and there has been considerable discussion concerning the interpretation of this experiment. It has now become of extreme importance in physics to perform additional experiments which can give information on the theory of general relativity especially with respect to a scalar component. The experiment suggested by Schiff is the first man made experiment which could accurately test

---

[3] P. G. Roll, R. Krotkov and R. H. Dicke, Annls of Phys. 26,442(1964).

[4] C. Brans and R. H. Dicke, Phys. Rev. 124, 925 (1961).

[5] R. H. Dicke and H. Mark Goldenberg, Phys. Rev. Letters 18,313(1967).

such effects without use of distant astronomical objects which can not be controlled in the laboratory. It has the further advantage that it avoids the necessity of calculating out large extraneous effects. Finally, the effect in the Schiff experiment due to the rotation of the earth depends on off diagonal terms in the metric never before checked by experiment. Thus we see that the gyroscope experiment proposed by Schiff is of extreme importance especially if it can be done with sufficient accuracy to check the scalar component suggestion of Brans and Dicke and if possible the much smaller earth rotational effect. Ideally it would be done accurately enough to see a possible scalar correction to the earth's rotational effect.

Thus it appears desirable to do the gyroscope experiment to an accuracy of about 2×10^{-3} arc/sec per year. Let us consider for a moment how small a drift of 2×10^{-3} arc/sec per year really is. It is many orders of magnitude smaller drift than the best available earth bound gyroscope. If a gyroscope sufficiently accurate to do the relativity experiment to 2×10^{-3} arc/sec accuracy had been available to Moses at the time he crossed the Red Sea then such a gyroscope would have precessed in the intervening thirty-two hundred years 6 seconds of arc, roughly equivalent to the drift in one hour of the best earth bound gyroscopes presently available. How can one possibly propose to do an experiment requiring such a dramatic improvement of existing technology? Furthermore, the experiment requires that the gyro readout be compared with the position of a fixed star by means of a telescope, at least at the beginning and the end of a year. How can one possibly achieve the required mechanical stability between the telescope and gyroscope since even one part in 10^8 of the solar radiation if allowed to fall on a quartz telescope would be sufficient to cause a thermal change in dimensions resulting in error of relative readout of the telescope and gyroscope in excess of the desired 2×10^{-3} arc/sec per year? This apparently hopelessly difficult experiment is achievable, we believe, by a combination of the zero g environment of a satellite and the unique properties of matter near the absolute zero of temperature.

At very low temperatures the coefficient of expansion becomes vanishingly small and the temperature of an object immersed in superfluid helium can be kept extremely constant. Thus it is possible, if the telescope and gyroscope are placed in an environment surrounded by superfluid helium at 1.2°K, to reduce the distortions due to thermal gradients to a negligible value. Furthermore since the satellite will be in approximately zero g environment, the mechanical distortions due to its own weight also vanish.

The largest torque on an earth bound gyro is due to the necessity of supporting the body against the gravitational pull of the earth. In space it is possible to reduce the gravitational pull sufficiently to reduce the support torques to the required level. However, the gravity-gradient torques which exist on any gyro which is not perfectly homogeneous and perfectly spherical remain as large in the satellite as they are on earth. To reduce the drift from this torque to less than .002 arc/sec per year it is necessary to have the ball homogeneous and spherical to the order of 1 part in 10^6. In addition one must eliminate magnetic and electric torques.

The requirement of nearly perfect sphericity poses a readout problem. Conventional readouts require knowing the position of the axis of rotation with respect to the ball. If the moments of inertia of all the axes of the ball are the same, it is not possible to anticipate about which axis the gyroscope will spin. Furthermore, the gyroscope must be allowed to spin freely in a vacuum. If the ball is to be kept at 1.2°K in a vacuum it can be cooled only by blackbody radiation and effectively no heat can be allowed to fall on the ball in the process of reading it out. These factors eliminate conventional readouts.

Superconductivity provides a solution for all these problems. Magnetic torques can be eliminated by placing the ball inside a superconducting shield from which the last quantum of flux has been eliminated. Electric torques are minimized by coating the ball with a superconductor. A perfect sphere which has zero resistance will experience no electric torques.

The spherical properties of superconductors lead to a unique solution of the readout problem. A superconductor spun up in zero field develops along its axis of spin a uniform magnetization of $10^{-7}\ \omega$ gauss[6,7] as a result of macroscopic flux quantization. With a gyroscope spinning at an ω of 2×10^3 radians/sec this would give 2×10^{-4} gauss along the axis of spin. The question then arises how can one detect to .002 seconds of arc the orientation of such a gyroscope by use of this very small magnetic field. Figure 2 shows the proposed readout which makes use of a superconducting loop. Shown in the figure is a spinning superconductor with a magnetic field as indicated along the axis of the spin. Around the spinning sphere is placed, as a method of readout, a superconducting loop. Since the resistance of the superconducting loop is zero, any change in the flux through the loop caused by a change in orientation of the gyro-sphere will cause a current to flow in the loop which exactly cancels this change in flux. If one could read out this current, one could determine the change in orientation of the

[6] F. London, Superfluids, (John Wiley and Sons, New York 1950 and Dover Publications, New York 1961) vol. 1 p. 78f.

[7] A.F. Hildebrandt, Phys. Rev. Letters 12,190(1964); M. Bol and W.M. Fairbank, Proc. IX int. Conf. on Low Temp. Phys.(Plenum Press 1965) p. 471· A King,Jr.,J.B.Hendricks,and H.E.Roschah,Jr. Ibid,p.471.

direction of the ball. Figure 3 shows the method we have developed to read out this current. In series with the first loop is placed a second superconducting loop indicated with an arrow through the loop. The current that flows in the two superconducting loops produces a cancelling flux which is distributed in the two loops instead of being confined to the one loop. The ratio of the flux in the two loops is equal to the ratio of the inductances of the two loops. Thus the change in flux through the first loop caused by the reorientation of the ball produces a cancelling flux distribution in the two loops. If the inductance of the second loop is changed, the current flowing in the two loops changes and the distribution of the cancelling flux in the two loops changes. If the inductance is changed 10^5 times/sec then a 10^5 cycle/sec. A.C. signal is produced which can be detected by a readout coil. Figure 4 shows such a circuit, including a readout and a nulling field. The modulator consists of a long superconducting lead evaporated on a flat surface. Adjacent to this long superconducting wire is a superconducting ground plane evaporated onto a quartz crystal. The crystal and surface are placed about 2000 Å apart and the crystal is driven such that the ground plane periodically approaches and recedes from the superconducting circuit. This modulates the inductance of the circuit and causes the flux to be pumped back and forth between the two loops. The oscillating current in the two loops flows through the coil as indicated and is read out through a transformer by an amplifier. It is possible to increase the sensitivity of this circuit by placing a condenser in the circuit as indicated on the diagram. The modulating current flows in and out of the condenser plates in such a way as to provide additional parametric amplification. John Pierce[8] has worked out in detail the sensitivity of such a circuit compared with the theoretical Johnson noise in an amplifier.

$$\Phi^2 \geq \frac{27\pi}{16} \frac{L}{Q} \left(\frac{L}{\Delta L}\right)^2 kT \frac{\Delta\nu}{\nu}$$

where L is the total inductance of the circuit, ΔL the inductance change, T the noise temperature of the circuit, Q the quality factor of the circuit, ν the frequency of modulation, and $\Delta\nu$ the bandwidth. We have verified the validity of this equation both with experiments and by model circuits on an analog computer. With the modulating crystal operating at 10^5 Hz and a Q of 1000 with room temperature amplifier noise the sensitivity is predicted to be 10^{-10} gauss which would allow an accuracy of readout of 0.1 arc/sec in 0.1 sec. of time. By averaging over a year or by reducing the noise temperature to below helium temperatures by feeding the signal to a Josephson junction amplifier the sensitivity can be increased to .002 sec of arc as required by the experiment.

---

[8] J.M. Pierce, Proc. of Symposium on Superconducting Devices, University of Virginia, April 28-29, 1967.

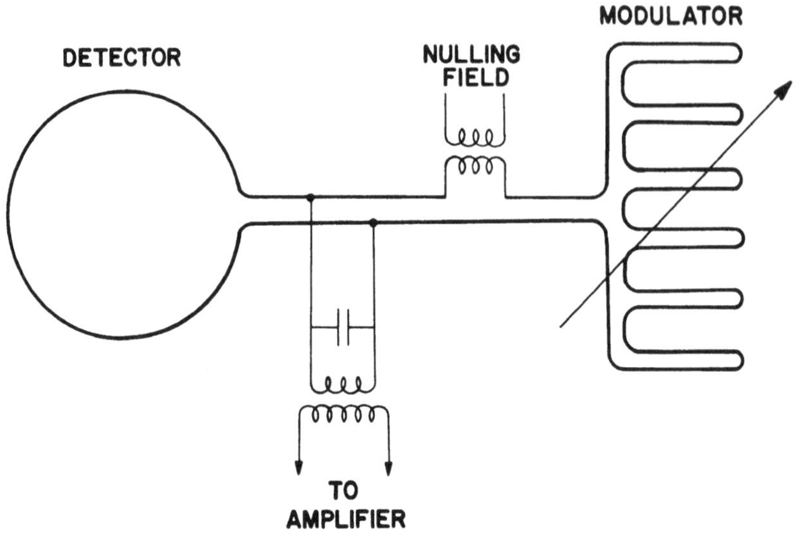

Figure 4

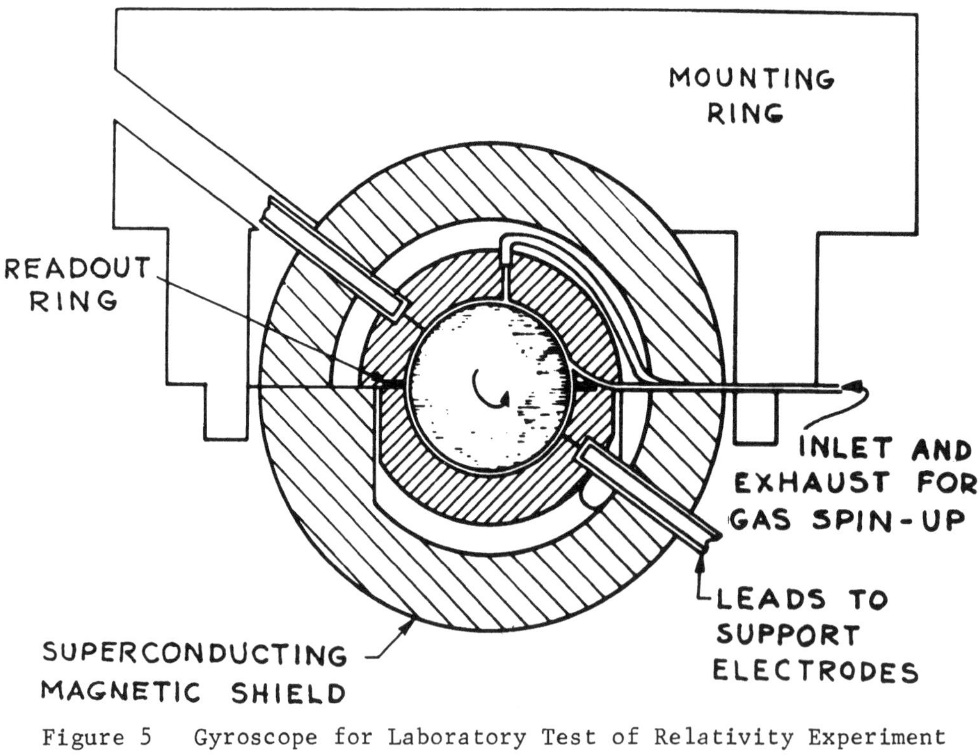

Figure 5 Gyroscope for Laboratory Test of Relativity Experiment

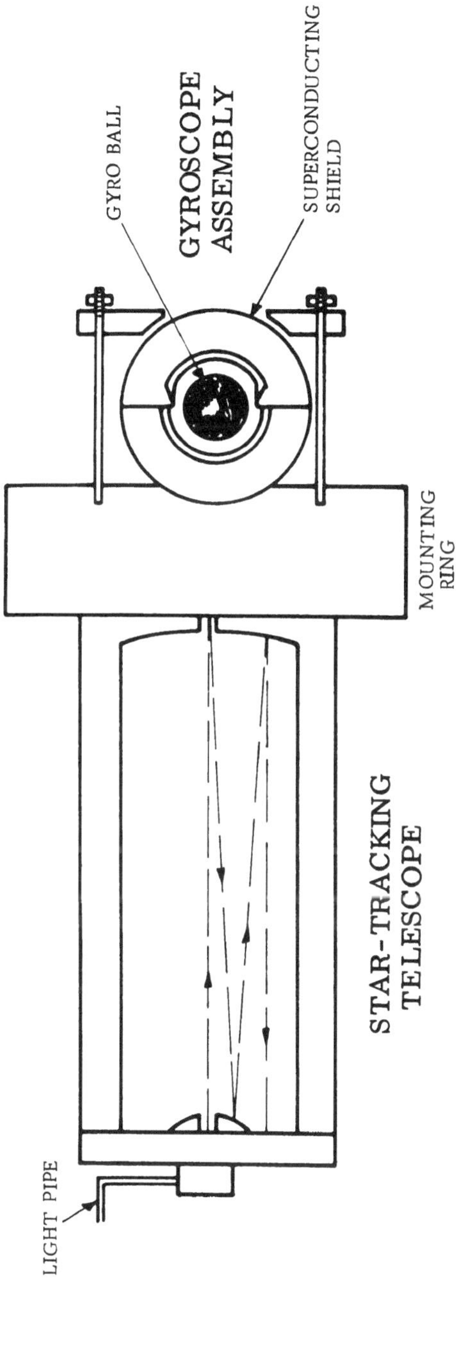

GENERAL VIEW OF APPARATUS FOR LABORATORY TEST OF RELATIVITY EXPERIMENT

Figure 6

Figure 5 is a diagram of the gyroscope. The gyroscope will consist of a very homogeneous quartz ball, 3/2 inches in diameter coated with a thin layer of niobium. The ball is made as spherical as possible. It is electrostatically supported across a supporting space of 3/2 thousandths of an inch. The ball is spun up in the superconducting state by helium gas spin up jets. The ball is normally operated in a vacuum with no friction. The readout system introducing essentially no losses in the ball. The ball is kept cold by black-body radiation to the walls. Surrounding the entire gyroscope is a superconducting 4 inch diameter spherical shield from which the last quantum of flux has been excluded. This is absolutely necessary if one is to make use of the sensitive magnetic readout discussed above. The readout loop is evaporated on an optically flat piece of quartz which is made an integral part of a telescope as shown in Figure 6. The telescope will be used to compare the axis of the gyroscope with the position of the star. Both the telescope, which is made of quartz and the gyroscope will be operated at liquid helium temperatures below the superfluid transition in helium. Thus the temperature will be kept very constant and since the coefficient of expansion of the materials is nearly zero there should be no change in the relative orientation of the telescope and gyro readouts to within 0.002 seconds of arc. We have calculated all of the known torques on the above gyro in a 500 mile polar orbit and find the net drift rate to be less than .002 seconds of arc per year. The gyroscope and telescope parts are being constructed by Honeywell and Davidson Optronics.

Figure 7 shows a proposed complete experiment including the helium dewar used to produce a low temperature environment. Included in the experiment are 4 gyroscopes, 2 checking the geodesic effect of 7 seconds per year, and 2 checking the motional effect of 0.05 seconds per year. The entire dewar contains about 75 lbs. of liquid helium which is calculated to keep the experiment cold for more than a year. The pressure and therefore the temperature of the liquid helium is kept constant by controlling the escape of the helium gas to the vacuum of space by means of a superfluid plug in the helium dewar and gas jets on the outside of the satellite. The flow of the escaping helium through these gas jets will be used to attitude control the satellite on a star to a calculated 1 second of arc and also zero g of the center of mass of the satellite to a proof mass shown in the diagram. An inner superconducting attitude control on the telescope is designed to keep the telescope on the star to better than 0.1 sec. of arc. This project is being supported by NASA and the air force.

Although the relativity experiment on earth requires a gyroscope millions of times more sensitive than any existing gyroscope it is interesting to contemplate whether there is any possibility at all of building an earth-bound gyroscope sensitive enough to do the experiment. For many years it has been realised that a spinning nucleus is a gyroscope which is unaffected by the usual torques caused by the necessity of supporting the gyroscope against the forces due to gravity. Schiff and Currin[9] have demonstrated theoretically that nuclei are subject to the same relativity precessions as regular gyros. The question arises, could nuclei be used as a gyroscope? Two kinds of nuclear gyroscopes are possible. One is a rate gyro, which would be slaved to an external magnetic field, the other is a free precession gyro. A rate nuclear gyro using water nuclei has been constructed. But in the relativity experiment we are interested only in a free precession gyro which will remain unaffected by all external torques.

There is, however, one overwhelming torque, which is the torque due to an external magnetic field. In contrast with the London-moment superconducting gyroscope which is very heavy and produces a very weak magnetic field along its axis, a nucleus is extremely tiny and produces, relatively speaking, a very large magnetic field along its axis. Thus a very weak external field oriented in any other direction than along the axis of spin will produce a torque. In fact, a magnetic field oriented at right angles to a He^3 nucleus will cause a precession of 2×10^4 radians per second per gauss. To detect the larger relativity effect with a He^3 nuclear gyro one would require in an earth bound laboratory at Palo Alto, California, a magnetic field less than 3×10^{-18} gauss in order that the magnetic field produce a precession of less than 7×10^{-14} radians per second as predicted by the relativity effect. Previous proposals for free nuclear gyroscopes have not been successful because of the impossibility of making the ambient field sufficiently small. With the discovery of quantized flux it has suddenly become possible to produce a truly zero magnetic field. When a solid perfect type I superconductor is cooled through the transition temperature into the superconducting state it excludes all the magnetic flux; this is known as the Meissner effect. However, all actual solid superconductors tend to trap some flux and all superconductors which contain a hole, such as a hollow cylinder or sphere trap approximately the flux originally contained in the hole. After the superconductor is in the superconducting state the amount of trapped flux remains constant even if the external field is changed. In the experiments of Deaver and Fairbank[10] it was found that the flux trapped in the hole in a small superconducting cylinder is quantized. In particular, they observed that if the cylinder is cooled down in a magnetic field of such a size that the flux in the hole

---

[9] J. D. Currin, Ph.D. Thesis, Stanford University, 1963 (under the direction of L. I. Schiff).

[10] B. S. Deaver, Jr. and W. M. Fairbank, Phys. Rev. Letters 7, 43 (1961); also see R. Doll and M. Nabauer, Ibid, 7, 51 (1961).

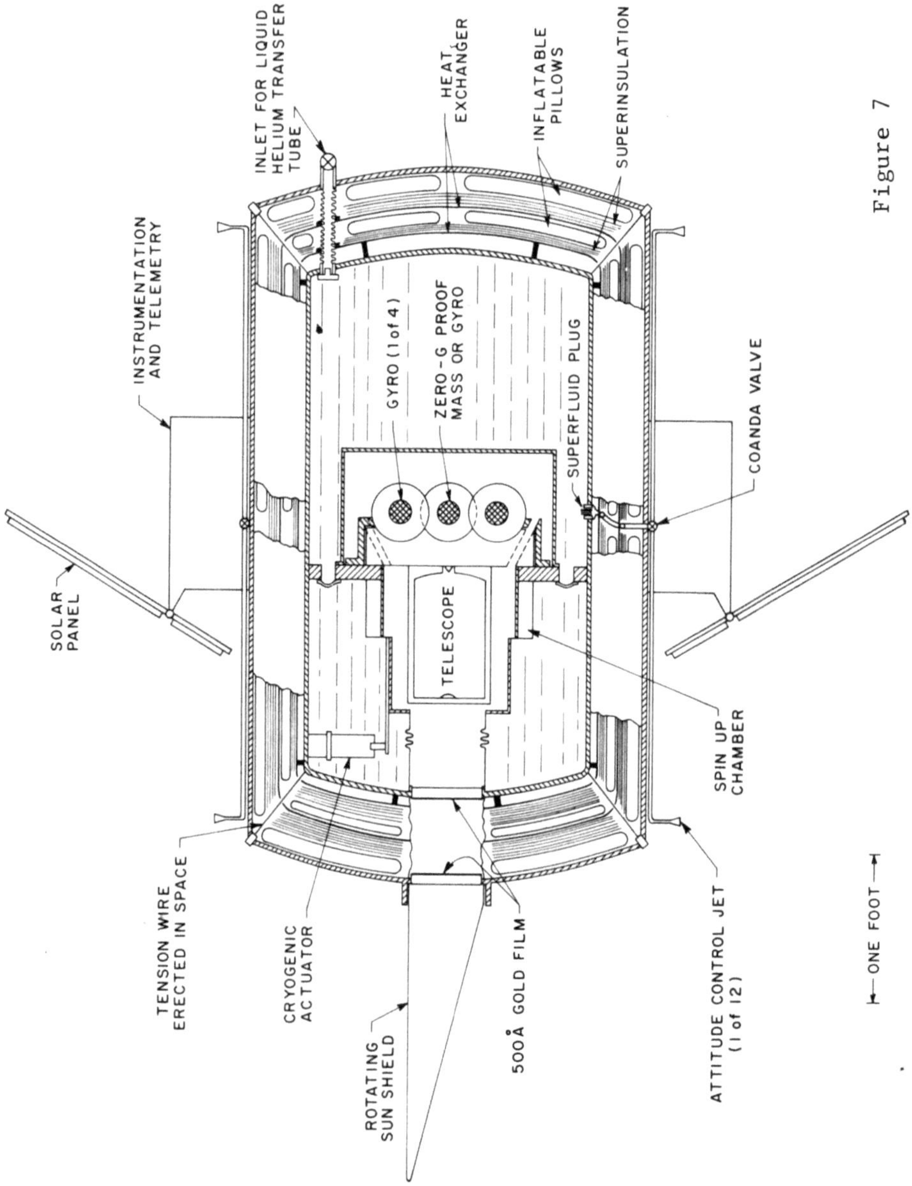
INLET FOR LIQUID HELIUM TRANSFER TUBE
HEAT EXCHANGER
INFLATABLE PILLOWS
SUPERINSULATION
INSTRUMENTATION AND TELEMETRY
GYRO (1 of 4)
ZERO-G PROOF MASS OR GYRO
SUPERFLUID PLUG
COANDA VALVE
SOLAR PANEL
TELESCOPE
SPIN UP CHAMBER
TENSION WIRE ERECTED IN SPACE
CRYOGENIC ACTUATOR
ROTATING SUN SHIELD
500Å GOLD FILM
ATTITUDE CONTROL JET (1 of 12)
ONE FOOT

Figure 7

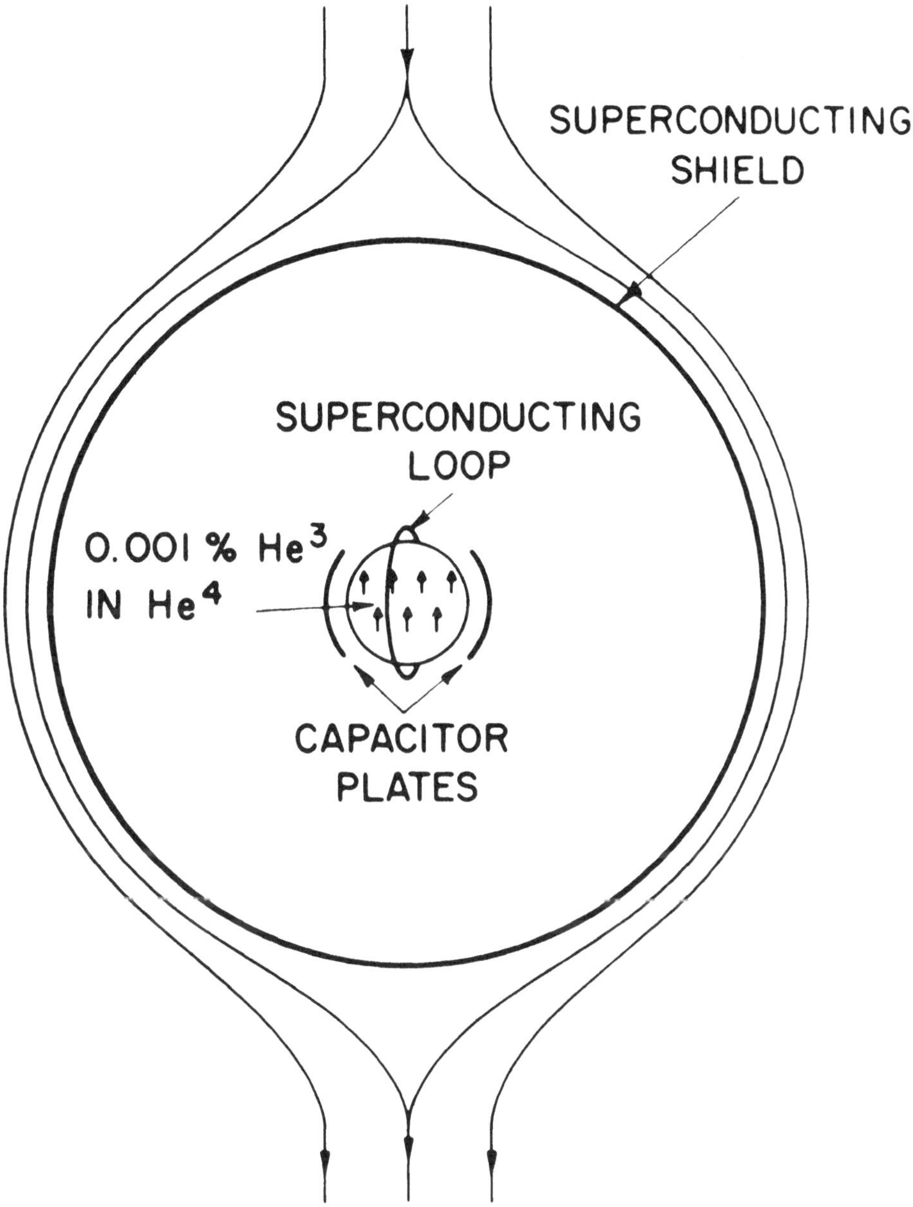
SUPERCONDUCTING
SHIELD
SUPERCONDUCTING
LOOP
0.001 % He3
IN He4
CAPACITOR
PLATES

Figure 8

of the cylinder is less than half a flux unit no flux at all is trapped in the hole. Thus it is possible to obtain a truly zero magnetic field inside a hollow superconducting shield provided the shield is cooled through its transition temperature in a magnetic field sufficiently small that the total flux passing through the shield is less than half a flux unit and the effects of thermal currents and fluctuations are eliminated. Experiments are underway at Stanford to produce large regions of zero magnetic field using flexible lead bladders.

Figure 8 shows a diagram of the proposed He^3 gyroscope. In order to have a free precession nuclear gyro that will be readable for a period of a year it is necessary that the relaxation times T_1 and T_2 of He^3 nuclei be longer than a year. The nuclei must be first polarized so that essentially all their spins point in the same direction. It must be a liquid or gas in order that the magnetic fields of the neighbors are effectively cancelled out giving a long T_2. It must have sufficiently small magnetic field due to the electrons that the relaxation time T_1 is as long as a year. Only one nucleus satisfies the requirements for such a free precession gyro. The He^3 nucleus has an inherent relaxation time in pure liquid He^3 of about 500 seconds. If the He^3 is diluted by the completely non-magnetic He^4, the relaxation time can be made longer linearly with the percent dilution. Thus a solution of 1 part of He^3 in 10^5 parts of He^4 would have relaxation times longer than a year. The He^3 nuclei can be polarized by optical pumping. Completely polarized He^3 nuclei of the desired concentration will produce along their axis of spin a magnetic field of about 10^{-5} gauss which is an order of magnitude smaller than the magnetic field expected from the superconducting gyro designed for space. The detection circuit for the He^3 gyro uses the same principle as that already discussed for the superconducting gyro in space.

To perform the relativity experiment on earth is extremely difficult. However long before the relativity experiment can be accomplished it is possible to perform another very fundamental experiment. By introducing an electric field at right angles to the nuclear magnetization and measuring any precession it is possible to measure an electric dipole moment if it exists on the He^3 nucleus. Such an electric dipole moment would violate parity conservation and time reversal invariance.

Another experiment underway at Stanford which is made possible by low temperature techniques is an experiment to measure the force of gravity on free falling electrons and positrons. The experiments on the electron have been completed[11] and the experiments on the

[11] F. C. Witteborn and W. M. Fairbank, Phys. Rev. Letters 19, 1049 (1967); Nature 220, 436 (1968).

positron are expected to give results during the next year. These will be followed by experiments on anti-protons. Morrison and Gold [12] have suggested that the separation of matter and antimatter in the universe might be caused by repulsive gravitational force on matter and antimatter, even though this would violate the equivalence principle of general relativity.

Professor Weber has discussed at this meeting his exciting evidence for the existence of gravitational waves.[13] As he has pointed out, the sensitivity of his gravitational detectors is limited by thermal fluctuations. To reduce these fluctuations by cooling to helium temperatures or below requires large scale refrigeration. We are developing at Stanford such large scale refrigeration as part of a program to build a 500 foot long superconducting accelerator. We are designing the apparatus to perform gravitational radiation experiments at low temperatures. Calculations indicate that a half wavelength resonator cooled to .003°K could detect a gravitational radiation of 10^{38} erg/sec from the Crab pulsar by phase coherent detection over a period of a day.

In this talk we described three relativity experiments in various stages of development which are made possible by low temperature techniques. The development of these techniques for use in space can make possible more sensitive measurements of the 3°K background black body radiation. New infrared detectors are also possible. The stable gyro reference system with the low temperature environment designed for the relativity experiment could be a useful addition to an orbiting observatory.

---

[12] P. Morrison and T. Gold, Essays on Gravity, 45 (Gravity Research Foundation, New Boston, New Hampshire 1957).

[13] J. Weber, Phys. Rev. Letters 22, 1320 (1969).

CONTRIBUTION TO THE HISTORY OF EINSTEIN'S GEOMETRY AS A BRANCH OF PHYSICS

Eugene Guth

Oak Ridge National Laboratory, Oak Ridge, Tenn., and

University of Tennessee, Knoxville, Tennessee

SECTION I.

INTRODUCTION

About a year ago in a short note[1] I outlined a "New Foundation of General Relativity." My main point was that the usual postulational approach to general relativity (GR) does not distinguish between the Nordström-Einstein-Fokker (NEF) scalar theory and Einstein's tensor theory. Thus a new postulate had to be added for which I have chosen the "derivability of the equations of motion from the field equations." Later, I remembered having seen a review (invited talk) of Einstein on the "Present Status of the Problem of Gravitation" which he delivered at the "85. Naturforscherversammlung zu Wien," in 1913 before a joint meeting of the Divisions of Physics, Mathematics, and Astronomy and which was published in the same year in the "Physikalische Zeitschrift,"[2] together with the lively discussion which Einstein's talk generated. In his review, Einstein also tried to distinguish between Nordström's second scalar theory (which in 1914 he and Fokker brought into a generally covariant form) and his own (preliminary) tensor theory. I noticed, with some surprise, that Einstein's argument for favoring his tensor theory was not correct. Nevertheless, his profound intuition led him to the right theory. As Goethe in his "Faust" said, "Der Mensch in seinem dunklen Drange ist sich des rechten Weges wohl bewusst." Applied to our case, "Einstein in his obscure drive was conscious of the right pathway."

Thus, I decided to study the historical development of Einstein's dream (or vision) to establish a union between geometry and physics so that geometry becomes a branch of physics[2a] or, conversely, physics a branch of geometry. This study turned out to be most fascinating.

We shall see that by distinguishing between the historically accidental and the logically essential elements in the progress of Einstein's ideas on gravity, the reconstructed past throws light upon present problems of gravity.

Some of the questions are:

a. Did Einstein distinguish between what Dicke later called weak and strong principles of equivalence? The answer is, he did this explicitly in his 1913 review but not so much in later papers where he essentially used only the strong principle of equivalence.

b.[69*] Did Einstein distinguish between the first and the second scalar theory of gravitation by Nordström? He used only the formalism of the second theory of Nordström. This (NEF) theory is based on the strong principle of equivalence; the first Nordström theory, by contrast, satisfies only the weak principle of equivalence.

c. Did Einstein distinguish sharply between what we shall call global and local Mach principles and their role in the first and second scalar theory of Nordström and in his own (preliminary) tensor theory of gravitation ? We shall call global Mach principle, Mach's original assumption of the (causal) relation between Newton's famous rotating bucket-of-water experiment and the mass of faraway stars (or galaxies). Under the local Mach principle, we understand Einstein's hypothesis of the "relativity of inertia," viz., "that inertia depends on a mutual action of the bodies." (Mach himself seemed to believe that the effect of nearby masses on the inertia of a test particle cancels out and emphasized directly only what we call the global Mach principle, though even that not in precise form.) Einstein refers qualitatively to the global Mach principle. Quantitatively, however, Einstein used only the local Mach principle; in particular, he stated already in his 1913 review and in all editions of his book "Meaning of Relativity" that in his tensor theory, "the inertia of a body must increase when ponderable masses are piled up in its neighborhood." He thought he proved this by quantitative calculations using the linearized form of (his provisional 1913 and) his final 1915 form of his tensor theory. He also is believed to have shown that this is not true for the second Nordström theory. In preparing the later editions of his cited book, he seemed to have overlooked that in a paper[3] in Rev. Mod. Phys. of 1945 (with E. G. Straus), he himself has shown that this effect does not exist in GR. On the contrary, it does exist in the first Nordström scalar theory. It also exists in the Brans-Dicke scalar-tensor theory.[4] The correct argument which Einstein could have used in 1913, but did not, in favor of his tensor theory, in contrast to the scalar theory of Nordström, would have been that it does not reduce completely to Newton's theory because it leads to zero value for the deflection of the light by the sun. Einstein could have used this argument in 1913 because already in 1911 he showed[5] that on the basis of the equivalence principle, one does get (Newtonian) light deflection, which, however, is only half of the value which later his final GR gave. Of course, in the 1911 paper, Einstein was somewhat apologetic about the necessary occurrence of light deflection because then he did not have yet a formal theory of gravitation.

d. The last argument applied to the Brans-Dicke theory shows that this theory at slow velocities reduces to the Newtonian form of the gravitational theory with a time-dependent gravitational constant. This time dependence causes a decrease in the light deflection as compared with the case of a time-independent gravitational constant. Thus, the decrease in the light deflection in the Brans-Dicke theory as compared with the Einstein value is due to this decrease in the Newtonian half of the light deflection.

e. All three famous tests of GR can heuristically be derived from the strong principle of equivalence augmented specifically by postulating the invariance of the Einsteinian gravitational constant, κ, and of the Planckian quantum constant, h; e.g., these last two constants are assumed to be independent of the absolute value of the gravitational potential. This "derivation" leads to the Schwarzschild line-element in second approximation. Alternatively, we shall also "derive" the full Schwarzschild line-element by generalized correspondence between Newton's theory and GR.

We shall discuss, briefly, also some milestones in the historical development of Newton's theory, which do not, however, seem to be widely known; e.g.,

A. Newton obtained first his gravitational law assuming circular planetary orbits. He combined Huygens' law of acceleration with Kepler's third law.

B. It took a hundred years for the general acceptance of Newton's theory of gravitation in contrast to his mechanics, which was almost immediately accepted. Laplace's two great memoirs of 1784 and 1787 turned the tide.

We shall also discuss briefly Soldner's early work on light deflection by the sun, and the pre-Einstein theories of Lorentz, Poincaré, Minkowski and again Lorentz.

We shall introduce the, apparently, new concepts of weak and strong correspondence of gravitational theories to Newton's. For weak (strong) correspondence any gravitational field equations have to reduce to Poisson's equation (and in addition, the line element to the Newtonian line element). Use of strong correspondence leads to a new, perhaps simplest, postulational foundation of GR.

SECTION II.[6]

HUYGENS → NEWTON → LAPLACE → POISSON → SOLDNER → LEVERRIER → LORENTZ

Birth of Universal Gravitation: Huygens and Newton

In a letter to Halley in 1686, Newton said that from Huygens' law of the acceleration in a circular orbit

$$a = v^2/r \;, \tag{II.1}$$

and Kepler's third law, it is easy to derive that the gravitational law varies with $1/r^2$. Since the eccentricities of the planets are small, it is quite permissible to consider their orbits as circular in the first approximation. Thus, Newton obtained for the gravitational force F_g between the earth and the sun

$$F_g = G \frac{m_E\, m_S}{r^2} \;, \tag{II.2}$$

where G is Newton's gravitational constant and m_E and m_S are the masses of the earth and of the sun, respectively. What cannot be derived and must be guessed by a stroke of insight and imagination is that this law is universal for all **macroscopic bodies**. Here is where Newton's real genius comes in. Therefore, the universal law of gravitation between any two bodies m and M is

$$F_g = G \frac{mM}{r^2} \;, \tag{II.3}$$

where G is the universal Newtonian gravitational constant.

Two Obstacles for Acceptance of Newtonian Gravitation: Cartesian Ideas and Planetary Inequalities

How did it come that Huygens did not make this same derivation of the gravitational law as did Newton? After all, Huygens published Eq. (II.1) in 1673 in his great book, "Horologium Oscillatorium." (This was much to the chagrin of Newton, who claims to have independently derived Eq. (II.1) around 1666, but did not publish it.) The reason must be that Huygens was committed to a type of explanation in the spirit, if not in the details, of the Cartesian theory which was en vogue at that time. As a matter of fact, even after the appearance of Newton's "Principia" in 1687, Huygens and the other great mathematicians and physicists in Europe, such as Leibnitz, Johann Bernoulli, and Cassini, rejected Newton's theory for the same or similar reasons as Huygens. It is then not surprising that even in Newton's own University of Cambridge, the textbook of physics in general use during the first quarter of the Eighteenth Century was still Cartesian.

While the great mathematicians and physicists named above were against the Newtonian doctrine of gravitation, mostly for emotional reasons culminating in their commitment to Cartesian or similar ideas, the astronomers had better reasons to deny the accuracy of Newton's law of gravitation for the explanation of the observed motions of the heavenly bodies. There was general agreement that it yielded well the first approximation to the planetary orbits, namely that these are ellipses with the sun in one focus. However, it was not clear how certain departures from elliptic motion, the so-called inequalities, can be obtained as a consequence of the Newtonian law.

Explanation of Planetary Inequalities: Laplace

The inequalities were of two types. The "periodic" inequalities righted themselves after some time and thus did not lead to cumulative effects. But more serious were the "secular" inequalities leading to a continuous magnification of the departures from the elliptic motion. The most serious of them was the "great inequality" of Jupiter and Saturn. The great mathematicians, Euler and Lagrange, wrote memoirs containing considerable advances of the Newtonian theory and received prizes from the French Academy of Sciences. Still, these two great men thought that the great inequalities of Jupiter and Saturn were of the secular type. However, Laplace in 1773 proved that the mean motions of the planets cannot have any secular accelerations whatever as a result of their mutual attractions. Laplace concluded therefore that the observed accelerations must really be periodic with a very long period. This theorem proved to be the key to the mystery. Laplace showed this in his great memoir of 1784. It turned out that a great inequality of a long period could be produced by a perturbation which was so small that Euler and Lagrange neglected it in their previous work. Another outstanding discrepancy between Newton's theory and observation was the secular acceleration of the mean motion of the moon. Euler and Lagrange published important papers in 1777, receiving additional prizes, but again it was Laplace in 1787 who found the solution and again the inequality was only apparently secular but in fact periodic (with an immensely long period of several million years). Thus, both great inequalities have been explained by Laplace in a striking vindication of the Newtonian theory, exactly one hundred years after the original publication of the "Principia." No wonder these great successes of Laplace induced him to become so to say the "Father of Determinism."

Field Theory of Gravity: Laplace and Poisson

Newtonian gravitation was originally a particle theory. The idea of a field theory of gravitation was brought up by Lagrange in

1773. He introduced a potential $V(r_{ik})$ whose directional derivative gives Newton's attractive force. The basic differential equation for V

$$\nabla^2 (V) = 0 , \tag{II.4}$$

was discussed in detail by Laplace in 1782 and is named after him. (However, Euler introduced this equation already before Laplace.) This is the beginning of the field theory of gravitation. Writing Newton's equation in the form

$$m_k \vec{r}_k = - \nabla . V , \tag{II.5}$$

$$V = G \sum_{i,k=1}^{N} \frac{m_i \, m_k}{r_{ik}} , \tag{II.6}$$

each term of the potential may be interpreted as a fundamental solution of the partial differential equation

$$\Delta V = 0 . \tag{II.7}$$

This field theory was generalized by Poisson in 1813 who, at that time, introduced a special case of the general equation named after him, which in our case takes on the form (ρ: density of matter)

$$\nabla^2 V = 4\pi \, G\rho . \tag{II.8}$$

The important feature of Newton's theory of gravitation is that it does not contain t explicitly, thus the gravitation is an instantaneous interaction and does not depend on retardation.

It does not seem to have been noticed in the literature that Kepler's third law can be obtained from the explicit form of the Eqs. (I.5) and (I.6), viz

$$m_k \vec{r}_k = G \sum_{i,k=1}^{N} m_i \, m_k \frac{\vec{r}_i - \vec{r}_k}{r_{ik}^3} , \tag{II.9}$$

by a symmetry argument. Eq. (II.9) is invariant under the conformal transformation

$$\left. \begin{aligned} x_k' &= \gamma^2 \, x_k \\ t' &= \gamma^3 \, t \end{aligned} \right\} . \tag{II.10}$$

Kepler's third law follows immediately from (II.10). (Incidentally, it is stated in the literature that Newton's general equation of motion (Eq. II.5) remains invariant under the transformations

$$\left.\begin{aligned} \vec{r}\,' &= \vec{r} + \vec{a} \\ V' &= V - \ddot{\vec{a}}\,\vec{r} \end{aligned}\right\} , \tag{II.11}$$

where the vector $\vec{a}$ depends on time, $\vec{a} = \vec{a}(t)$.)

Early Work on the Gravitational Deflection of Newton's Light Corpuscles: Soldner

It is of historical interest that in 1801 Soldner had the idea that Newton's light corpuscles interact with the gravitational field. This was the same year in which Thomas Young published his fundamental paper proving the wave theory of light. If Soldner had read this paper previously, he would probably never had gotten his idea. The main assumptions in Soldner's work were that (a) the inertial mass of the light corpuscle is equal to its gravitational mass, and (b) obeys Newton's nonrelativistic equations of motion.[6a]

With assumption (a), m, the mass of the light corpuscle, cancels out in the equation of motion. Therefore, Soldner did not have to worry what value of m he should assign to a light corpuscle. The geometry of the passage of light particles into the vicinity of a larger mass, M, (for example, the sun) is shown in Fig. 1. The impact parameter is designated by R. Introducing rectangular coordinates which give the path of light parallel to the x axis, the deflection of light occurs in the x-y plane.

The equation of motion is

$$m \frac{d^2y}{dt^2} = - \frac{GMm}{r^2} \frac{y}{r} , \tag{II.12}$$

where

$$r^2 = x^2 + y^2 \; ; \tag{II.13}$$

if the deflection is small, then $y \approx R$. We write $x = ct$. With these substitutions, Eq. (II.12) may immediately be integrated to give

$$\frac{dy}{dx} = \frac{GMx}{c^2R} \frac{1}{(x^2 + R^2)^{1/2}} \; . \tag{II.14}$$

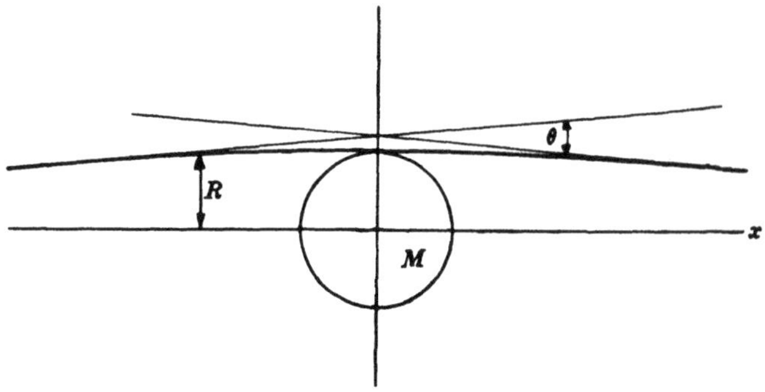

Figure 1. The path of a light beam bent by the gravitational field of the sun.

The angle of deflection of the path of light is

$$\theta \cong \left(\frac{dy}{dx}\right)_{x=-\infty} - \left(\frac{dv}{dx}\right)_{x+\infty} = -\frac{2Gm}{Rc^2} \qquad (II.15)$$

This formula is identical with Einstein's which was derived on essentially the same basis in 1911 without knowing of Soldner's earlier work.

After Einstein obtained his final gravitational field equation which gives twice the deflection given by Eq. (II.15), Soldner's work has been rediscovered and his paper republished in Ann. der Phys. (1921).

Inequality in the Perihelion of Mercury: Leverrier and Newcomb

We discussed briefly in the previous section how Laplace explained the puzzling inequalities, e.g., certain apparent discrepancies between Newton's theory and planetary observations. However, Leverrier, in 1845, showed that a remainder is present in the perihelion motion of Mercury which cannot be explained on the basis of Newton's theory. "It may be of interest to quote from Simon Newcomb's article on Mercury in the 11th edition of the "Encyclopaedia Britannica," published in 1911:

> A perplexing problem is offered by the secular motion of the perihelion of Mercury. In 1845 Leverrier found that this motion, as derived from observations of the transits, was greater by 35" per century than it should be from the gravitation of all the other planets. This conclusion has been fully confirmed by subsequent investigations, a recent discussion showing an excess of motion to be 43" per century.

> It follows from this either that Mercury is acted upon by some unknown masses of matter, or that the intensity of gravitation does not precisely follow Newton's law....

Clemence's most recent value hardly changes this figure, while decreasing the estimate of probable error. It would appear that Newcomb's alternative – that gravitation does not precisely follow the Newtonian law – is realized in Einstein's theory."[7]

Vector Theory of Gravitation: Lorentz

In 1906 H. A. Lorentz gave lectures at Columbia University on "The Theory of Electrons" which were later published (First Edition in 1909, Second Edition in 1915).[8] R. H. Dicke, in a charming scenario,[9] depicts the possibility that Lorentz would have discussed a scalar theory of gravitation and developed it further in his lectures. Then if at the same time observations of the three famous effects have been undertaken, and they would have checked Lorentz' theory, perhaps the young Einstein would have never gotten the idea to work on his geometric theory of gravitation. The Lorentz-invariant theory of gravitation would have been in a more or less complete analogy to the only field theory which was known then, viz., Maxwell's theory of electrodynamics.

Well, Lorentz did discuss such a theory under the title "Considerations on Gravitation"[10] even before 1906, e.g., in 1900. Among several other Maxwell-type theories of gravitation, the theory by the master of electrodynamics, Lorentz, was of most interest -- it was a vector theory of gravitation. As Maxwell already pointed out the energy of a static gravitational field is negative in this type of theory. The same is true for a non-static gravitational field. There are more recent vector theories of gravitation in which this difficulty is (at least partly) avoided. However, in a recent comparison of Lorentz-invariant theories of gravitation it is stated "Generally speaking it appears that vector theories similar in character to Maxwell's electromagnetic theory are particularly poor for the study of gravitational phenomena, if the comparative success of general relativity in Einstein's three classic tests is taken as a reliable guide."[11]

From a psychological point of view it would have been difficult, if not impossible, for Lorentz to develop in 1906 a Lorentz-invariant scalar theory of gravitation. At this time, as was clear from his previously cited book, Lorentz was not at all convinced that Einstein's theory of special relativity was superior to his own theory of 1904 (with which Einstein was not familar in 1905). Lorentz did not have then the correct (completely symmetric in space and time) form of the special relativity (SR) theory.

First Lorentz-Invariant Theory of Gravity: Poincaré, (Minkowski, Sommerfeld, and Lorentz)

Newton's theory of gravity, based on instantaneous action-at-a-distance, is clearly not in accord with SR. The latter requests c as the velocity of propagation also for gravity. Poincaré, in his great paper of 1906,[12] where he developed SR independently of Einstein, already set up a Lorentz-invariant theory of gravity. In his theory, the force between two particles depends on positions which are apart by a time interval t = r/c as well as on their velocities and, possibly, also on their accelerations. Minkowski[13] essentially used a special case of Poincaré's theory. Strangely, he could not have read Poincaré's paper carefully! He claimed that his theory "was quite different from Poincaré's". The only differences were that Minkowski used four-vector calculus and introduced acceleration. Poincaré was the greatest mathematician at that time. Therefore, one would think that Minkowski (also a great mathematician) would study any paper of Poincaré relevant to his own (Minkowski's) work with great care. Sommerfeld,[14] who also used four-vector calculus, recognized that Minkowski's theory was just a special case of Poincaré's, but he apparently did not notice Minkowski's claim to the contrary, e.g., he also did not read Minkowski's paper carefully!

Lorentz also discussed[15] a special case of Poincaré's theory.

It is interesting to note the attitude toward action-at-a-distance theories in those days. Pauli, in his 1921 article on relativity, published first in the "Mathematical Encyclopedia", wrote "The objection to all these considerations is that they take as their starting point a fundamental law of force, instead of Poisson's equation. Once the finite propagation of an effect has been demonstrated, one can only expect to arrive at simple, generally valid laws if one describes it in terms of continuously varying functions of position and time (a field) and looks for the differential equations of this field. The problem thus consists in modifying the Poisson equation (II.8), and the Newton equation of motion of the particle (II.1), in such a way that they become Lorentz-invariant." (cf. ref. 42).

Today, our attitude is more flexible as shown by the very interesting work of Wheeler and Feynman.[16a] More specifically, a Lorentz-invariant theory of interacting particles has been developed by van Dam and Wigner[16] which, in fact, is very closely related to Poincaré's theory of gravity, although the authors were unaware of his work.

SECTION III.

EINSTEIN, ABRAHAM, NORDSTRÖM, MIE

Einstein's First Work on Gravity: Equivalence Principle and Red Shift

Almost simultaneously, but independently of Poincaré's Lorentz-invariant theory of gravitation, Einstein turned his attention into another direction. In a review article[17] on SR which appeared in 1907, Einstein already tried to extend the relativity principle of uniform motion to frames in non-uniform motion. He postulated that the laws of physics should retain their form even in non-Galilean frames. For this purpose, Einstein for the first time introduced what he later called equivalence principle.

In Newtonian theory, a frame in a homogeneous gravitational field is completely equivalent to a uniformly accelerated frame from the point of view of mechanics. Now, the equivalence principle postulates that, in addition, all physical processes should take place in the same way in both systems. At this early stage, apparently, Einstein did not distinguish between strong and weak equivalence principles which he did explicitly in 1913 as we shall see later.

The equivalence principle makes it possible to calculate the effect of homogeneous gravitational fields on arbitrary processes. Using this principle, Einstein was able to derive the red shift. He also pointed out that the velocity of light is not constant in a gravitational field so that light rays may curve. He also showed that both inertial and gravitational mass $m = E/c^2$ has to be ascribed to any energy E. We shall see later, that equivalence principle and red shift are already "implied" by Newton's theory of gravity, if we generalize his gravitostatics to gravitodynamics.

Einstein's First Derivation of (Newtonian) Light Deflection

Four years later in 1911, Einstein[5] derived Soldner's formula for light deflection (11.15) by another method. In retrospect, Einstein's derivation implied a line element which in the Newtonian approximation to GR has the form

$$ds^2 = c^{*2}dt^2 - dx^2 - dy^2 - dt^2 \qquad \text{(III.1a)}$$

where the effective light velocity c^* is given by

$$c^* \cong c(1 - \frac{GM}{c^2r}) \qquad \text{(III.1b)}$$

Incidentally, Eq. (III.1a) implies in general, a Riemannian space-time. Only for the case of a homogeneous gravitational field does Eq. (III.1a) imply a Euclidean space-time. Einstein does not give equation (III.1a) explicitly. However, he does state explicitly equation (III.1b.) This last equation is all he needed to derive the light deflection using Huygens' principle. Thus, it must have been at least "tacit knowledge" (using a concept introduced by Polanyi) for Einstein that equivalence principle, red shift, and light deflection are implied by Newton's theory of gravitation. The agreement of the value for the light deflection obtained by Einstein's method as compared to that of Soldner's method is a proof for the (at least partial) "consistency" of the line element (III.1a) with the usual formulation, e.g., equation of motion. In fact, the equation of motion follows from the line element using the variational principle

$$\partial \int ds = 0 . \qquad \text{(III.1c)}$$

The Newtonian line element (III.1a) implies that Newtonian theory applied to phenomena involving the light also implies the replacement of Laplace's equation for the gravitational potential (II.4) by the wave motion

$$\nabla^2 V - \frac{1}{c^{*2}} - \frac{\partial^2 V}{\partial t^2} = 0 \qquad \text{(III.1d)}$$

and a similar change in Poisson's equation (II.8). Therefore, (approximate) Lorentz-invariance is also implied in application of Newtonian gravity to light phenomena. Clearly, the treatment of light phenomena by Newtonian mechanics is not self-consistent because light doesn't move slowly. For example, according to Newton, the kinetic energy of light corpuscles would be

$$E = \frac{mc^2}{2} , \qquad \text{(III.1e)}$$

and this expression will give double of the radiation pressure. It is interesting that Planck even in the latest edition of his "Theory of Heat Radiation" uses equation (III.1e) as he did in the first edition of his book, and does not emphasize that instead of Newtonian mechanics, SR should be applied because the photons do not move slowly. In both Soldner's and Einstein's derivation (III.1e) does not enter.

In GR, the light deflection turned out to have twice the Newtonian value, the other half being due to Riemannian character, (curvature) of space.

Weak and Strong Correspondence

Under classical correspondence, we understand the postulate that any theory of gravitation should reduce to Newtonian theory for slow motion. We shall distinguish between two types of classical correspondence. In one type which we shall dub weak classical correspondence, we postulate that for slow motion of a test body the gravitational field equation should reduce to Poisson's equation. Strong classical correspondence postulates that the line element of any theory of gravity should reduce to the Newtonian equation (III.1a) in addition to the reduction of the field equation to Poisson's equation.

Preliminary Scalar Theories by Abraham and Einstein[18]

The next problem was how to set up a theory based on the equivalence principle which is also applicable to non-homogeneous gravitational fields. This problem was attacked first by Abraham and then by Einstein in a somewhat different manner. However, both tried to characterize the general static gravitational field by the value of the light velocity c at each space-time point. Thus, c would play the role of a gravitational potential and both Einstein and Abraham tried to construct differential equations which c had to satisfy.

Mie's Theory[19]

Mie also developed a theory of gravity based on the principle of SR. However, in this theory even the weak equivalence principle is not satisfied rigorously. Therefore, Mie's theory never had much chance to succeed.

Of greater influence was Mie's general idea of developing a complete theory of physics based on the principle that electric fields, charges and currents, and magnetic fields yield a complete description of the world -- leading to an electromagnetic "world-picture". Mie also introduced the concept of a single "world-function", which yields, using a variational principle, the laws of all physics. This concept was later developed by Hilbert and influenced also Born's (and Infeld's) non-linear electrodynamics as we shall see later.

Nordström's Two Theories of Gravity[20]; Einstein on First Scalar Theory

The difficulties implied by these theories lead Nordström to develop two scalar theories of gravitation. In both these theories, the principle of SR is strictly valid. The velocity of light is constant, and light is not deflected in a gravitational field. The lack of coupling between light and gravitational field is related to the vanishing of the trace of the electromagnetic stress-energy tensor because this is the only scalar of the electromagnetic field to which the gravitational field can be coupled. Incidentally for photons, the only available scalar, the norm of the four-momentum, also vanishes.

The first theory of Nordström cannot be derived from a Lagrangean and obeys only the weak principle of equivalence. It seems that Einstein has also considered a scalar theory of gravitation similar to Nordström's first theory. However, he soon abandoned his effort. He believed to have proven that a rotating object would fall with less acceleration than a nonrotating object. However, this argument is not correct.

The second Nordström theory can be derived from a Lagrangean and satisfies the strong equivalence principle. This theory brings, in a logically satisfactory way, the Poisson equation into a Lorentz-covariant form. It guarantees conservation of energy-momentum. Moreover, Einstein and Fokker in 1914 showed that it can be reformulated in a generally covariant form.[21] There are three main objections against both scalar theories of Nordström.

First, their line element does not reduce to Eqs. (III.1a, 1b), the Newtonian line element. (We shall show this later in conjunction with the transition of GR to Newtonian theory.) This is related, of course, to the vanishing of the light deflection in both Nordström scalar theories.

Second, even the second theory of Nordström is formally more complicated than Einstein's tensor theory. At first sight, this statement doesn't look right. As a matter of fact, Nordström himself and his contemporaries like Abraham and Mie and even Einstein thought that a theory using one gravitation potential is simpler than one using ten gravitational potentials. Only the one scalar $T = T^{\mu}_{\mu}$ occurs in Nordström's second theory and it is proportional to the curvature invariant R. The remaining equations permit writing the line element in the form

$$ds^2 = \Phi(\chi)\delta_{\mu\nu}d\chi^{\mu}d\chi^{\nu} \qquad \text{(III.1)}$$

and imply the constancy of the velocity of light. From the point of view of general tensor calculus, Nordström's field equations look much more artificial and complicated than the field equation of Einstein's tensor theory in which all components $T^{\mu\nu}$ appear on equal footing.

Third, the equations of motion do not follow from the field equations, again in contrast to GR. The adherence of Abraham and Nordström to scalar theories can only be psychologically explained. They have not imbued the spirit of the general tensor calculus of Ricci and Levi-Civita and they did not like Einstein's geometrization of gravity. However, Nordström, after Einstein discovered GR, participated in its further development (cf. Reissner-Nordström line element).

Although Nordström's scalar theories ultimately had to be abandoned, his was the most sustained and logical effort before Einstein to develop a consistent (Lorentz-invariant) theory of gravity. His work has just now more than historical interest. A scalar-tensor theory like that of Brans and Dicke[4] is the most natural generalization of Einstein's tensor theory should a need arise for such a generalization. Let me remark right here that at present I cannot see any compelling reason for such a generalization. We shall discuss this point in more detail later.

Einstein, around 1912, seems to have developed simultaneously with and independently of Nordström a scalar theory of gravity similar to and perhaps even identical with the first scalar theory of Nordström. There are three sources for this information. First close to the end of Nordström's first paper on his first scalar theory of gravity, he mentions that he received a letter from Einstein in which the latter indicated that he developed a scalar theory similar to Nordström's first scalar theory. Second, in an essay he wrote later "Notes on the Origin of the General Theory of Relativity",[22] he talks of a scalar theory he advanced. The third indication comes from his Autobiography.[23]

According to Nordström, Einstein, in his letter to him mentions as chief objection against this scalar theory: it predicts that a rotating object will fall more slowly than a nonrotating object. It is instructive to quote verbatim from his "Notes" and Autobiography. In his "Notes" he states: "According to classical mechanics the vertical acceleration of a body in the vertical gravitational field is independent of the horizontal component of velocity. Hence in such a gravitational field the vertical acceleration of a mechanical system or of its center of gravity works out independently of its internal kinetic energy. But in the theory I advanced the acceleration of a falling body was not independent of the horizontal velocity or the internal energy of a system."

In his Autobiography he says, "In classical mechanics, interpreted in terms of the field, the potential of gravitation appears as a scalar field (the simplest theoretical possibility of a field with a single component). Such a scalar theory of the gravitational field can easily be made invariant under the group of Lorentz-transformations. The following program appears natural, therefore: The total physical field consists of a scalar field (gravitation) and a vector field (electromagnetic field); later insights may eventually make necessary the introduction of still more complicated types of fields; but to begin with one did not need to bother about this.

The possibility of the realization of this program was, however, dubious from the very first, because the theory had to combine the following things:

(1) From the general considerations of special relativity theory it was clear that the inert mass of a physical system increases with the total energy (therefore, e.g., with the kinetic energy).

(2) From very accurate experiments (specially from the torsion balance experiments of Eötvös) it was empirically known with very high accuracy that the gravitational mass of a body is exactly equal to its inert mass.

It followed from (1) and (2) that the weight of a system depends in a precisely known manner on its total energy. If the theory did not accomplish this or could not do it naturally, it was to be rejected. The condition is most naturally expressed as follows: the acceleration of a system falling freely in a given gravitational field is independent of the nature of the falling system (specially therefore also of its energy content).

It then appeared that, in the framework of the program sketched, this elementary state of affairs could not at all or at any rate not in any natural fashion, be represented in a satisfactory way. This convinced me that, within the frame of the special theory of relativity, there is no room for a satisfactory theory of gravitation."

All these, essentially identical, arguments of Einstein refer to the first Nordström (Einstein) scalar theory. It seems that Einstein did not develop a scalar theory corresponding to Nordström's second scalar theory, but turned already in the latter half of 1912 to a tensor theory, after learning enough about the mathematics of tensors. (cf. subsection: Einstein's First Attempt for Geometrization of Gravity); there we shall discuss Einstein's Machian arguments against the second scalar theory of Nordström.

Kottler's Generally Covariant Formulation of the Electromagnetic Field Equations in Vacuum

Without the direct connection with the theory of gravitation, Kottler[24] in his 100 page thesis, "Space-Time Lines of the Minkowski-World", made the first extensive application in physics of the work of Goursat[25] on integral forms and of Ricci and Levi-Civita's "Absolute Differential Calculus".[26] Whittaker,[27] in his "History, etc.", Vol Il, l.c., p. 154, 156, gives Bateman[28] credit for the first use of the general quadratic form of the line element in physics; cf. his eq. (1) on p. 155. He also says "Bateman realized the connection of his work with the tensor calculus of Ricci and Levi-Civita." This statement and the credit given by Whittaker to Bateman is misleading. In connection with two theorems on the transformation of integral forms, Bateman says in a footnote "they appear to be closely connected with two theorems used by Ricci and Levi-Civita." Kottler in his paper on p. 1686 refers to Bateman's paper in a footnote in connection with integral forms of Maxwell's equation which were introduced by Hargreaves.[29] This is one of the many examples where Whittaker, who was close to 80 when he published his book, must have forgotten and/or mislaid some of his notes. (Whittaker says that the only reason at his age he was able to publish his two books was because whenever at earlier days he read papers, he made notes and comments.) I am sure that Whittaker had a copy of Kottler's paper. Still he did not mention him in this connection and gave a lot of unjust credit to Bateman. Kottler essentially solved the problem how a field-law, given in the terminology of SR, can be transferred to the case of a Riemann metric.

Einstein's First Attempt for Geometrization of Gravity

For the historical development, it was fortunate that Einstein was not satisfied with the scalar theory of gravity. His first argument against the scalar theory[30] was incorrect and was taken back by Einstein himself[31] but the second argument[32] was never taken back by Einstein even though, as we will see later, he disproved himself without realizing (or perhaps forgetting) the connection with his early arguments against the scalar theory. At any rate, these early difficulties of the scalar theory seemingly strengthened Einstein in his endeavor: to develop the laws of physics, in general, and the laws of gravity, in particular, in such a form that they be covariant under the widest possible group of transformations. Until about 1912, Einstein learned and used mathematics only to the extent to which it was absolutely necessary for the treatment of the physical problems he was interested in. It is described vividly in his autobiography that although he was very enthused about Euclidean geometry in his early school days, he still didn't turn to mathematics because he felt he could not judge well which field in mathematics was the most important one (at that time). In contrast to mathematics, Einstein says that he was soon able to learn to scent out what is important in physics. Although he had excellent teachers in mathematics, like Minkowski and Hurwitz, he neglected to profit from them. This explains why Minkowski, when he heard of Einstein's epochal work on SR, said that he was surprised that Einstein could do this work because he was not such a good student in Zurich!

Around 1912, however, he realized that he had to learn more mathematics for his efforts to geometrize gravity. Therefore, he asked his old friend, Marcel Grossmann, of Zurich student days, for help in mathematics. Both became professors at their old Alma Mater, ETH, (Federal Institute of Technology), Einstein in physics and Grossmann in mathematics. (Incidentally, Grossmann in his student days knew Haller, the director of the Swiss federal patent office in Bern, and through his connection got Einstein his first job as a "technical help of third class".) Grossmann helped him to assimilate the work of Riemann and Christoffel as generalized and extended to an "absolute differential calculus" by Ricci and Levi-Civita.

In 1913 Einstein and Grossmann[33] published a brochure with the title "Outline of a General Relativity and of a Theory of Gravitation" in which Einstein wrote the physical part and Grossmann, the mathematical part. Here they transform the square of the line element into an arbitrary curvilinear space-time coordinate system resulting in a quadratic form in the coordinate differentials with ten coefficients, $g_{\mu\nu}$. The gravitational field is now determined by this ten-component tensor of the $g_{\mu\nu}$. This was the great leap forward. It abandoned the time-honored belief that the gravitational field can be described by a single scalar potential. This paper for the first time put forward the idea that straight-line motion of a free particle is determined by the equation

$$\delta\int ds = 0 \tag{III.2}$$

where

$$ds^2 = c^2dt^2 - dx^2 - dy^2 - dz^2. \tag{III.3}$$

The motion of a free particle in a gravitational field is determined by the same equation. But now Eq. (III.3) has to be replaced by

$$ds^2 = g_{\mu\nu}\, dx^\mu\, dx^\nu. \tag{III.4}$$

Einstein and Grossmann also put the energy-momentum law and the electromagnetic field equations for the vacuum into a generally covariant form, giving proper credit for the last accomplishment to Kottler. They got acquainted with Kottler's work while they were preparing their work for publication. However, the field equations which were written in the form

$$\Gamma_{\mu\nu} = \kappa T_{\mu\nu}, \tag{III.5}$$

(where $\Gamma_{\mu\nu}$ was a complicated tensor) were neither invariant nor self-consistent).

[In a letter to Sommerfeld, dated October 12, 1912, (cf. ref. 37, l.c.p. 26), Einstein said, he works now exclusively on the problem of gravitation and believes to have mastered, with the help of his mathematical friend, Grossmann, all difficulties. He adds, that he never worked so hard all his life and developed high regards for mathematics which, in its subtler parts, he previously considered to be pure luxury.]

Post-Script

After this manuscript was finished Lawrence Dresner kindly provided me with a copy of Helle Zeit - Dunkle Zeit - In Memoriam Albert Einstein, (C. Seelig, ed.), Europa Verlag, Zurich (1956). This book contains an "autobiographic sketch" (l.c. page 9) which Einstein wrote about a month before he died in March, 1955. The sketch contains a more detailed description of his psychological makeup from his student days up to the time when he made his greatest contributions to physics and, particularly, his friendship and collaboration with Marcel Grossmann. He says, among other things, that while he was anxious for more profound understanding, he had little talent for retention and a poor memory. (This of course is in keeping with his well-known modesty.) We would interpret this to mean that he was only interested in learning and retaining that which was important in connection with his own thoughts. In light of this fact, we can better understand why, as stated in this paper, Einstein did not read and remember certain books and papers. In general, he was inclined much less to reading than to thinking.

Weak and Strong Equivalence

In Sept. 21, 1913, the Austrian-German equivalent of our AAAS, the "Naturforscherversammlung" held its 85th meeting in Vienna. Einstein was invited by the joint sections of Physics, Mathematics, and Astronomy to review the present (1913) status of the problem.

In his review article,[39] Einstein starts by comparing the transition of Newtonian theory to GR with the similar transition from

Electrostatics to Electrodynamics. He emphasizes that the large number of possible generalizations are limited by the great formal insight of Minkowski into the symmetry between space and time coordinates. He proceeds to state some general postulates which a theory of gravitation can but must not satisfy. These are:

1. Validity of conservation of energy and momentum.
2. Equality of inertial and gravitational mass of closed systems.
3. Validity of SR: equations invariant under linear orthogonal substitutions (generalized Lorentz transformations).
4. The observable laws of nature do not depend on the absolute values of the gravitational potentials. This has the following physical significance: All relations between observable quantities, which can be found in a laboratory will not be changed by bringing the whole laboratory in a region of a different (but constant in space and time) gravitational potential.

Einstein emphasizes that while all theoreticians will agree about the necessity of postulate 1, there might be arguments about 2 and 3. However, he personally feels that their validity should be assumed.

Postumate 4 cannot be based on experience, but only on the faith in the simplicity of the laws of nature. Einstein is conscious of the fact that the postulates 2 and 4 express more a scientific faith than an established truth. He also does not claim that Nordström's scalar theory and his (provisional) tensor theory are the only possible generalizations of Newton's theory; but they are the most natural ones corresponding to the present (1913) knowledge.

Clearly, postulate 2 is the weak equivalence principle. On the other hand, postulate 4 corresponds to the strong equivalence principle. In his later papers, Einstein did not make a sharp distinction between these two equivalence principles, but implied mostly the strong equivalence principle. The distinction made here corresponds exactly to that made (much later but independently) by Dicke.

Einstein emphasizes that only Nordström's second scalar theory and his own (provisional) tensor theory satisfy all four postulates (1, 2, 3, and 4). He does not point out in his review that Nordström developed two different scalar theories. The first one obeys only the weak equivalence principle and cannot be formulated in Lagrangean form. The second scalar theory of Nordström does satisfy the strong equivalence principle and can be put into Lagrangean form, which Einstein presents in his review. None of the other theories of gravitation, for example by Abraham or by Mie, satisfy all four postulates and Einstein does feel justified not to discuss them in his review although he did have some discussion about them with Abraham and Mie. We must remember that at this time (1913), Einstein wasn't able yet to predict the three famous tests. He did predict (half of) the light deflection in 1911, but he did say that it would be interesting if someone would check his prediction even when his derivation appeared to have an insufficient basis or even appeared much too fantastic. In this situation, Einstein had only theoretical arguments to decide between Nordström's second scalar theory and his own preliminary tensor theory. For this decision, he used what he later called Mach's principle .

Global and Local Mach Principles

The young Einstein was greatly impressed by Mach's philosophy of nature, in general, and his discussion of the rotating-bucket-of-water experiment of Newton, in particular. We have now to distinguish sharply what we shall call global and local Mach principles. We shall call global Mach principle, Mach's original assumption of the (causal) relation between Newton's famous rotating bucket-of-water experiment and the mass of faraway stars (or galaxies). Under the local Mach principle, we understand Einstein's hypothesis of the "relativity of inertia," viz., "that inertia depends on a mutual interaction of the bodies." (Mach himself seemed to believe that the effect of nearby masses on the inertia of a test particle cancels out and emphasized directly only what we call the global Mach principle, though not in a precise form.) Qualitatively, Einstein did refer to the global Mach principle. Quantitatively, however, Einstein used only the local Mach principle.

On pages 226-228 of the Einstein-Grossmann paper, Einstein seems to have for the first time believed to have proven that, as a consequence of the local Mach principle:

(1) "The inertia of a body must increase when ponderable masses are piled up in its neighborhood." He mentions this conclusion also in a letter which he wrote to Mach on the 25th of June, 1913, referring to a paper which he sent to Mach before. He also mentions, there, two further effects:

(2) "A body must experience an accelerating force when neighboring masses are accelerated."

(3) "A rotating hollow body must generate inside of itself a Coriolis field which deflects moving bodies in the sense of rotation and a centrifugal field as well."

Einstein must have come to the conclusions 2. and 3. after he finished the paper he sent to Mach.

In his 1913 review, § 9, pp. 1260-1262, Einstein again attempts to prove statement 1., and he also tries to prove statements 2. and 3. as we shall see later in the section Later Developments. Einstein's "proofs" of statement 1. are based on the use of particular coordinate systems, therefore they cannot have invariant significance. The truth is that Nordström's first scalar theory which obeys only weak equivalence and weak classical correspondence does satisfy the local Mach principle. On the other hand, Einstein's (provisional and his) or his final tensor theory satisfy both strong equivalence and strong classical correspondence, but do not obey the local Mach principle.

Einstein's Round-About Pathway to the Final GR Field Equations

In a footnote on p. 1257 of his 1913 review, Einstein says that during the last days, he found a proof that generally covariant field equations cannot exist. This "proof" is indicated at the end of Einstein and Grossmann's paper, l.c., p. 260. In a second joint paper of Einstein and Grossman,[35] this "proof" is somewhat generalized.

Shortly thereafter, Einstein said that in his theory of gravitation (which he developed in part with his friend Grossmann) the heuristic tools were a colorful mixture of physical and mathematical postulates. Therefore, it was not easy to see how the theory is related to a formal mathematical point of view. Consequently, in this paper he tried to simplify the basic laws of physics using tensor calculus. He also used some physical problems to illustrate the uses of the mathematical methods. Finally, in a short physical section, he discusses reduction of the theory to Newton's for slow motion and obtains red shift and Newtonian light deflection. Clearly, his field equations were neither generally covariant nor consistent. However, in spite of the inconsistency of the theory, Einstein correctly ascribes half light deflection and red shift as characteristic properties of a Newtonian (static) gravitational field. This paper was presented to the Berlin Academy of Science on the 29th of October, 1914.[36]

Einstein stated later that in addition to these somewhat complicated and not generally covariant field equations, he and Grossmann, using Riemann's curvature, had already considered the right field equations. However, they were unable to see how it could be used in physics. On the contrary, he felt sure that they both did not agree with Nature.[37] Moreover, as we discussed before, he thought that he could show on general grounds that covariant field equations contradicted the principle of causality. "These were errors of thought which cost me two years of excessively hard work, until I finally recognized them as such at the end of 1915 and succeeded in linking the question up with the facts of astronomical experience, after which I ruefully returned to the Riemannian curvature.

In the light of knowledge attained, the happy achievement seems almost a matter of course, and any intelligent student can grasp it without too much trouble. But the years of anxious searching in the dark, with their intense longing, their alternations of confidence and exhaustion, and the final emergence into the light; only those who have experienced it can understand that."

On the way to the correct field equations, he again postulated general covariance which he had previously "abandoned only with a heavy heart." Still his first attempt presented to the Berlin Academy November 4, 1915, proposing as field equations [38]

$$R_{\mu\nu} = -\kappa T_{\mu\nu} , \qquad \text{(III.6)}$$

was not quite correct. However, this equation was sufficient to obtain the correct expressions for the three famous effects.[39] He finally realized, that since the covariant derivative of the matter-energy tensor vanishes, therefore, the covariant derivative of the left hand side should vanish also. Thus, he arrived at the final field equations of GR presented to the Berlin Academy on Nov. 25, 1915.[40]

$$R_{\mu\nu} -1/2\, g_{\mu\nu} R = -\kappa T_{\mu\nu} . \qquad \text{(III.7)}$$

Einstein and Hilbert

Apparently, Weyl, in the first textbook of general relativity (first edition published in 1918),[41] originated the story that the field equations of general relativity have been established also by Hilbert simultaneously with and independently of Einstein. This remark was taken over also by Pauli in his well-known article on relativity[42] and later on by several other text book writers. The remark of Weyl and Pauli does not correspond with the historical truth. In the first place, these two great experts did not look up carefully the references in Hilbert's paper. Otherwise, they would have noticed that Hilbert quotes all of Einstein's communications of November, 1915, in the Proceedings of the Berlin Academy. In particular, he quotes the paper on p. 844 which contains the final form of the GR field equations.[43]

It is interesting, however, how this "myth" has arisen. For the following remarks I am indebted to Professor P. P. Ewald who was, in 1915, Hilbert's "assistant for physics." His story is reinforced by the recently published exchange of letters between Einstein and Sommerfeld.[37] In a letter dated July 15, 1915, Einstein says about his visit in Göttingen that he had great joy there; everything he said was understood in detail. He was very enthusiastic about Hilbert. Clearly Einstein gave a talk in Göttingen. Hilbert, whose absent-mindedness was legendary, started to think about the problem on the basis of what Einstein said, who at that time did not have yet the correct form of his field equations. I heard from Ewald that Hilbert, very likely in the fall of 1915, gave a talk in Göttingen presenting the correct equations without referring to Einstein. However, by that time, Einstein must have had also the correct equations, certainly before Hilbert did. Sommerfeld heard about Hilbert's talk in Göttingen and suggested that Hilbert write a letter apologizing to Einstein, which Hilbert of course did. At any rate, he never claimed having been the independent discoverer of the field equations of GR.

It is interesting for the change of fashions and/or different tastes that Pauli states Hilbert's "presentation though would not seem to be acceptable to physicists for two reasons. First, the existence of a variational principle is introduced as an axiom. Secondly, of more importance, the field equations are not derived for an arbitrary system of matter, but are specifically based on Mie's theory of matter." On the other hand, Whittaker[44] states that Hilbert's procedure introducing a world-function to be used in conjunction with a variational principle represents "a distinct advance on Einstein's methods." At any rate, at present we are inclined to prefer or at least not look down upon the derivation of any physical law from a variational principle. Perhaps it is interesting to point out that historically already Ricci and Levi-Civita mentioned what turned out to be later the Lagrangean of GR as of some possible physical interest, but, of course, they did not think of the connection with gravity.[45]

Even though Hilbert's work was not independent of Einstein's, it was very clarifying. For example, Hilbert recognized first that the general solutions of the field equations must contain four arbitrary functions. Thus, there must be four identities between the ten field equations for the ten unknown $g_{\mu\nu}$. The apparent contradiction with the causality principle is resolved by the remark that the many possible solutions of the field equations are physically completely equivalent and only formally different. Later on, Hilbert gave lectures on GR; Laue, in his textbook of GR, thanks Hilbert for putting his lecture notes at Laue's disposal. Hilbert also emphasized that not only must the general laws of physics be covariant, but every result of the theory which has a physical significance must be covariant with respect to arbitrary transformations of the coordinate system. Therefore, f.e. a statement that the particle is at rest for an interval of time of definite duration cannot have physical significance because this statement is not covariant.[46]

SECTION IV

LATER DEVELOPMENTS[48]

Reception of GR

In the fall of 1915 Einstein arrived, after many false starts, at his final GR. The only test of the theory he had was the explanation of the perihelion motion of Mercury as determined by Newcomb (c.f. the quotation in Section II, Subsection: Inequality in the Perihelion of Mercury: Leverrier and Newcomb). For some time, a few astronomers tried to disprove Newcomb's value of 43" per century, and thus throw doubts on the significance of the agreement between GR and planatary observations. Newcomb's value, however, was fully confirmed by later (and more accurate) work by Clemence and Wayman. The recent ingenious measurement of solar oblateness by Dicke and Goldenberg might change the situation only if Dicke's interpretation of his experiment is the right one, but this is a controversial subject. Many astrophysicists disagree with Dicke's intrepretation.) Mathematicians like Hilbert, Klein and Weyl (who wrote the first book on GR), astronomers like Schwarzschild and Eddington and theoretical physicists like Lorentz, who either were familiar with tensor calculus or had no difficulties in learning its application, have very fast recognized the significance of Einstein's GR.

On the other hand, v. Laue, a close friend of Einstein who wrote the first textbook on SR in 1911, was skeptical about GR for some time. More than a year after the appearance of Einstein's first paper on the final form of his GR, he wrote a review article[49] on Nordström's scalar theories, although these have been discussed in some detail in a review article by Abraham[49] which was written in 1914, three years before v. Laue's review article. In 1917, Laue still thought that the Lorentz-invariant scalar theories of Nordström are <u>simpler</u> than the generally covariant tensor theory of Einstein with its basic modifications (Riemannian geometry) of Newtonian ideas on space and time. He even cast doubt on the significance of the agreement between GR and observation concerning the perihelion motion of Mercury, saying that the GR calculation might apply to mass points but not to extended bodies. Of course, we know now that this objection is not valid.

It wasn't until the triumphant confirmation of Einstein's prediction of the solar light deflection by the British expeditions to Africa in 1919 that GR was more generally accepted. Still even then, there had been stories that only half a dozen people really understood relativity (perhaps the half dozen mentioned before). There was another story according to which Eddington was asked: Was it true that only three people really understood relativity and Eddington supposedly asked, "Who is the third?".

Certainly, Eddington was convinced of the validity of GR even before its confirmation by the observations on light deflection. Sir Frank Dyson, the Astronomer Royal, was asked by Crommelin (one of the astronomers who went on the African expedition) what would happen if observations give double of Einstein's prediction? He supposedly said, "Then Eddington would go mad, and you would have to come home alone." We won't discuss here the somewhat ridiculous arguments by physicists who did not understand GR like Lenard and Stark. In 1920, there was a "Naturforscherversammlung" at Bad Nauheim where the many friends and the few opponents of GR clashed.

In 1921 in the preface of his book on GR, v. Laue says that GR is much admired and much criticized. He also says that because of his initial skepticism of GR, he started comparatively late to work on it, but the more he thought about it the more he felt convinced by its inherent logical consistency, although he was still of the opinion that GR is neither with regard to agreement with experiments nor with regard to finality comparable to SR at the time (1911) when he wrote his book on SR.

Einstein emphasized on several occasions that he didn't consider his GR "the final theory." Lanczos,[50] around 1922 when he first met Einstein, told him about a mathematical method to solve the strongly non-linear field equations of GR through a method of successive integrations. To Lanczos' surprise, Einstein said ... "why should one make efforts to solve these equations rigorously if they have only an ephemeral significance." In the same vein, Infeld quotes Einstein saying that the left hand side of his field equation (III.7) is solid as rock, but the right hand side is weak as straw.[51]

In his Autobiography he says, "The right side (Eq.III.7) is a formal condensation of all things whose comprehension in the sense of a field-theory is still problematic. Not for a moment, of course, did I doubt that this formulation was merely a makeshift in order to give the general principle of relativity a preliminary closed expression. For it was essentially not anything more than a theory of the gravitational field, which was somewhat artifically isolated from a total field of as yet unknown structure.

If anything in the theory as sketched -- apart from the demand of the invariance of the equations under the group of the continuous co-ordinate-transformations -- can possibly make the claim to final significance, then it is the theory of the limiting case of pure gravitational field and its relation to the metric structure of space."

The situation did not change much since 1921. GR is still very much admired and much criticized. We quote from a recent (1966) excellent review of GR by Trautman:[52] "Many physicists are very emotional about the Einstein theory. Most of them admit that the general theory of relativity is a beautiful theory, but then they add that because of the weakness of gravitational forces the theory of gravitation is on the sidelines relative to the rest of physics. Some physicists go so far as to say that GR should not be regarded as a physical theory.

A less extreme position consists in the assertion that those who are at present developing the Einstein theory are mathematicians rather than physicists. In certain circles relativists are regarded as "socially undesirable elements." Most discussions begin or end with the reproach that GR has been subjected to too few tests -- as if this were a fault of the relativists. There are many reasons for these misunderstandings. On the one hand, certain relativists regard Einstein's theory as standing in some higher relation to other theories, and believe that in the last analysis it will combine all the others (hence the search for "unified theories"); some of them are inclined to ignore quantum physics. Such a position had its beginning with Einstein who regarded the quantum theory, with its statistical interpretation, with mistrust. On the other hand, many physicists working in other fields are not inclined to study the fundamentals of Riemannian geometry that are necessary for understanding GR, and blame the theory because it is not included among the Lorentz-invariant quantum fields.

At the present time the general theory of relativity is the best of the existing classical theories of space, time, and gravitation. It is an exceptional example of a theory having a good logical foundation, but confirmed by a small number of experimental data. Under such conditions it is natural to develop the theory as far as possible."

Cosmology [53]

The cosmological problem in Einstein's formulation is based on the hypothesis (suggested by experience) "that there exists an average density of matter in space which differs from zero". Relativistic cosmology, as initiated by Einstein in 1917, is an attempt to reconcile this hypothesis with GR. For this purpose Einstein introduced a new universal constant (cosmological constant) by adding a term $\lambda g_{\mu\nu}$ on the left hand side of Equation (III.7). Einstein was aware of the fact that the introduction of this term constitutes a complication of the theory which reduces its logical simplicity.

In 1922, however, the Russian meteorologist and flying champion, A. Friedmann, showed that the field equations of GR can yield a solution for which there is finite density in the "whole three-dimensional space, without enlarging these field equations ad hoc". Einstein first (1923) doubted the validity of Friedmann's solution, but a few months later, after correspondence with Friedmann, he admitted that his criticism was based on an error in his calculations and agreed that Friedmann's was a correct solution of the field equations. Still, Einstein did not take up at that time a study of cosmology using Friedmann's solution. The reasons for this were probably three-fold: first, at that time Einstein was already actively pursuing unified field theories; second, he was still working actively on quantum theory; and third, and perhaps more importantly, before 1929 there were no observations which would have shown agreement with the main prediction following from Friedmann's solution, namely the possible expansion of our universe. Incidentally, the application of Friedmann's solution to stellar evolution was made in 1927 by Lemaître. Therefore, it is not surprising that Einstein became interested in the work of Friedmann and Lemaître only after Hubble's discovery of the expansion of our universe in 1929. Einstein always insisted that the main postulates of physical theory must be freely invented, but he also insisted that only a thorough comparison of the theoretical conclusions with experiment can prove whether a particular theory is physically significant or is only a figment of imagination.

Unified Field Theory

Soon after (1915) Einstein established GR, Weyl developed a unified field theory based on a generalization of Riemannian geometry. This theory attempts to describe all of physics in terms of gravitation and electromagnetism and these in turn of the world metric. This theory was the first in which the whole physics is presented as a branch of geometry or geometry as a branch of physics. Unfortunately, as first pointed out by Einstein, this theory led to contradiction with experience.

Starting in the twenties, Einstein in his later years more and more turned to various forms of the generalization of the original GR. For a short account we refer to Supplementary Note 23., pp. 224-232 of Pauli's book. These attempts by Einstein and numerous other authors of unifying physics have led to interesting formal developments but not to any physically interesting results. It is easy to understand, psychologically, why Einstein insisted in his unification of gravitation and electromagnetism. Gravitation is universal and macroscopic. When Einstein started on unified field theories, there were only two elementary particles, namely, electron and proton. Thus, it was natural to consider electromagnetism also as universal. Moreover, electrodynamics was originally also a macroscopic theory. Furthermore, both gravitating and charged particles interact via long range forces. On the other hand even before elementary particles had been discovered showing that electromagnetism cannot be universal, it was not clear that gravitation is not just a macroscopic property; while already in the twenties, it was clear that electromagnetism is valid even in nuclear distances. As a matter of fact, this has been clear since Rutherford's basic scattering theory.

Another argument for unified field theory was that GR with its non-linear field equation, from which the equations of motion follow, might be a model for a non-linear theory of all matter. Perhaps another reason for Einstein was that Einstein did not believe that statistical quantum theory was the last word. He said, "God doesn't play dice." This attitude was another reason why he hoped that perhaps quantum phenomena can also be obtained from a unified field theory. Furthermore, as Einstein stated in an obituary of Ehrenfest in 1933, for any scientist above 50 it is very difficult to keep abreast of new developments. (In 1933 Einstein was 53 years old.) In fact while Einstein in Berlin usually attended seminars, he seldom did this in Princeton.

The "already unified field theory" of Rainich, Wheeler, and Misner leading to Wheeler's Geometrodynamics is a topic of current research.

Very recently and closely related to geometrodynamics, the Hamilton-Jacobi formulation ("Superspace") of GR is considered as an interesting and hopeful field to connect gravitation with other fields of physics. Quantization of the gravitational field, in analogy to the electromagnetic field is a main objective of these generalizations and/or reformulations of GR. However, experimentally it has not been established, that gravity has a meaning at subnuclear distances down to the "Planck length" (10^{-33}). This was discussed in several talks at this conference.

Recently, the universality of gravity has been established, with high accuracy, also for elementary particles. Electrons and positrons, kaons and antikaons, etc. behave the same way in gravitational fields.

The Many Faces of Mach

We borrowed this subheading from the title of a very stimulating article of Dicke.[54] In Section III under the subheading Global and and Local Mach Principles, we discussed only "two faces of Mach". However, Dicke is quite right. In the many articles about Machian principles many faces of Mach are presented. Here, we cannot do justice to all these faces but restrict ourselves essentially to recent developments concerning the local Mach principle.

In May, 1921, Einstein delivered four Stafford Little Lectures at Princeton University. The text of his lectures has been published in book form under the title, The Meaning of Relativity.[55] The original text has not been changed in four subsequent editions. In the second edition an appendix was added on cosmology and in the fifth edition another appendix was added on unified field theory. On p. 100 of his book Einstein says "... the theory of relativity makes it appear probable that Mach was on the right side in his thought that inertia depends upon a mutual action of matter. For we shall show in the following that, according to our equations, inert masses do act upon each other in the sense of the relativity of inertia, even if only very feebly. What is to be expected along the line of Mach's thought?

(1) The inertia of a body must increase when ponderable masses are piled up in its neighbourhood.

(2) A body must experience an accelerating force when neighbouring masses are accelerated, and, in fact, the force must be in the same direction as the acceleration.

(3) A rotating hollow body must generate inside of itself a "Coriolis field," which deflects moving bodies in the sense of the rotation, and a radial centrifugal field as well.

We shall now show that these three effects, which are to be expected in accordance with Mach's ideas, are actually present according to our theory, although their magnitude is so small that confirmation of them by laboratory experiments is not to be thought of."

Einstein then goes on to derive these effects from the linear approximation to his theory. His result is the following in slightly changed notation: $\ell = ct$,

$$T^{\mu\nu} = \sigma \frac{dx\mu}{ds} \frac{dx\nu}{ds} \; ; \; ds^2 = g_{\mu\nu} \, dx^\mu \, dx^\nu \; . \qquad \text{(IV.1a*)}$$

$$\frac{d}{dl}\left[(1 + \bar{\sigma})\vec{v}\right] = \text{grad}\ \bar{\sigma} + \frac{\partial \vec{A}}{dl} + [\text{curl}\ \vec{A} \times \vec{v}] \quad \text{(IV.1a)}$$

$$\bar{\sigma} = \frac{\kappa}{8\pi}\int \frac{\sigma dV_0}{r} \quad \text{(IV.1b)}$$

$$\vec{A} = \frac{\kappa}{2}\int \frac{\sigma \frac{dx_\alpha}{dl}\ dV_0}{r} \quad \text{(IV.1c)}$$

The equations of motion, show now, in fact, that

(1) The inert mass is proportional to $1 + \bar{\sigma}$, and therefore increases when ponderable masses approach the test body.

(2) There is an inductive action of accelerated masses, of the same sign, upon the test body. This is the term $\frac{\partial \bar{\vec{A}}}{\partial l}$.

(3) A material particle, moving perpendicularly to the axis of rotation inside a rotating hollow body, is deflected in the sense of the rotation (Coriolis field). The centrifugal action, mentioned above, inside a rotating hollow body, also follows from the theory, as has been shown by Thirring.

That the centrifugal action must be inseparably connected with the existence of the Coriolis field may be recognized, even without calculation, in the special case of a co-ordinate system rotating uniformly relatively to an inertial system; our general co-variant equations naturally must apply to such a case."

From now on we are concerned with effect 1. which corresponds to the local Mach principle. The form of the left hand side of the equation (IV.1a) invites the introduction of an effective mass

$$m^* = m\ (1 + \bar{\sigma})\ . \quad \text{(IV.2)}$$

Unfortunately, Einstein seems to have overlooked that his conclusion 1. is valid only in a particular coordinate system he has chosen and that this effect is simply a cordinate effect which happens for that particular coordinate system only. Since the choice of his particular coordinate system is arbitrary, the result can be transformed away by the choice of another coordinate system and therefore, his effect doesn't exist and naturally cannot be observed.

It is interesting that without reference to Einstein's book which at that time had not appeared, v. Laue in the first edition of his book on GR (the preface to which was written in the same month (May) and year (1921) when Einstein gave his lectures at Princeton) states[56] very clearly on pp. 186-187 that a relation like equation (IV.2) can be valid only for a particular system and does not have invariant significance. It is somewhat strange that Einstein who must have at least looked through the book of his close friend, v. Laue, did not notice the contradiction to his own considerations.

In 1945 Einstein and Straus wrote a paper on The Influence of the Expansion of Space on the Gravitational Fields Surrounding the Individual Stars.[57] We quote the "statement of the problem" and the "conclusions" from the original paper:

"Statement of Problem

In the theory of relativity one is used to representing the gravitation field in the neighborhood of a single star by the centrally symmetric static solution of the field equations, which were first stated by Schwarzschild. This field goes over asympototically with increasing distance from the generating mass into the Euclidean (or rather, Minkowskian) space. That is to say, it is embedded in a "flat" space. On the other hand, we know that real space is expanding, and that, for the existence of a non-vanishing average density of matter, the field equations will imply such an expansion.

The boundary conditions on which the Schwarzschild solution is based are, therefore, not valid for a real star. In particular the boundary conditions which are valid for the expanding space are dependent on time. One has to expect, therefore a priori, that the field surrounding a single star is essentially dependent on time.

The problem of this time dependence is of particular interest, since such a time-dependent behavior could be of essential importance for the theory of matter. The assumption has been voiced in this connection that there may exist connecting relations between the cosmic and the molecular constants.

The investigation below yields that the expansion of space has no influence on the structure of the field surrounding an individual star, that it is a static field -- if only for an exactly delimited neighborhood."

"Conclusion

The field of the mass point in the interior of (a region) G, which is imbedded in an expanding space is, considered in "local coordinates," a static field given by the Schwarzschild solution. The time dependence implied by the expansion does not make the solution time-dependent. What becomes time-dependent is the boundary of G where the Schwarzschild field goes over into the field generated by homogeneously distributed matter."

From this result, it followed clearly that Einstein's effect 1. did not exist. Again, it is somewhat strange that Einstein didn't think of the connection with the statement in his book on the occasion of the second and later editions of his book.

The next episode in this story is a paper by W. Davidson[58] who reconsidered Einstein's derivation of equation (IV.1a) and pointed out that a more consistent approximation leads to $3\bar{\sigma}$ instead of $\bar{\sigma}$ in the equation (IV.1a) for the effective mass.

$$m^* = m\,(1 + 3\bar{\sigma}). \qquad \text{(IV.2a)}$$

However, he also did <u>not</u> notice that equation (IV.2a) has no invariant significance.

In the early 60's Dicke noticed the non-invariant nature of Einstein's effect 1. At Dicke's and Misner's suggestion, Brans[59] made an extensive study of effect 1. which again established its non-invariant character. Brans[60] has also studied the occurrence of the same effect in the Brans-Dicke scalar-tensor theory and concluded that there this effect exists. It also exists in Nordström's first scalar theory which obeys the principle of <u>weak equivalence</u> and <u>weak</u> correspondence.

In GR which obeys <u>strong equivalence</u> and <u>strong correspondence</u> and in Nordström's second scalar (NEF) theory which obeys <u>strong equivalence</u> but <u>weak correspondence</u>, this effect does not exist.[61]

Einstein's Argument Concerning the First Nordström Scalar Theory

In section III under the subheading <u>Nordström's Scalar Theories of Gravitation</u>, we stated Einstein's counter-argument with regard to Nordström's first scalar theory: Rotating objects fall slower than non-rotating objects so that <u>the gravitational acceleration would become dependent on the internal structure of the object</u>.

Recently, a lively discussion developed concerning this argument. Harvey[60a] reconstructed Einstein's argument in detail. However, Dicke[60b] and Sexl[60c] showed that, in fact, <u>the gravitational acceleration turns out to be independent of the internal structure of the object</u>.

New Foundation of GR: Derivability of Equations of Motion from Field Equations

In Section III we discussed Einstein's 1913 paper in which he attempted to distinguish between the second scalar theory of Nordström and his tensor theory using what we call the local Mach principle. It is mentioned there and described later in detail in this section (subsection: The Many Faces of Mach) that Einstein's arguments in favor of his tensor theory were not correct.

In a recent note[2], I indicated another postulate to distinguish between GR and the NEF theory: The derivability of the equation of motion from the field equation.

Einstein gave, perhaps, the clearest and most concise formulation of his basis principles of GR in Ann. Physik 55, 241 (1918), [answer to Kretschmann's criticism of the physical implications of general covariance; similar criticism has been lately voiced by Fock]. The three basic postulates which are not "independent of each other" he formulates as follows[(b) and (c) constitute "strong" equivalence]: (a) Principle of relativity: Laws of nature are only statements about coincidences in space-time; therefore, the one and only natural formulation must be in terms of generally covariant equations. (b) Principle of equivalence: Inertia and gravity are completely equivalent. From this principle and from the results of the special theory of relativity (SRT) follows necessarily that the symmetric fundamental tensor ($g_{\mu\nu}$) determines the metrical properties of space, the inertial behavior of bodies in space, and the gravitational interaction. The state of space described by the fundamental tensor, we call G field. (c) Mach's principle: G field is completely determined by the masses of the bodies. Since mass and energy according to the results of SRT are the same and since the energy is, formally, described through the symmetric energy tensor $T_{\mu\nu}$, this says, in effect, that the G field is caused and determined by the energy tensor of matter." [Einstein adds to (c) "Till now, I didn't distinguish between (a) and (c), and this caused confusion. The name Mach's principle I have chosen because this principle is a generalization of Mach's postulate, that the inertia is due to an interaction between the bodies.]

Following Einstein, GR is, usually, based on two postulated principles: (I) general covariance and (II) strong equivalence. In addition to the postulates I and II one has, of course, to assume all the other postulates which Einstein, more or less tacitly, assumed in his development, based on I and II. These are as follows: (1) The field equations should reduce to Newton's (e.g., in the nonrelativistic approximation). (2) No new constant should be introduced (in addition to Newton's). (3) Special relativity should be valid locally. (4) The field equations should be the simplest possible ones (second order, Riemannian geometry, etc.).

Postulate 1 was meant in the sense of what we call in Section III weak correspondence. Postulate 2 excludes the cosmological term and also theories of the Brans-Dicke type. Postulate 4 excludes quadratic or higher order invariant Lagrangeans which would all lead to four identities. Quadratic Lagrangeans were first considered by Eddington.[62] These 6 postulates are not independent of each other; for example, it is only sufficient to use I, and 1, 2, 3, and 4. Now, my postulate of the derivability of the equation of motion in addition to the previously enumerated postulates excludes NEF scalar theory (and incidentally any vector theory) and thus, GR is uniquely established. I would like to emphasize that under the tensor theory of Einstein I mean not only the usual tensor form of GR but also any mathematical form of GR which is mathematically and physically equivalent to GR. This includes, for example, the tetrad formulation of GR which is equivalent to the usual tensor formulation of GR.

In an interesting paper Havas[63] has shown that in some sense the linearized field equations of GR, considered as an exact theory also lead to "equations of motion". However, these equations (yielding free motion) would be considered trivial by Einstein and Infeld. Only non-linear equations can lead to non-trivial equations of motion.

Einstein says: "The group of the general relativity is the first one which demands that the simplest invariant law be no longer linear or homogeneous in the field-variables and in their differential quotients. This is of fundamental importance for the following reason. If the field-law is linear (and homogeneous), then the sum of two solutions is again a solution; as, for example: in Maxwell's field-equations for the vacuum. In such a theory it is impossible to deduce from the field equations alone an interaction between bodies, which can be described separately by means of solutions of the system. For this reason all theories up to now required, in addition to the field equations, special equations for the motion of material bodies under the influence of the fields. In the relativistic theory of gravitation, it is true, the law of motion (geodetic line) was originally postulated independently in addition to the field-law equations. Afterwards, however, it became apparent that the law of motion need not (and must not) be assumed independently, but that it is already implicitly contained within the law of the gravitational field."

We might add that to some extent Maxwell's theory does yield conclusions about equations of motion. We quote an interesting discussion by Weber:[64]

"The fact that the equations of motion are contained within the field equations has always been considered a most attractive feature of General Relativity. I wish to invite attention of the fact that other theories in a certain sense also contain equations of motion within their field equations. Consider electrodynamics alone. Suppose we have an electric circuit consisting of an inductance and a capacitance, but with an open switch. Suppose the capacitor is initially charged, then the switch is closed. Maxwell's field equations alone give a description of the future motion of the electrical energy. This is complete for all practical purposes because the inertial effects of the electron mass are small. If we have energy in the form of a wave packet of light, Maxwell's equations again predict the future evolution. It is clear that Maxwell's equations contain the equations of motion for mass (energy) of purely electrical origin, if other (i.e., gravitational) interactions are unimportant.

The field equations of General Relativity contain in some degree the equations of motion of mass elements made up of all kinds of energy. For certain distributions of energy, the equations of motion for both gravitational and non-gravitational forces follow from Einstein's field equations alone. This is the case we are going to discuss. In the general case, the coupled equations of the gravitational and non-gravitational fields must be considered."

It is not sufficient just to postulate that the field equations should give the conservation laws

$$T^{\mu\nu}_{;\nu} = 0, \qquad \text{(IV.3)}$$

because these conservation laws might be trivally satisfied, for example, by the dipole field model considered by Einstein and Infeld. In order to see that the conservation law (Eq. IV.3) is not trivally satisfied we have to derive physical conclusions from it and the best and simplest way is to derive non-trival equations of motion from it. It has been pointed out by Peter Bergmann[65] that the equation of motion problem is not ideally solved in GR:

"The original belief that the field equations of general relativity determine the equations of motion of the ponderable bodies has been borne out only in part; it has been confirmed to the extent that the development in time of either a point-like or an extended region with non-vanishing energy-stress is constrained by the requirement that the covariant divergences of this source tensor must vanish." Still in GR we have the best possible "unification" of field equations and equations of motion."

I am indebted to Peter Havas for calling my attention to a paper by A. E. Scheidegger, Helv. Phys. Acta. 23, 740 (1950). It is shown there that a fairly general invariant Lagrangean, introduced by Peter Bergmann, Phys. Rev. 75, 680 (1949) does lead for tensor fields of at least rank two, but not for a scalar or vector fields to meaningful equations of motion. This is related to our approach in ref. 2, but it does not postulate derivability of the equations of motion from the field equations. Furthermore, it does not point out that the postulates I, II and 1, 2, 3, 4 are also satisfied by the NEF theory.

We have to refer to ref. 2 for further literature on the equations of motion problem.

New Foundations of GR: Strong Correspondence

In Section III, we distinguish between strong and weak correspondence of GR with classical (Newtonian) gravitational theory. From our discussions it follows that strong correspondence can be used in a new foundation of GR because it distinguishes between GR and NEF theory. As a matter of fact, the postulates enumerated in the last subsection: I, the postulate of general covariance, and postulates 1, 2 3, and 4 are sufficient to characterize uniquely GR if postulate 2 is understood in the sense of strong correspondence. This seems to us the simplest postulational foundation of GR. The three famous effects and any other directly observable consequence of GR and the derivability of the equations of motion from the field equations are then all "bonuses" or "surplus values" of GR.

GR and Experiment

Regarding comparison of GR with experiment cf. B. Bertotti, D. Brill and R. Krotkov in *Gravitation* (L. Witten, ed.) Wiley, N. Y. (1962). More recent reviews (preprints) are: R. H. Dicke, *General Relativity, Survey and Experimental Tests*, presented at International Symp. of Contemporary Physics, ICTP, Trieste (1968); L. I. Schiff, Some Experiments on Gravitation, presented at the 5th International Conference on Gravitation and Relativity, Tbilisi, USSR (1968).

Heuristic Derivation of the Schwarzschild Line Element

Under a heuristic derivation of the Schwarzchild line element, we understand a derivation which is not based on the explicit form of the field equations. We only assume that the field equations have the form

$$G_{\mu\nu} = \kappa\, T_{\mu\nu} \qquad \text{(IV.4)}$$

and that $G_{\mu\nu}$ linearly depends on the second derivatives of the (dimensionless) $g_{\mu\nu}$. Herefrom follows that the dimensions of $G_{\mu\nu}$ must be (indicating cgs dimensions by ℓ, m, t)

$$[G_{\mu\nu}] = \ell^{-2}\ . \qquad \text{(IV.5)}$$

Moreover, we assume that the Newtonian line element is given by (III.1a) and (III.1b). From (III.1b) it follows immediately, that we can introduce an effective time t*.

$$t^* = t\ (1 + \bar{\sigma}). \qquad \text{(IV.6)}$$

Since we know that the dimension of $T_{\mu\nu}$ must be

$$[T_{\mu\nu}] = K\ell^{-1}\, mt^{-2}\ , \qquad \text{(IV.7)}$$

from (IV.5) and (IV.7) follows:

$$[\kappa] = \frac{[G_{\mu\nu}]}{[T_{\mu\nu}]} = \ell^{-1}m^{-1}t^{2}\ . \qquad \text{(IV.8)}$$

We can now further assume that κ is "invariant" in the sense that it does not depend on the Newtonian gravitational potential, e.g., $\bar{\sigma}$; then we obtain the relation

$$\ell m \to t^2 \to 1 + 2\bar{\sigma} \qquad \text{(IV.9)}$$

If we further assume that, in the same sense, Planck's constant h, with dimension

$$[h) = \ell^2\, m\, t^{-1}\ , \qquad \text{(IV.10)}$$

is also "invariant", then we obtain

$$\ell^2\, m \to t \to 1 + \bar{\sigma}\ . \qquad \text{(IV.11)}$$

From equations (IV.9, 11), we obtain immediately for the effective length ℓ^* and effective mass m^*

$$\ell^* = \ell\ (1 + \bar{\sigma}) \qquad \text{(IV.12)}$$

$$m^* = m\ (1 + 3\bar{\sigma}) \qquad \text{(IV.13)}$$

It is clear that all these effective quantities do not have invariant significance. We notice that the effective mass is the same which follows from Davidson's paper. From the expressions (IV.12) and (IV.13) for the effective time and for the effective length, we obtain immediately the Schwarzschild line element in first approximation

$$ds^2 = c^{*2}\, dt^2 - \ell^{*2}\, (dx^2 + dy^2 + d_z{}^2). \qquad \text{(IV.14)}$$

The line element (IV.14) gives the Einsteinian light deflection, in addition to the red shift, which is already given by the Newtonian line element (IV.1a) and (IV.1b). The motion of the Mercury perihelion can be, again heuristically, obtained, by introducing the effective mass into Newton's equation of motion. While the expressions for ℓ^*, m^*, t^* depend on the particular coordinate system used, the resulting observable consequences of the line element (IV.14) do have invariant significance.

It is clear that without knowing the explicit form of $G_{\mu\nu}$ we cannot decide whether the assumptions made for the heuristic derivation of the first approximation of the line element are self-consistent or not. Thus, our derivation is mostly of a pedagogical interest only.

We shall outline now a heuristic derivation of the exact Schwarzschild line element. Again, we start out from the Newtonian line element (IV.1a, 1b). We also assume that

$$g_{00}\, g_{11} = -c^2;\ g_{11} = g_{22} = g_{33};\ g_{\mu\nu} = 0\ (\mu \neq \nu) \qquad \text{(IV.15)}$$

This last assumption can be considered as a choice of a particular coordinate system or a normalization condition. Using the condition (IV.15), we obtain immediately the exact Schwarzschild line element.

This type of derivation has been first given in a paper by Kottler[66] in 1918 which is hardly, if ever, mentioned in the literature. Recently, independently, a derivation essentially equivalent to Kottler's, has been presented by Tangherlini.[67] We have simplified and clarified the derivation by starting out from the Newtonian line element (III.1a,1b).[68]

Pauli's Remarks on the Position of Relativity Theory in the Development of Physics

"There is a point of view according to which relativity theory is the end-point of "classical physics", which means physics in the style of Newton-Faraday-Maxwell, governed by the "deterministic" form of causality in space and time, while afterwards the new quantum-mechanical style of the laws of Nature came into play. This point of view seems to me only partly true, and does not sufficiently do justice to the great influence of Einstein, the creator of the theory of relativity, on the general way of thinking of the physicists of today. By its epistemological analysis of the consequences of the finiteness of the velocity of light (and with it, of all signal-velocities), the theory of special relativity was the first step away from naive visualization. The concept of the state of motion of the "luminiferous aether", as the hypothetical medium was called earlier, had to be given up, not only because it turned out to be unobservable, but because it became superfluous as an element of a mathematical formalism, the group-theoretical properties of which would only be disturbed by it.

By the widening of the transformation group in general relativity the idea of distinguished inertial coordinate systems could also be eliminated by Einstein as inconsistent with the group-theoretical properties of the theory. Without this general critical attitude, which abandoned naive visualizations in favour of a conceptual analysis of the correspondence between observational data and the mathematical quantities in a theoretical formalism, the establishment of the modern form of quantum theory would not have been possible. In the "complementary" quantum theory, the epistemological analysis of the finiteness of the quantum of action led to further steps away from naive visualizations. In this case it was both the classical field concept, and the concept of orbits of particles (electrons) in space and time, which had to be given up in favour of rational generalizations. Again, these concepts were rejected, not only because the orbits are unobservable, but also because they became superfluous and would disturb the symmetry inherent in the general transformation group underlying the mathematical formalism of the theory.

I consider the theory of relativity to be an example showing how a fundamental scientific discovery, sometimes even against the resistance of its creator, gives birth to further fruitful developments, following its own autonomous course." (cf. ref. 42, Preface)

REFERENCES

The subject of history of GR is so large that it was impossible to quote all pertinent literature. I am quoting most important papers of Einstein, but had to be very selective with regard to other references. Sometimes I hope to write a more detailed history of GR, in which I plan to include more references. Certainly, all four Sections could easily be expanded into separate historical papers.

An expanded form of the two subsections of Section IV on "New Foundation of GR" based on (a) derivability of the equations of motion from the field equation, (b) strong correspondence, and of our heuristic derivations of the Schwarzschild line element we plan to publish separately. These two topics, of some current interest, were "spin-offs" of our reconstruction of the past.

1. E. Guth, Phys. Rev. Letters, 21, 106 (1968).

2. A. Einstein, Phys. Z. 14, 1249 (1913) including discussion by Mie, et al; subsequent discussion by Mie, Einstein and Nordström in Phys. Z. 15, 115, 169, 176 and 375.

2a. "Geometry as a Branch of Physics" was the title of a very interesting article by H. P. Robertson in "Albert Einstein, Philosopher-Scientist, (P. A. Schilpp, ed.), Tudor, N. Y. (1949), l.c. p. 313.

3. A. Einstein and E. G. Straus, Rev. Mod. Phys. 17, 120 (1945).

4. C. A. Brans and R. H. Dicke, Phys. Rev. 124, 925 (1961).

5. A. Einstein, Ann. d Phys. 35, 898 (1911); also contained in collection: The Principle of Relativity, Dover, N. Y. (1923).

6. In this Section, covering sketchily more than two centuries, I have omitted any detailed references. General references are: (a) J. Zenneck, Encykl. Math. Wiss, V. 2 and S. Oppenheim (ibid, VI, 2, 22, in particular Part V). For more details, particularly of papers after 1900, cf. also F. Kottler, (ibid VI 2, 22); (b) E. T. Whittaker, A History of the Theories of Aether and Electricity, Vol. II, Chap. V, Nelson, London (1953).

6a. For the following concise derivation, cf. J. Weber, in Gravitation and Relativity (H-Y, Chiu and W. F. Hoffmann, eds.) Benjamin, N. Y. (1964), Chap. 11.

7. Quoted from, H. P. Robertson, Space-Time Astronomy (A. J. Deutsch and W. B. Klemperer, eds.) Acad. Press, New York (1962) p. 228.

8. H. A. Lorentz, The Theory of Electrons, Teubner, 1st ed. (1909), 2nd ed (1915).

9. R. H. Dicke, in Evidence for Gravitational Theories, (C. Møller, ed) Acad. Press, N. Y., (1962).

10. H. A. Lorentz, Proc. Amst. Acad. p. 559 (1900).

11. G. J. Whitrow and G. E. Morduch, Nature, 188, 790 (1960), and Vistas in Astronomy, (A. Beer, ed.), Pergamon Press, N. Y., Vol. 6, (1965); l.c. p.l.

12. H. Poincaré, Rend. Circ. Mat. Paleramo, 21, 129 (1906); cf. also W. de Sitter, M. N. R. A. S., 71, 388 (1911).

13. H. Minkowski, Gött. Nachr, p. 53 (1908); Phys. Z. 10, 104, 1909 (1908).

14. A. Sommerfeld, Ann. d. Phys. 33, 649 (1910).

15. H. A. Lorentz, Phys. Z. 11, 1234 (1910).

16a. J. A. Wheeler and R. P. Feynman, Rev. Mod. Phys. 21, 425 (1949); cf. also J. W. Dettman and A. Shild. Phys. Rev. 95, 1059 (1959).

16. H. van Dam and E. P. Wigner, Phys. Rev., 138, 1576 (1965); 142, 838 (1960).

17. A. Einstein, Jb. Radioakt, 4, 440 (1907); l.c. Chap. V.

18. A. Einstein, Ann. d. Phys. 38, 355, 443 (1912); M. Abraham, Phys. Z, 13, 1, 4, 794; discussion between Einstein and Abraham, Ann. d. Phys. 38, 1056, 1059 (1912), ibid, 39, 444, 704 (1912).

19. G. Mie, Ann. d. Phys. 40, 1, (1913), l.c. Chap. V; Elster-Geitel Festschrift, p. 251 (1915).

20. For a modern brief review of Lorentz-invariant scalar theories cf. A. L. Harvey, Amer. J. Phys. 33, 449 (1965). Here Nordström's first theory and its modifications by O. Bergmann and by Whitrow and Morduch, field theoretic approach of W. Thirring and Nordström's second theory are discussed. Included also are the theory of Poincaré (our ref. 12), its presentation by Kottler (our ref. 6) and by Whitrow and Morduch (our ref. 11). Scalar theories are also discussed in the review by M. Abraham, Jb. Radioakt, 11, 470 (1914). Nordström's 2nd theory, in particular, is reviewed in detail by M. v. Laue, Jb. Radioakt. 14, 263 (1917) G. Nordström, Phys. Z. 13, 1126 (1912); Ann. d. Phys. 42, 533 (1913).

21. A. Einstein and A. D. Fokker, Ann. d. Phys. 44, 321 (1914).

22. A. Einstein, Essays in Science, Wisdom (Philosophical) Library, N. Y. (1934), p. 78, l.c. p. 80.

23. cf. ref. 2a. p. l., l.c., p. 63-65.

24. F. Kottler, S. B. Akad. Wiss, Wien, 121, 1659 (1912).

25. E. Goursat, Liouville's J. p. 331 (1908).

26. G. Ricci and T. Levi-Civita, Math. Ann. 54, 135 (1901); Kottler, ref. 24 quotes also J. E. Wright, Invariants of Quadratic Differential Forms, Cambridge U. Press (1908), perhaps the first book dealing with the calculus of Ricci and Levi-Civita, which, in turn, was based mainly on the work of Gauss, Riemann and Christoffel.

27. E. T. Whittaker, l.c. ref. 6.

28. H. Bateman, Proc. London Math. Soc. 8, 223 (1910); Phil. Mag. 37, 2196 (1919).

29. R. Hargreaves, Camb. Phil. Trans., 21, 107 (1908). Whittaker, ref. 6. l.c.f. 64 comments that the treatment of the time coordinate on the same level as the space coordinates was introduced and developed by Hargreaves simultaneously with and independently of Minkowski's paper: his work suggests the use of space-time vector's just as Minkowski's does. For comments on this point, Whittaker refers to: H. Bateman, Phys. Rev. 12, 459 (1918). (Bateman thinks (1918!) that Hargreaves' paper might turn out to be of greater significance than Minkowski's!? According to Whittaker most important in Minkowski's paper was his formulation of physics in terms of a space-time manifold, the use of tensors in this manifold, and the discovery of some of the more important tensors in this manifold, the most important being the stress-energy tensor for an electromagnetic field $S_{\mu\nu}$ unifying the Poynting-Heaviside vector (1894) and the Maxwell stress tensor (1873). Here Whittaker gives a better appraisal of the significance of Minkowski's work than Bateman did. These questions will be discussed in detail in an article on the origins of SR which I am preparing.

30. Einstein must have meant the second theory of Nordström. The argument is developed in §7 of ref. 33 . Strangely, in this paper Nordström's work is not mentioned.

31. c.f. ref. 2, p. 1253.

32. Einstein implied "Machian" arguments, c.f. sub-section: Global and Local Mach Principles.

33. A. Einstein and M. Grossmann, Z. Math. Phys. 62, 225 (1913).

34. Ref. 2.

35. A. Einstein and M. Grossmann, Z. Math. Phys. 63, 215 (1914); this paper also attempts to prove the covariance of the Eq. (III.5) under a class of particular ("manifold-fitting") transformations. It still maintains that generally covariant equations for the determination of the $g_{\mu\nu}$ do not exist.

36. A. Einstein, S. B. Berlin Akad. Wiss, 1030 (1914).

37. Einstein had written his "Notes" and his Autobiography many years after 1915. The recently published exchange of letters between Einstein and Sommerfeld [Briefwechsel Einstein/Sommerfeld (edited with comments by A. Hermann) Schwabe et Co., Basel/Stuttgart (1969); l.c. p. 32], contains a letter Einstein wrote to Sommerfeld on the 28th of November, 1915. There, he gives detailed arguments why his previous field equations were not correct:

(1) He proved that the gravitational field on a uniformly rotating system does not obey the field equations.

(2) The motion of the perihelion of the Mercury turns out to be 18" instead of 45" per century.

(3) The covariance consideration in his paper from last year (1914) does not yield the Hamilton function H. It admits, properly generalized, an arbitrary H. Therefore, the covariance with regards to a particular class of cordinate systems turned out to be useless.

Einstein also says that he has considered with Grossmann three years ago [cf. also ref. 38, l.c. p. 778; (not two as stated in his "Notes")], the equation (III.7) (without the second term on the left hand side). However, he concluded, erroneously, that it doesn't reduce to the Newtonian approximation.

38. A. Einstein, S. B. Berlin, Akad. Wiss. p. 778, 779 (1915).

39. Ibid. p. 831.

40. Ibid. p. 844, Summarizing paper: Ann. d. Phys. 49, 769 (1916).

41. H. Weyl, Raum, Zeit, Materie, Springer, Berlin, 1st edition (1918), 5th edition (1923).

42. W. Pauli: Theory of Relativity, Pergamon, London (1958), l.c. p. 145, ref. 277, originally in: Encykl . Math. Wiss, V 19, Teubner, Leipzig (1921).

43. D. Hilbert, Nachr. Ges. Wiss. Göttingen, p. 395 (1915); presented Nov. 20. Perhaps the presentation date of this paper which was five days before November 25, the presentation date of Einstein's final paper, ref. 40, confused Weyl and Pauli.

44. E. T. Whittaker, ref. 6, l.c. p. 171.

45. I am indebted to P. Ruffini for this remark.

46. D. Hilbert, Math. Ann. 92, 1 (1924).

47. This section is, necessarily, rather sketchy. Many important (formal and physical) developments had to be omitted. For a brief but excellent summary of the present status of GR cf. A. Trautman, Sov. Phys. Uspekhi, (1966). For still more recent important developments, cf. the other articles in this book.

48. M. v. Laue, Relativitätstheorie, Vol. II, Vieweg, Braunschwwig (1921).

49. cf. ref. 20; Laue discusses mostly N's second scalar theory.

50. C. Lanczos, Einstein-Symposium, 1965, Akademic-Verlag, Berlin (1966); l.c. p. 40.

51. L. Infeld, Quest, Doubleday, Doran (1941), l.c. p. 93.

52. A. Trautman, ref. 47, l. c. p. 334-335.

53. For references, cf. H. P. Robertson and T. W. Noonan, Relativity and Cosmology, W. B. Saunders, Philadelphia (1968). Robertson made many important contributions to relativistic cosmology.

54. R. H. Dicke, ref. 6a, l.c. Chap. 7.

55. A. Einstein, The Meaning of Relativity, Princeton U. Press, 1st ed. (1921); 5th ed. (1955).

56. M. v. Laue, ref. 48.

57. A. Einstein and E. G. Straus, ref. 3., I am indebted to Charles W. Misner for pointing out to me the relevance of this paper in the present context.

58. W. Davidson, M. N. R. A. S. 117, 212 (1957); c.f. also D. W. Sciama, ibid. 113, 34 (1953); F. Hund , Z. Phys. 124, 742 (1948).

59. C. A. Brans, Phys. Rev. 125, 388 (1962).

60. C. A. Brans, ibid, 125, 2194 (1962).

60a. A. L. Harvey, Ann. of Phys. 31, 240 (1965).

60b. R. H. Dicke, ibid, 31, 235 (1965).

60c. R. U. Sexl, Fortschritte der Phys. 15, 269 (1967), l.c. p. 305-306.

61. For more references and newer developments cf. the review, H. Goenner, Mach's Principle and Einstein's Theory of Gravitation, Temple U. preprint (1968). The most recent development is based on integral formulations of GR. cf. D. Lynden-Bell, M.N.R.A.S. 135, 413 (1967); B. L. Altshuler, JETP, 24, 766 (1967); D. A. Sciama, P. C. Waylen and R. C. Gilman (Cambridge preprint; B. Bertotti (personal communication). I am indebted to Roger Penrose and Bruno Bertotti for calling my attention to these latest developments. (see also our ref. 70*).

62. A. S. Eddington, The Mathematical Theory of Relativity, Cambridge U. Press, 2nd ed. (1924), l.c. p. 140-144; c.f. also G. Stephenson, N. C. 9, 263 (1958).

63. P. Havas, in Recent Developments in General Relativity, Pergamon, London (1962), l.c. p. 259.

64. J. Weber, in Relativity, Groups and Topology, C.B.S. and C. M. DeWitt, eds.) Gordon and Breach, N. Y. (1964); Chap. Gravitational Radiation Experiments.

65. P. G. Bergmann, Encyclopedia of Physics, (S. Flüge , ed.) Springer, Berlin (1962), Vol. IV, p. 203, l.c. p. 245.

66. F. Kottler, Ann. d. Phys. 56 (1918), l.c. III p. 411 ff.

67. F. R. Tangherlini, N. C. 25, 1081 (1962).

68. For other heuristic derivations, c.f. W. M. Sacks and J. A. Ball, Amer. J. Phys. 36, 240 (1968).

69*. In Section II, sub-section: Nordström's Two Theories of Gravity; Einstein on First Scalar Theory, we discuss in detail Einstein's argument against the first Nordström scalar theory, which theory Einstein seems to have advanced simultaneously with and independently of Nordström.

70*. In connection with GR in our reference 61, it is interesting that Einstein in 1946 considered, in general, a formulation of physics in terms of integral equations but abandoned this effort because he did not see how to restrict such a formulation so that solutions with non-physical properties could be excluded. This is stated in an article by E. G. Strauss in the book quoted in Post-Script (after Eq. (III.5) in which he tells about his assistanship to Einstein from 1946-50 when he worked with him on this problem.

GRAVITATIONAL RADIATION DAMPING*

William L. Burke and Kip S. Thorne

California Institute of Technology

Pasadena, California

ABSTRACT

Descriptions are presented of two different analyses in which systems which radiate gravitational waves are damped.

The first analysis shows how the back-reaction of gravitational radiation can be incorporated with ease into any slow-motion expansion of general relativity. This analysis reveals that the dominant effects of general-relativistic radiation resistance can be incorporated into the Newtonian theory of gravity by a simple change in the boundary condition on the Newtonian potential at $r = \infty$. The rate of energy loss from a radiating system, as calculated by this analysis, agrees with the power carried in the gravitational waves, as calculated by the Isaacson stress-energy tensor or the Landau-Lifschitz pseudotensor.

The second analysis treats the small-amplitude, nonradial pulsations of fully relativistic stellar models, which may be arbitrarily close to their Schwarzschild radii. The slow-motion and weak-field approximations, which are crucial to the first analysis, are not made here. As a consequence, the radiation damping appears at first order in the amplitude of pulsation. This analysis reveals that a neutron star formed by gravitational collapse should emit a burst of gravitational waves with frequencies of $\sim 10^3$ hz, energy $\sim 10^{52}$ ergs, and damping time ~ 1 second.

*Supported in part by the National Science Foundation [GP-9433, GP-9114] and the Office of Naval Research [Nonr-220(47)].

In the slow-motion, weak-field limit, the second analysis reduces to the first.

I. SLOW-MOTION, WEAK-FIELD EXPANSIONS

In this section we discuss some work being carried out at Caltech on the problem of how to incorporate radiation resistance into slow-motion expansions of general relativity theory.

Slow-motion expansions, such as that of Einstein, Infeld, and Hoffman (1938) for "point masses" and that of Chandrasekhar (1965, 1969) and Chandrasekhar and Nutku (1969) for a perfect fluid, have provided us in the past with our best insights into the problem of motion in general relativity theory. And at present they are beginning to provide us with powerful tools for theoretical studies of astrophysical systems (the solar system, dense white dwarfs, pulsars, supermassive stars, etc.).

In any slow-motion expansion one begins by introducing a nearly-inertial frame throughout the system. (If the curvature of spacetime were so great as to prohibit nearly inertial frames, then it would produce velocities so high that the slow-motion assumption would break down.) One then assumes that the typical velocities, $\langle v\rangle$, of the matter relative to this nearly-inertial frame are small compared to the speed of light; and one expands the equations of general relativity theory in powers of the resultant small, dimensionless parameter

$$\epsilon \equiv \langle v/c\rangle = \text{"}1/c\text{" in the notation of Chandrasekhar (1965).} \qquad (1)$$

One also assumes (often without stating it explicitly) that the size of the system, L, is small compared with the mean wavelength, $\langle\lambda\rangle$ of the waves which it emits, $L/\langle\lambda\rangle \ll 1$; if this is not true, then retardation effects will come in at the lowest order. In addition, so as to prevent large unbalanced forces from producing high velocities, one assumes that the stresses in the system are small compared to the density of total mass-energy, $T^{ij}/T^{oo} \ll 1$, and that the Newtonian potential is small throughout the system, $U/c^2 \ll 1$.

Precisely how the expansion proceeds depends on the relative sizes of the various dimensionless parameters. All expansions which have been performed to date deal with systems for which

$$L/\langle\lambda\rangle \sim (T^{ij}/T^{oo})^{1/2} \sim \epsilon \sim \langle v/c\rangle. \qquad (2)$$

In fact it is hard to conceive of a system which could live for any length of time without satisfying these conditions. Systems of interest then divide themselves into two classes, depending on the strength of the gravitational field relative to the expansion parameter ϵ. One class consists of "very-weak-field" systems, for which

$$U/c^2 \ll \epsilon^2 \tag{3}$$

so that the forces in the system are primarily non-gravitational (example: laboratory-sized elastic bodies). The other class consists of "gravitationally-bound systems", for which

$$U/c^2 \sim \epsilon^2 \tag{4}$$

so that gravitational forces are as significant as material forces (examples: the solar system and most other astrophysical systems of interest).

The expansion schemes used by EIH and Chandrasekhar deal with the gravitationally-bound case. This case is more difficult in practice than the very-weak-field case because gravitational nonlinearities come in immediately after the Newtonian order and might contribute significantly to the radiation resistance.

In this discussion of slow-motion expansions we, like almost everyone else, take spacetime to be flat in lowest order, not only in and near the system, but also far away in the wave zone. In this case the corrections of $O(\epsilon^2)$ to flat spacetime produce the usual Newtonian theory of gravity. When one extends the expansion on to higher, post-Newtonian orders, one finds that only terms of even order (ϵ^4, ϵ^6, ...) are generated by the Newtonian terms. The field equations never mix odd and even terms.

The lack of any source for the odd terms is very disturbing since radiation resistance can occur only in the odd terms — only the odd terms are sensitive to the direction of time! Does this mean that radiation resistance is absent from general relativity theory? Some workers have thought so. Others have sought ways to generate odd terms from Newtonian terms by means of the outgoing-wave boundary condition, but they have <u>not</u> been markedly successful. For example, in the problem of radiation damping for a binary-star system Trautman (1958) and Carmeli (1964, 1965) found damping; Hu (1947), Peres (1959), and Havas and Goldberg (1962) found "antidamping"; and Infeld and Plebanski (1960) (see also work of Infeld and Scheidegger cited therein) found no radiation reaction at all. All of these calculations are so complicated that we have not tried to follow or reproduce them in detail.

Many workers have remarked that their slow-motion expansions were not valid far from the system, where radiation fields dominate, but they have not realized that such behavior is indicative of a "singular perturbation problem". Singular perturbation theory has been the focus of intense interest in applied mathematics for many decades, but for some strange reason nobody seems to have noticed that this theory is precisely the mathematical technique which is needed for slow-motion, radiative problems like ours. A major purpose of this paper is to call singular perturbation theory to the attention of relativity theorists who wish to do slow-motion expansions. For a clear exposition of the subject see Julian Cole's (1968) recent book.

Here we will give a model problem illustrating how one of the techniques of singular perturbation theory -- the method of "matched asymptotic expansions" -- can be used to incorporate radiation and irreversibility into a slow-motion expansion. In linear theories, such as electromagnetism or acoustics, where the exact solutions are easily accessible, such fancy techniques result in little or no saving of effort. For nonlinear problems, such as our gravitationally bound systems, the use of matched asymptotic expansions simplifies the calculations enormously and provides a consistent and systematic framework for the approximations.

The crucial feature of slow-motion expansions in any radiating theory, which makes it necessary to use the technique of matched asymptotic expansions, is that they are not uniformly valid for large distances. More particularly, they are not valid in the wave zone, where the outgoing-wave boundary conditions are to be imposed. Terms which were small and ignored in the inner zone are not small and not ignorable in the wave zone. A different asymptotic expansion which does not ignore these terms must be used to represent the solution in the wave zone. On the other hand, the wave-zone fields are much weaker than the near-zone fields, so even in the case of gravitationally-bound systems, they are more nearly linear. Thus at each stage in the calculation the inner expansion keeps many nonlinear terms and drops many time derivatives, while the outer expansion drops many of the near-zone nonlinearities but keeps all time derivatives. The alternative to using two expansions is to use one expansion which keeps all nonlinearities and time derivatives simultaneously -- this would lead to the same answers but by needlessly complicated intermediate steps.

To see how a matched asymptotic expansion for a radiating system proceeds, let us examine a trivial model problem -- a scalar field φ, which is generated by a sinusoidally oscillating system of period λ/c, size L, and spherical-harmonic angular shape

$$\Box\varphi \equiv -\frac{1}{c^2}\frac{\partial\varphi}{\partial\varphi^2} + \nabla^2\varphi = -4\pi\rho_0 Y^{\ell}{}_{m}(\theta,\varphi)\exp(i\omega t), \qquad (5a)$$

$$\rho_0(r) = 0 \text{ for } r > L, \qquad \omega \equiv 2\pi c/\lambda. \qquad (5b,c)$$

We will not specify here how the φ field acts on its sources, but will content ourselves with finding the way in which the time-asymmetric, outgoing-wave, boundary condition generates time-asymmetric terms in the inner expansion of the solution.

A person unfamiliar with singular perturbations might try to analyze this system by working entirely in the near zone. He would notice that $(1/c^2)\partial^2\varphi/\partial t^2$ is two orders smaller in ϵ than is $\nabla^2\varphi$, so upon expanding φ in powers of ϵ, he would get equations in which even terms couple to even and odd terms couple to odd:

$$\begin{aligned}
\varphi &= \varphi_0 + \varphi_1 + \varphi_2 + \cdots; \quad \varphi_n = O(\epsilon^n);\\
\nabla^2\varphi_0 &= -4\pi\rho_0(r)\, Y^\ell{}_m(\theta,\varphi)\exp(i\omega t)\\
\nabla^2\varphi_1 &= 0,\\
\nabla^2\varphi_2 &= (1/c^2)\partial^2\varphi_0/\partial t^2,\\
\nabla^2\varphi_3 &= (1/c^2)\partial^2\varphi_1/\partial t^2,\\
\nabla^2\varphi_4 &= (1/c^2)\partial^2\varphi_2/\partial t^2, \;\ldots\;.
\end{aligned} \tag{6}$$

He would then solve these equations to find, outside the source,

$$\begin{aligned}
\varphi_0 &= [Q\, r^{-(\ell+1)} + E_0 r^\ell]\, Y^\ell{}_m\, e^{i\omega t},\\
\varphi_1 &= E_1 r^\ell\, Y^\ell{}_m\, e^{i\omega t},\\
\varphi_2 &= \left[\frac{(\omega/c)^2 Q}{2(2\ell-1)}\, r^{-(\ell-1)} - \frac{(\omega/c)^2 E_0}{2(2\ell+3)}\, r^{\ell+2} + E_2 r^\ell\right] Y^\ell{}_m\, e^{i\omega t},\\
\varphi_3 &= \left[-\frac{(\omega/c)^2 E_1}{2(2\ell+3)}\, r^{\ell+2} + E_3 r^\ell\right] Y^\ell{}_m\, e^{i\omega t}, \;\ldots;
\end{aligned} \tag{7a}$$

where

$$\begin{aligned}
Q &= [4\pi/(2\ell+1)] \int \rho_0 r^{\ell+2}\, dr\\
E_k &\text{ are arbitrary.}
\end{aligned} \tag{7b}$$

At this point our near-zone friend might be tempted to make a disasterous mistake. He might argue that all of the E_k must vanish because (i) the field φ must not blow up at infinity; and (ii) one should restrict himself solely to terms which are generated explicitly by the lowest-order, "Newtonian" solution

$$\varphi_N = Q\, r^{-(\ell+1)}\, Y^{\ell}_{m}\, e^{i\omega t}. \tag{8}$$

If he so argued, our near-zone friend would find that the solution for φ would remain always time-symmetric. At no point would he be able to insert the information that in the wave zone the waves were purely outgoing. He would never find damping in his equations of motion.

An expert on matched asymptotic expansions would be more cautious. He would realize that without the terms in E_k he might not be able to match a wave-zone solution with purely outgoing waves to his near-zone solution, so he would temporarily leave the E_k arbitrary.

In the wave zone the $(1/c^2)\partial^2\varphi/\partial t^2$ term is the same size as $\nabla^2\varphi$; on the other hand, the sources are confined to a region whose size is $L \sim \epsilon\lambda$, so as $\epsilon \to 0$, from the standpoint of someone focusing on the radiation, the sources shrink into what appears to be a singularity at the origin. The terms in an outer expansion for φ,

$$\varphi \sim \psi_0 + \psi_1 + \cdots \tag{9}$$

are thus uncoupled from each other and obey the sourceless equations

$$\Box\psi_0 = 0, \quad \Box\psi_1 = 0, \quad \text{etc.} \tag{10}$$

Our expert would focus his attention on the general outgoing-wave solutions to these equations with sinusoidal time behavior and Y^{ℓ}_{m} angular dependence:

$$\psi_n = A_n\, [j_\ell(\omega r/c) - i n_\ell(\omega r/c)]\, Y^{\ell}_{m}\, e^{i\omega t}. \tag{11}$$

Here j_ℓ is the spherical Bessel function, n_ℓ is the spherical Neumann function, and A_n is a constant.

Our expert would then begin the matching process, the goal of which is to determine the constants A_n and E_k in terms of the ℓ-pole moment, Q, of the source.

In this model problem the matching is trivial; but in the gravitational problem it is not. Here our expert need only expand the wave-zone solution (11) in powers of ωr (which is small in the matching domain at the inner edge of the near zone), obtaining, e.g., for ψ_0

$$\psi_0 \to \left\{ A_0 \frac{(\omega r/c)^\ell}{(2\ell+1)!!} + \cdots + iA_0 \frac{(2\ell-1)!!}{(\omega r/c)^{\ell+1}} \right\} Y^{\ell}_{m}\, e^{i\omega t}. \tag{12}$$

(Note that near the source $\omega r/c \sim \langle L/\lambda\rangle = \epsilon$.) He would then match the near-zone term φ_0 (eq. [7a]) to the dominant term of ψ_0, obtaining

$$\varphi_0 = [Q r^{-(\ell+1)} + E_0 r^{\ell}] Y^{\ell}_{m} e^{i\omega t} = iA_0(2\ell-1)!!\,(\omega r/c)^{-(\ell+1)} Y^{\ell}_{m} e^{i\omega t}; \quad (13)$$

so that

$$A_0 = [-i/(2\ell-1)!!]\,(\omega/c)^{\ell+1} Q, \qquad E_0 = 0. \quad (14)$$

He would then notice that all of the remaining terms in the real part of the series (12) for the wave-zone field ψ_0 match perfectly to the even terms φ_2, φ_4, ... in the near-zone field if and only if the unknown near-zone constants E_2, E_4, ... vanish

$$E_2 = E_4 = \dots = 0. \quad (15)$$

That would leave the imaginary part of the wave-zone field,

$$(\psi_0)_{imag} \to -iQ(\omega/c)^{2\ell+1} r^{\ell} Y^{\ell}_{m} e^{i\omega t}/[(2\ell-1)!!(2\ell+1)!!] + \dots \quad (16)$$

(cf. eqs. [12] and [14]) still to be matched. Recall that in the inner zone $\omega r/c \sim \epsilon$. This means that the dominant imaginary term in ψ_0 must match to φ at the $\epsilon^{2\ell+1}$ order. Since $Y^{\ell}_{m} r^{\ell}$ is a solution of the homogeneous near-zone equation (not just the limit of such a solution, which was all we might have expected) we have in fact

$$E_{2\ell+1} = \frac{(-1)^{\ell+1}(i\omega)^{2\ell+1} Q}{(2\ell-1)!!\,(2\ell+1)!!} \quad (17)$$

so that

$$\varphi_{2\ell+1} = \frac{(-1)^{\ell+1}(i\omega)^{2\ell+1} Q}{(2\ell-1)!!\,(2\ell+1)!!} r^{\ell} Y^{\ell}_{m} e^{i\omega t}. \quad (18)$$

This is the dominant odd term that our expert was seeking! Because the dominant odd term in the near zone is $\varphi_{2\ell+1}$, our expert must set

$$E_3 = E_5 = \dots = E_{2\ell-1} = 0. \quad (19)$$

Our expert could then wind up his analysis by noticing that, for appropriate choices of $E_{2\ell+3}$, $E_{2\ell+5}$, ... the imaginary part of the wave-zone field ψ_0 matches perfectly to $\varphi_{2\ell+1} + \varphi_{2\ell+3} + \dots$. As a consequence, he would conclude that in this model problem, unlike the gravitational case, only the first-order wave-zone field ψ_0 is needed to perform a complete match to the near zone:

$$\Psi_1 = \Psi_2 = \dots = 0. \tag{20}$$

In words, the results of the matching process are these: The solution (7) to the near-zone, slow-motion equations (6) contains an ℓ-pole moment of the source, Q, which is uniquely determined by what the source is doing; and it also contains an infinite number of arbitrary constants, E_k (arbitrary functions of time in the non-sinusoidal case). By matching the near-zone solution to the general wave-zone solutions (11), (12), with outgoing waves, one learns the value of the amplitude, A_0, of the lowest-order waves $\Psi_0(t,r,\theta,\varphi)$. By then matching Ψ_0 back into the near zone one learns the values of the constants, E_k. For even k, the E_k turn out to vanish, while for odd k they do not. The resultant terms containing the odd E_k change sign under time reversal and lead to radiation damping.

Notice also a very important fact: The dominant odd term, $\varphi_{2\ell+1}$, is determined uniquely in terms of the source multipole moment Q and satisfies the sourceless Laplace equation, $\nabla^2\varphi_{2\ell+1} = 0$. This means that, if we will content ourselves with only the lowest-order radiation-reaction effects, and if we are willing to ignore all of the "post-Newtonian" (φ_2, φ_4, ...) effects, we can build a self-consistent near-zone theory that will do this by simply absorbing $\varphi_{2\ell+1}$ into φ_0 and ignoring $\varphi_2, \varphi_4, \dots, \varphi_{2\ell}, \varphi_{2\ell+2}, \varphi_{2\ell+3}, \varphi_{2\ell+4}, \varphi_{2\ell+5}, \dots$. The resultant modified "Newtonian" theory is identical to the original "Newtonian" theory except that the potential has a small, new, radiation-resistance term, $\varphi_{2\ell+1}$ ($\varphi = \varphi_0 + \varphi_{2\ell+1}$), which diverges as r^ℓ at $r = \infty$.

Burke (1969) has solved a variety of slow-motion mechanical, electromagnetic, and gravitational problems using the techniques illustrated by the above model problem. For general relativity he has carried the expansions far enough in the very-weak-field case to show conclusively that the rate at which energy is damped from the system agrees, to accuracy ϵ^2, with the famous formula [Landau and Lifschitz 1962, eq. (104.12)]

$$\langle dE/dt \rangle = -\frac{G}{45\, c^5}\left\langle \sum_{i,j} \left(\frac{\partial^3 D^{ij}}{\partial t^3}\right)^2 \right\rangle \tag{21}$$

for the rate at which energy is carried off by the gravitational waves.

For the self-gravitating case the calculations are more difficult. However, Burke (1969), Thorne (1970), and Chandrasekhar (work reported at this conference) have all carried the calculations far enough to discover that, when the near-zone Newtonian terms are matched out into the radiation zone and then back into the near zone, they produce radiation resistance (terms of order

ϵ^5 smaller than Newtonian) which agrees with formula (21). But this is not enough to give one confidence in equation (21) for self-gravitating systems. One must check that the reaction, produced by matching the post-Newtonian and post-post-Newtonian near-zone solutions outward and then back in, is of smaller order than the Newtonian-produced reaction (cf. Figure 1). This has not been done yet, but the techniques for doing it are perfectly straightforward, thanks to the applied mathematicians who have developed in great detail the technique of matched asymptotic expansions.

Just as in our model problem, so also in general relativity, the matching calculations yield a modified version of Newtonian theory which automatically incorporates all radiation reaction effects (to accuracy ϵ^2). This modified Newtonian theory can be regarded as rigorously derived from general relativity theory for the very-weak-field case, but as not yet satisfactorily derived (see preceding paragraph) for the gravitationally bound case. The modified theory is as follows:

Modified Newtonian Theory of Gravity, which gives correctly to accuracy ϵ^2 all general-relativistic radiation-reaction effects: Gravity is described by the usual Newtonian potential, $U(\underset{\sim}{x},t)$, which produces forces on bodies in the usual way

$$\underset{\sim}{F} = m\underset{\sim}{\nabla} U . \tag{22}$$

The gravitational potential satisfies the usual source equation

$$\nabla^2 U = -4\pi G\rho , \tag{23}$$

and as usual it must be nonsingular throughout the system. However, by contrast with the usual Newtonian theory, the boundary condition at $r = \infty$ is not $U(\infty) = 0$. Rather, at a particular moment of time t, when the multipole moments of U are

$$D^\ell_m(t) \equiv (4\pi/2\ell+1) \int_0^R \rho r^\ell Y^{\ell\,*}_m \, d\text{ volume} \tag{24}$$

($* \equiv$ complex conjugate), the form of U at large r must be

$$U = \frac{M}{r} + \sum_{\ell=2}^{\infty} \sum_{m=-\ell}^{\ell} \left[\frac{D^\ell_m(t)}{r^{\ell+1}} + \frac{(-1)^{\ell+1}(\ell+1)(\ell+2)}{\ell(\ell-1)(2\ell+1)[(2\ell-1)!!]^2} \frac{d^{2\ell+1}D^\ell_m}{d(ct)^{2\ell+1}} r^\ell \right] Y^\ell_m . \tag{25}$$

As a consequence, anywhere in space the potential of the modified theory is related to that of the usual theory by

$$U = U_{usual} + U_{reaction}, \tag{26a}$$

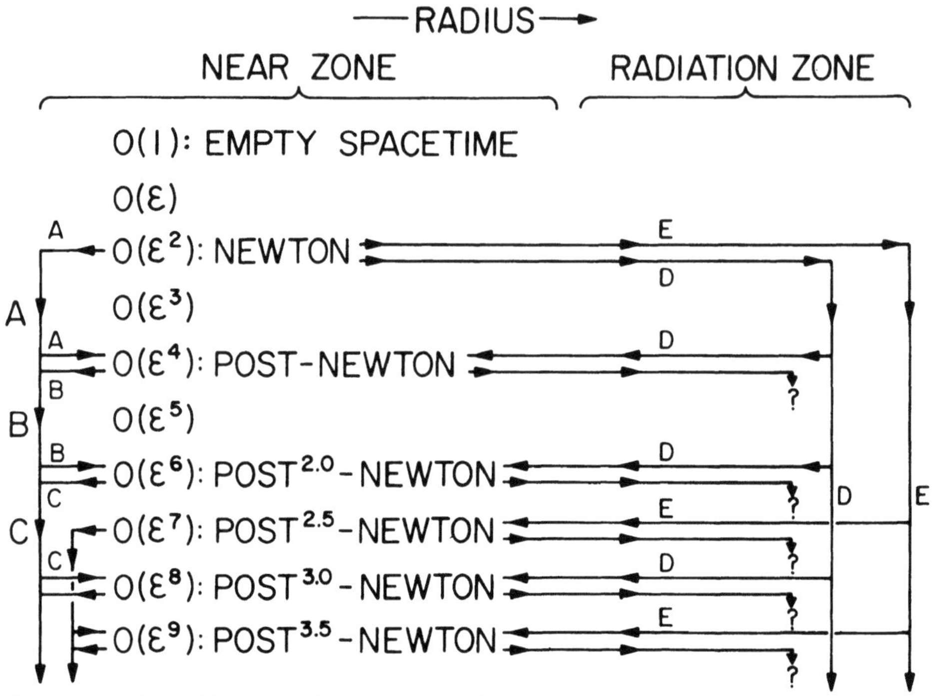

Fig. 1. Flow diagram for slow-motion expansions of general relativity. (Here and throughout the text, the near-zone orders refer to the equations of motion rather than to terms in the metric.) The near-zone terms of order ϵ^n generate, by means of the slow-motion field equations, terms of order ϵ^{n+2}, ϵ^{n+4}, ... Of this type of coupling, step A has been formulated by Chandrasekhar (1965), and step B has been formulated by Chandrasekhar and Nutku (1969). The Newtonian terms couple out into the radiation zone at $O(\epsilon^2)$; and then, via the outgoing-wave boundary condition, they couple back into the near zone at $O(\epsilon^4)$, $O(\epsilon^6)$, $O(\epsilon^8)$, etc. (step D) and at $O(\epsilon^7)$, $O(\epsilon^9)$, $O(\epsilon^{11})$, etc. (step E). The coupling to odd orders (step E) produces radiation reaction in the source, whereas the coupling to even orders (step D) does not -- at least not at $O(\epsilon^4)$ or $O(\epsilon^6)$. This wave-zone coupling is the subject of Burke's (1969, 1970) work and of § III of Thorne (1970). The treatment of the gravitationally-bound case will not be fully satisfactory until all wave-zone coupling has been carried out down through $O(\epsilon^7)$. [This figure is taken from Thorne (1970).]

where

$$U_{\text{reaction}} = \sum_{\ell=2}^{\infty} \sum_{m=-\ell}^{\ell} \frac{(-1)^{\ell+1}(\ell+1)(\ell+2)}{\ell(\ell-1)(2\ell+1)[(2\ell-1)!!]^2} \frac{d^{2\ell+1}D^{\ell}{}_m}{d(ct)^{2\ell+1}} r^{\ell} Y^{\ell}{}_m . \quad (26b)$$

The term $m\nabla U_{\text{reaction}}$ in the force law represents radiation reaction. But notice that the radiation reaction force on a given particle is caused by the behavior of the system as a whole, rather than by the particle's behavior alone.

Several warnings must be made about this modified Newtonian theory of gravity: (i) It has not yet been fully proved for the gravitationally-bound case (see above). (ii) Only over the long run, as its effects become cumulative, is the resultant radiation reaction larger than other, post-Newtonian effects (e.g., the perihelion shift of Mercury), which are neglected in this theory. (iii) The multipole terms with $\ell \geq 3$ in the radiation reaction potential (26b) are smaller than the $\ell = 2$ terms by a factor of order ϵ^2. Since the reaction given by this theory is guaranteed to be accurate only to order ϵ^2, these $\ell \geq 3$ terms are "frosting on the cake" which have no real significance. (iv) The dominant "magnetic-parity" radiation, which is generated by the mass currents and is smaller by $\sim \epsilon^2$ than our dominant "electric-parity" radiation, has been ignored here, but has been worked out by Burke (1969).

II. PULSATION OF FULLY RELATIVISTIC STARS

Two possible astrophysical sources of gravitational waves which cannot be analyzed by means of slow-motion, weak-field expansions are neutron stars and "black holes". In currently accepted models of neutron stars [see, e.g., tables 1-4 of Hartle and Thorne (1968)], the weak-field expansion parameter is as large as

$$\epsilon^2 \sim \left(\frac{\text{Gravitational Radius}}{\text{Actual Radius}}\right) \equiv \frac{2M}{R} \approx 0.6, \quad (27a)$$

$$\epsilon^2 \sim \left(\frac{\text{Central Pressure}}{\text{Central Density of Mass-Energy}}\right) \equiv \frac{p_c}{\rho_c} \approx 0.5, \quad (27b)$$

$$\epsilon^2 \sim \left(\begin{matrix}\text{Gravitational Redshift of Neutrino}\\ \text{Emitted from Center of Star}\end{matrix}\right) \equiv \left[\frac{g_{00}(\infty)}{g_{00}(0)}\right]^{1/2} - 1 \approx 2.2. \quad (27c)$$

(Here and throughout this section we set $c = G = 1$.)
For black holes, the relevant expansion parameters are even larger.

Under these circumstances, no weak-field, slow-motion expansion can be trusted at all!

At present no reliable strong-field analysis has been developed for the gravitational waves emitted by a star as it collapses to form a black hole, or for the waves emitted by a star which collides with and is destroyed by an already-existent black hole. However, at Caltech we have recently carried out an analysis of the waves emitted by a newly formed, pulsating neutron star. In this section we describe that analysis, which is being presented in detail in a series of 5 papers in the Astrophysical Journal [Thorne and Campolattaro (1967), Paper I; Price and Thorne (1969), Paper II; Thorne (1969a), Paper III; Thorne (1969b), Paper IV; Campolattaro and Thorne (1970), Paper V].

The analysis is based on a perturbation-series expansion about a fully relativistic, equilibrium stellar model, which is composed of a perfect fluid (no shear stresses possible). The equilibrium model is described by the usual spherically-symmetric line element

$$ds^2 = (ds^2)_0 \equiv e^{\nu(r)}dt^2 - e^{\lambda(r)}dr^2 - r^2\,(d\theta^2 + \sin^2\theta d\varphi^2), \qquad (28)$$

and by the radial distributions of the density of total mass-energy, $\rho(r)$, isotropic pressure, $p(r)$, and adiabatic index, $\gamma(r)$. [For the theory of the structure of such an equilibrium model, see, e.g., chapter 3 of Thorne (1967).]

The pulsations of the equilibrium model are analyzed only to first order in the amplitude of pulsation. Because the equilibrium model is fully relativistic and no slow-motion assumption is made, radiation reaction is present in this first-order analysis. By contrast, in the weak-field, slow-motion expansions of § I, radiation reaction first appears 5 orders beyond Newtonian theory.

The perturbation functions (perturbations of the metric, stress-energy tensor, etc.) describing the stellar pulsations are expanded in spherical harmonics, which are characterized by the usual integers ℓ and m, and by the parity π. Because the unperturbed stellar model is spherically symmetric, there is no coupling between first-order pulsational modes with different values of (ℓ, m, π).

The spherically symmetric pulsations ($\ell = 0$), which have been treated in detail by S. Chandrasekhar (1964), produce no gravitational waves (Birkhoff's theorem). The dipole pulsations ($\ell = 1$) are treated in detail in Paper V; like the spherical pulsations, they produce no gravitational waves. In fact, despite the finite-amplitude motion of the star's surface in a dipole pulsation, the geometry of spacetime outside the star is completely unaffected by the pulsation (to first order); it retains its unperturbed

Schwarzschild form.

The quadrupole and higher-order pulsations ($\ell \geq 2$), which can produce gravitational waves, are treated in Papers I-IV.

The modes with $\ell \geq 2$ split into two classes -- "even-parity", or "electric-type" modes, which have $\pi = (-1)^{\ell}$; and "odd-parity", or "magnetic-type" modes, which have $\pi = (-1)^{\ell+1}$.

For a star made of perfect fluid, the odd-parity modes are not pulsations at all; rather, they represent differential rotations which do not radiate (cf. Paper I). This fact is easily seen from the nonexistence of scalar spherical harmonics with parity $\pi = (-1)^{\ell+1}$, which implies that all scalar quantities, including the star's pressure and density, are unaffected by an odd-parity perturbation.

Shortly after a neutron star has formed by collapse, and after its initial pulsations have been damped away, its cooling matter may crystalize at densities between 10^8 g/cm^3 and 10^{14} g/cm^3 (see Ruderman 1968). Since the resultant crystalized star can support shear stresses, it can undergo torsional oscillations with odd parity, that radiate. However, such torsional oscillations are idealized away by our simplifying assumption that the star is made of perfect fluid.

Papers I-IV concentrate almost exclusively on even-parity pulsations, since they are the only ones that radiate in our idealized, perfect-fluid case.

Paper I derives the even-parity equations of motion for the star and its spacetime geometry, using a precisely fixed gauge -- that of Regge and Wheeler (1957). In this gauge the line element for perturbations with a particular choice of ℓ and m takes the form

$$ds^2 = e^{\nu}(1 + H_0 Y^{\ell}{}_m)\,dt^2 + 2\,H_1 Y^{\ell}{}_m\,dtdr - e^{\lambda}(1 - H_2 Y^{\ell}{}_m)\,dr^2$$
$$- r^2(1 - K Y^{\ell}{}_m)(d\theta^2 + \sin^2\theta d\varphi^2). \tag{29}$$

Here $Y^{\ell}{}_m(\theta,\varphi)$ is the usual scalar spherical harmonic, and $H_0(t,r)$, $H_1(t,r)$, $H_2(t,r)$, $K(t,r)$ are functions of t and r. The stellar pulsation is described not only by the gravitational amplitudes, H_0, H_1, H_2, and K, but also by amplitudes $W(t,r)$ and $V(t,r)$ for the radial and tangential displacements of the fluid:

$$\delta r = \frac{e^{-\lambda/2}}{r^2} W Y^{\ell}{}_m, \quad \delta\theta = -\frac{V}{r^2}\partial_\theta Y^{\ell}{}_m, \quad \delta\varphi = -\frac{V}{r^2}\frac{\partial_\varphi Y^{\ell}{}_m}{\sin^2\theta}. \tag{30}$$

Of the six perturbation functions only K, W, V represent true dynamical degrees of freedom. H_0, H_1, H_2 are fixed in terms of K, W, V by certain initial-value equations. The stellar pulsation and the emission of gravitational waves are governed by hyperbolic differential equations (the perturbed, dynamical Einstein field equations) of the form

$$(\partial^2/\partial t^2)\,\{K, W, V\} = \mathscr{L}\,\{K, W, V\}\ , \tag{31}$$

plus certain boundary conditions. Here $\mathscr{L}$ is a particular third-order, linear, integro-differential operator in r (cf. Paper I, equations (8a) and (9) where the integral aspect of the operator $\mathscr{L}$ is hidden in the function H_0).

For a solution of the perturbation equations (31) to be physically acceptable, it must be composed, in the wave zone, of outgoing gravitational waves. In § I we could impose the outgoing-wave condition only by performing independent asymptotic expansions in the near zone and wave zone, and by matching them in the intermediate region. Here, however, because no slow-motion assumption is made, a single expansion (of which we study only the first order) is valid everywhere. However, the dynamical equations (31) are much more complicated here than in the slow-motion expansion, so the imposition of the outgoing-wave condition is not easy.

In § IV of Paper I a complex-eigenfrequency technique is devised, which singles out pulsations with purely outgoing waves. This technique is closely related to S-matrix techniques of quantum-mechanical scattering theory. (See Thorne 1968 for additional discussion.)

Paper III and Appendix C of Paper II develop, and use in numerical work, an alternative, closely related technique for applying the outgoing-wave condition. We shall describe that technique here:

The basic idea of the technique is to first study standing-wave normal modes for the star, and then to build outgoing-wave modes as linear superpositions of the standing-wave modes.

The standing-wave modes are pulsations which have sinusoidal time dependences with real angular frequencies, ω, and real eigenfunctions, $K_\omega(r)$, $W_\omega(r)$, $V_\omega(r)$

$$\{K, W, V\} = \{K_\omega(r),\ W_\omega(r),\ V_\omega(r)\}\, e^{i\omega t}\ . \tag{32}$$

For such modes the dynamical equations (31), together with the boundary conditions of smooth behavior at the star's center and surface, determine the eigenfunctions $K_\omega(r)$, $W_\omega(r)$, $V_\omega(r)$ uniquely (aside from normalization), for each choice of ℓ, m, and ω. Thus, for each choice of ℓ and m there is a continuous, one-parameter spectrum of standing-wave modes. As one expects from group-theoretic considerations, the eigenfunctions K_ω, W_ω, V_ω are independent of m.

The standing-wave modes can be thought of physically as follows: Construct a spherical cavity of infinite radius, whose wall is a perfect reflector of gravitational waves. Place the star at the center of the cavity, and into the cavity introduce standing gravitational waves with a particular but arbitrary angular frequency ω and spherical-harmonic indices ℓ and m. The gravitational waves will couple to the star, driving it into an undamped, sinusoidal pulsation. These coupled star-wave pulsations constitute a standing-wave mode.

Such a standing-wave mode is an exact (numerical) solution to the linearized Einstein equations (31) (linearized about a fully relativistic star; not about flat spacetime!). However, it is not a good approximation to the full, nonlinear Einstein equations, because it contains a finite energy density throughout an infinite region of space. This need not worry us, however, because the outgoing-wave modes built by superposing standing-wave modes will contain finite, arbitrarily small total energy, and will (presumably) be good approximations to exact solutions of Einstein's equations.

For a particular choice of ℓ and m one can probe the properties of the star by subjecting it to standing gravitational waves with various frequencies ω. When ω is very close to a resonant vibration frequency, σ_n, of the star, small-amplitude waves will excite large-amplitude fluid motions; but when ω is far from any resonant frequencies, the fluid will be excited hardly at all.

For a given, realistic stellar model [in the form of a numerical table of $\nu(r)$, $\lambda(r)$, $\rho(r)$, $p(r)$, $\gamma(r)$] one can discover the resonant frequencies by the following procedure: Calculate and plot the ratio of pulsation energy of the matter, E_M, to energy in one wavelength of the standing gravitational waves, E_W, as a function of angular frequency, ω (cf. Figure 2). This plot will exhibit resonances at the characteristic resonant frequencies, σ_0, σ_1, ..., of the star. The numerical techniques for calculating E_M/E_W as a function of ω are spelled out in the appendix to Paper III.

The resonances in the E_M/E_W plot obey the Breit-Wigner resonance formula

$$\left.\frac{E_M}{E_W}\right|_{\text{Standing Wave}} = \frac{\sigma_n/2\pi\tau_n}{(\omega - \sigma_n)^2 + (1/\tau_n)^2} ; \qquad (33)$$

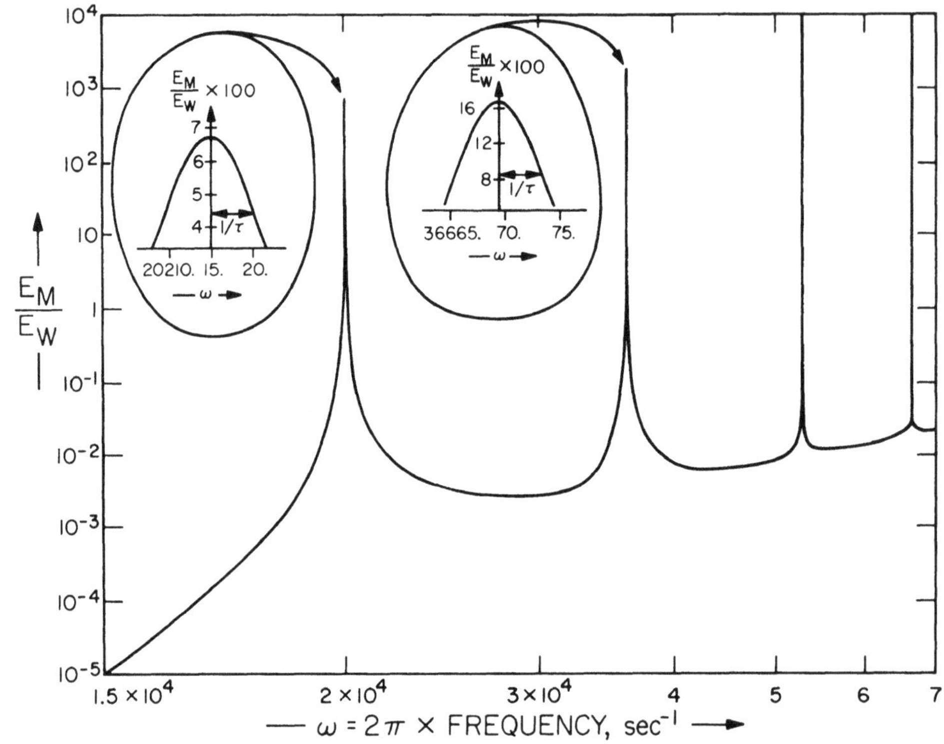

Fig. 2. Resonances in the quadrupole ($\ell = 2$) standing-wave normal modes for the H-W-W neutron-star model of maximum mass ($\rho_c = 6 \times 10^{15}$ g/cm^3, $M = 0.682\ M_\odot$). Plotted vertically is the ratio of the oscillation energy of the stellar matter, E_M, to the energy in one wavelength of the standing gravitational waves, E_W. Plotted horizontally is the angular frequency, ω, of the normal mode. Corresponding to each value of ω there are 5 independent standing-wave modes ($m = -2, -1, 0, +1, +2$), which all have the same eigenfunctions and energies, but different angular dependences. [This figure is taken from Thorne (1969a).]

and as one passes through resonance the phase, δ_ω, of the gravitational-wave function $K_\omega(r)$ far from the star changes by π

$$\delta_\omega = \delta_{\sigma_n} + \tan^{-1}[(\omega - \sigma_n)\tau_n]. \tag{34}$$

Here σ_n is the resonant frequency, and $1/\tau_n$ is the half-width that characterizes the resonant behavior of the star. Formulae (33) and (34) are derived analytically in Paper III and are compared with numerical calculations to determine the resonant frequencies and half-widths of the lowest few quadrupole modes of a variety of neutron-star models.

Corresponding to each standing-wave resonance of a stellar model there is a pulsational mode (called a "quasi-normal mode") with purely outgoing radiation and finite, arbitrarily small total energy. The quasi-normal mode has a sharp wave front, in front of which spacetime is unperturbed, and behind which the waves travel. Aside from transients near its wave front, and normalization of its amplitude, the quasi-normal mode is completely determined by the properties of the corresponding standing-wave resonance. This is proved in Appendix C of Paper II and is discussed further in Paper III.

Each quasi-normal mode can be constructed as a linear superposition ("wave packet") of standing-wave modes with frequencies in the vicinity of the corresponding resonance. This construction procedure is completely analogous to the way in which one constructs, from standing-wave states, the wave function for a particle leaking out of a quantum-mechanical potential well (cf. Breit and Yost 1935).

At a fixed point in the interior of the star, after the wave front has departed, the amplitude of the fluid motion for a quasi-normal mode undergoes damped sinusoidal oscillations

$$W \sim V \sim e^{i\sigma_n t}\, e^{-t/\tau_n}. \tag{35}$$

The angular frequency of the oscillations is σ_n, the angular frequency of the corresponding resonance; and the decay rate is $1/\tau_n$, the half-width of the resonance. (This was to be expected from the quantum-mechanical analogue; it is proved in Appendix C of Paper II and is discussed further in Paper III.)

In the radiation zone the outgoing gravitational waves of a quasi-normal mode have the expected radial and time dependence, in the canonical, transverse gauge

$$\begin{aligned} h_{\theta\theta}/r^2 &\sim h_{\theta\varphi}/r^2 \sin\theta \sim h_{\varphi\varphi}/r^2 \sin^2\theta \\ &\sim r^{-1} \exp\{[i\sigma_n - 1/\tau_n][t - r - 2M\ln(r - 2M)]\}. \end{aligned} \tag{36}$$

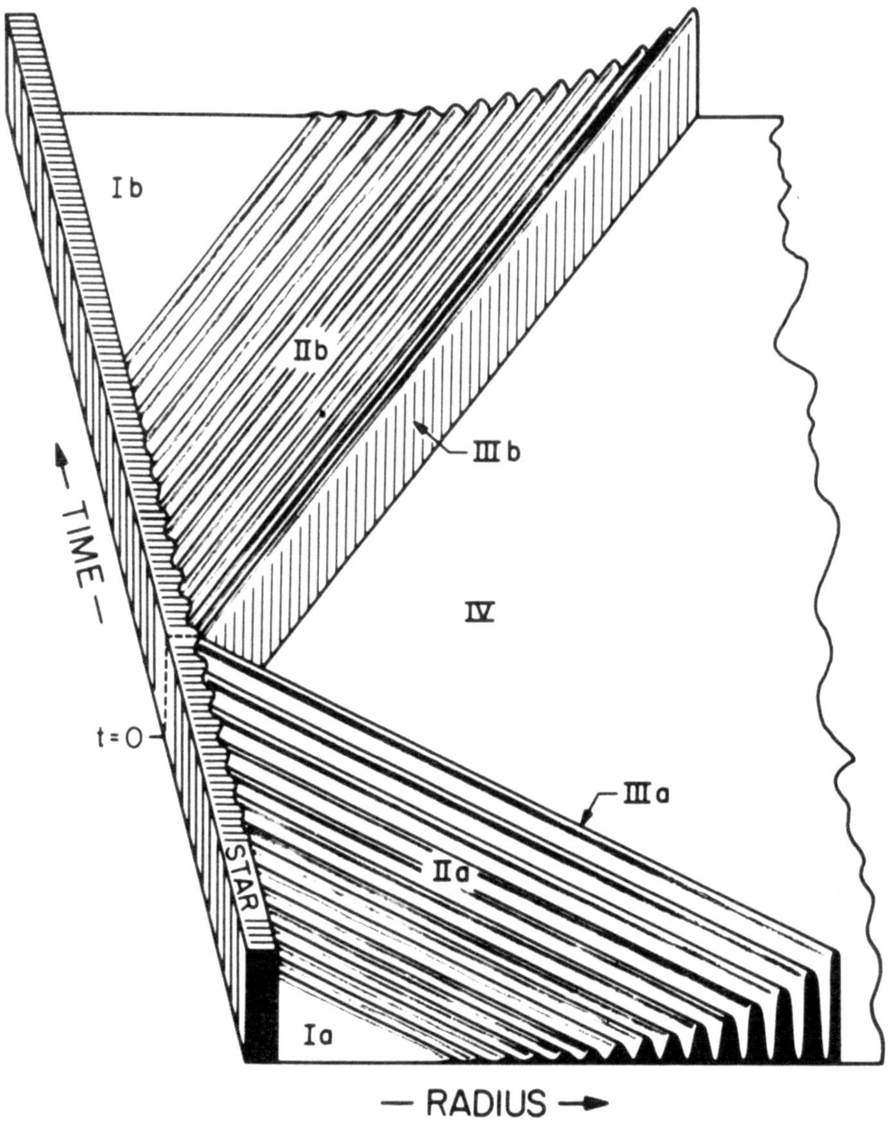

Fig. 3. Schematic spacetime diagram for a quasi-normal mode of pulsation. In such a quasi-normal mode a train of gravitational waves impinges upon the star from infinity, exciting it into pulsation. The star then reradiates a train of identical waves, thereby damping its pulsations. The wave trains have sharp fronts and carry with them finite, small amounts of energy and angular momentum. [This figure is taken from Price and Thorne (1969).]

Here M is the mass of the star; and the term $2M\ln(r-2M)$ guarantees that the waves travel along null radial geodesics of the external Schwarzschild geometry, and that the locally measured wavelength undergoes the usual gravitational redshift. These properties of the waves, as well as their polarization, energy, and angular momentum, are derived and discussed in Paper II. In Paper III it is shown that the energy carried off by the waves is precisely balanced by the pulsation energy damped out of the fluid.

An important problem of principle is the manner by which the star is excited into its quasi-normal pulsation. In practice a neutron star will be excited by the collapse in which it is formed, and the pulsations will approximate a superposition of quasi-normal modes only after the pulsation amplitude has dropped to $\delta r/r < 0.1$. However, the analysis described here has not attempted to treat such a realistic process for exciting the star. Instead, it has yielded -- via the superposition of standing-wave modes of Appendix C of Paper II -- a quasi-normal mode which is excited by an incoming train of gravitational waves. In fact, for our particular solution the excitation process is the time-reversal of the radiation process (see Figure 3).

By studying numerically the standing-wave modes of a realistic neutron-star model one can learn all of the properties of its quasi-normal modes. This is done in Paper III for the quadrupole modes of several realistic neutron-star models. The numerical analyses yield results which agree surprisingly well (to within a factor of 2 or 3) with rough weak-field calculations: The typical pulsation periods for massive neutron stars are $T = 2\pi/\sigma \sim 3 \times 10^{-4}$ sec, the typical damping times are $\tau \sim 1$ sec, and the total energy that might be radiated when a neutron star is formed by collapse is $E_M \sim 10^{52}$ erg.

Paper IV examines the slow-motion, weak-field limit of this nonradial-pulsation analysis. In the weak-field, slow-motion limit, analytic formulae can be derived for both standing-wave modes and quasi-normal modes in terms of the Newtonian pulsations of the stellar model. These same formulae can be derived independently from the weak-field, slow-motion expansion described in § I of this paper. The relationship between the two analyses is spelled out in detail in Paper IV.

REFERENCES

Breit, G. and Yost, F. L. 1935, Phys. Rev., 48, 203.
Burke, W. L. 1969, Ph.D. Thesis, California Institute of Technology (available from University Microfilms, Ann Arbor, Michigan).
1970, paper in preparation.

Campolattaro, A., and Thorne, K. S. 1970, Ap. J., in press. Paper V.
Carmeli, M. 1964, Phys. Letters, **9**, 132.
__________ 1965, Nuovo Cim., **37**, 842.
Chandrasekhar, S. 1964, Ap. J., **138**, 185.
__________ 1965, Ap. J., **142**, 1488.
__________ 1969a, Ap. J., in press.
Chandrasekhar, S. and Nutku, Y. 1969, Ap. J., in press.
Cole, J. 1968, Perturbation Methods in Applied Mathematics (Waltham, Mass.: Ginn-Blaisdell).
Einstein, A., Infeld, L., and Hoffman, B. 1938, Ann. Math., **39**, 66.
Hartle, J. B., and Thorne, K. S. 1968, Ap. J., **153**, 807.
Havas, P., and Goldberg, J. N. 1962, Phys. Rev., **128**, 398.
Hu, N. 1947, Proc. Roy. Irish Acad., **A51**, 87.
Infeld, L., and Plebanski, J. 1960, Motion and Relativity (New York: Pergamon Press).
Landau, L. D., and Lifschitz, E. M. 1962, The Classical Theory of Fields (Reading, Mass.: Addison Wesley).
Peres, A. 1959, Nuovo Cim., **11**, 644.
Price, R., and Thorne, K. S. 1969, Ap. J., **155**, 163. Paper II.
Regge, T. and Wheeler, J. A. 1957, Phys. Rev., **108**, 1063.
Ruderman, M. 1968, Nature, **218**, 1128.
Thorne, K. S. 1967, in High Energy Astrophysics, vol. 3, eds. C. DeWitt, P. Véron, and E. Schatzman (New York: Gordon and Breach).
__________ 1968, Phys. Rev. Letters, **21**, 320.
__________ 1969a, Ap. J., in press. Paper III.
__________ 1969b, Ap. J., in press. Paper IV.
Thorne, K. S., and Campolattaro, A. 1967, Ap. J., **149**, 591 and **152**, 673. Paper I.
Trautman, A. 1958, Bull. Acad. Polon. Sci. III, **6**, 627.

THE NATURE OF THE SCHWARZSCHILD SINGULARITY

Nathan Rosen

The Weizmann Institute of Science, Rehovot, Israel
and
Technion-Israel Institute of Technology, Haifa, Israel*

I. THE SCHWARZSCHILD SINGULARITY

Let us define the Schwarzschild solution of the Einstein gravitational field equations for empty space as the static, spherically symmetric solution for which space-time is flat at spatial infinity. By "static" is meant that, if we have coordinates x^1, x^2, x^3, x^4, where $x^4 = t$ is the time, then

$$\frac{\partial g_{\mu\nu}}{\partial t} = 0, \tag{1}$$

and

$$g_{4k} = 0 \qquad (k = 1, 2, 3). \tag{2}$$

The solution can be written in polar coordinates $(x^1, x^2, x^3) = (r, \theta, \phi)$ in the form (1)

$$ds^2 = -\frac{1}{1-2m/r}dr^2 - r^2 d\theta^2 - r^2 \sin^2\theta\, d\phi^2 + (1-2m/r)\,dt^2, \tag{3}$$

so that, as $r \to \infty$,

$$ds^2 = -dr^2 - r^2 d\theta^2 - r^2 \sin^2\theta\, d\phi^2 + dt^2, \tag{4}$$

---

* Permanent address

corresponding to flat space-time. In (3) m is the mass of the particle (in units of general relativity theory) which is at the origin and is the source of the field.

Let us work with polar coordinates, and let us therefore require that the line element have the form of Eq. (4) at infinity. The coordinate transformations under which the general form of the Schwarzschild solution is maintained are: (a) transformations of θ and ϕ corresponding to rotations and reflections of the coordinate system about the origin, (b) transformations of t of the form, $\bar{t} = t + \text{const.}$, or $\bar{t} = -t + \text{const.}$, and (c) transformations of r of the form

$$\bar{r} = f(r), \tag{5}$$

where $f(r) \to r$ as $r \to \infty$.

The form of the solution given by Eq. (3) is characterized by the fact that it contains the terms $- r^2 d\theta^2 - r^2 \sin^2\theta\, d\phi^2$ which are also present in the line element for flat space, Eq. (4). Hence if one takes a circle having its center at the origin and specified by a value of r, its circumference will be $2\pi r$. One can say that this is what defines the coordinate r used in (3). Alternatively, one can consider a sphere specified by a value of r, in which case the surface of the sphere will have an area $4\pi r^2$.

We see that in the form of the line element given by (3) there is a singularity for r = 2m, the Schwarzschild radius, for g_{11} is infinite and g_{44} vanishes. The question is whether this is a real singularity in the geometry of the space, or only an apparent one associated with the particular set of coordinates being used.

If one calculates the Riemann-Christoffel tensor, one finds that the non-vanishing components $R^{\mu\nu}{}_{\sigma\tau}$ are given by

$$\left.\begin{aligned} R^{12}{}_{12} = R^{13}{}_{13} = R^{24}{}_{24} = R^{34}{}_{34} = \frac{m}{r^3}, \\ R^{14}{}_{14} = R^{23}{}_{23} = -\frac{2m}{r^3}, \end{aligned}\right\} \tag{6}$$

together with those related to the above on the basis of the symmetry properties of the tensor. One can show that these components are invariant under the transformations that are allowed for the Schwarzschild solutions, e.g., that of Eq. (5). Actually, the components are the same as the so-called physical components, those obtained at a given point by the use of a local Galilean coordinate system having this point as origin and having its axes in the directions of the above coordinate lines. Since the components of (6) are singular at r = 0, and not at $r = 2m$, one concludes that the singularity at $r = 2m$ is an apparent one, arising from an unfortunate choice of coordinates.

Indeed, several writers have shown that, by a suitable coordinate transformation, one can get rid of the above singularity. Thus, Finkelstein (2) rediscovered the form of the solution that had been found a long time ago by Eddington (3), obtained from (3) by writing

$$t = t' + 2m \ln(r-2m). \tag{7}$$

This is given by

$$ds^2 = -\left(1+\frac{2m}{r}\right)dr^2 - r^2(d\theta^2 + \sin^2\theta\, d\phi^2) + \frac{4m}{r}\,dr\,dt' + \left(1-\frac{2m}{r}\right)dt'^2, \tag{8}$$

and does not have any singularity at $r = 2m$. However, in this case Eq. (2) is not satisfied, so that we do not have here a static solution. This solution is not invariant under a change of sign of t, whereas one would expect such invariance in the case of a static solution.

Kruskal (4) has found a transformation from the coordinates (r, t) to coordinates (u, v) given by

$$\left.\begin{aligned} u &= (2\alpha r - 1)^{\frac{1}{2}} e^{\alpha r} \cosh \alpha t, \\ v &= (2\alpha r - 1)^{\frac{1}{2}} e^{\alpha r} \sinh \alpha t, \end{aligned}\right\} \tag{9}$$

with $\alpha = 1/4m$, which leads to

$$ds^2 = f^2(-du^2 + dv^2) - r^2(d\theta^2 + \sin^2\theta\, d\phi^2). \tag{10}$$

Here

$$f^2 = (32m^3/r)\, e^{-r/2m}, \tag{11}$$

and r is given by the relation

$$\left(\frac{r}{2m} - 1\right) e^{r/2m} = u^2 - v^2. \tag{12}$$

One sees that the singularity for $r = 2m$ has disappeared. However the new coordinate system corresponds to an accelerated frame of reference, and the metric is no longer static. Similarly, the solution given recently by Israel (5) is non-static.

If we restrict ourselves to static solutions, obtained from (3) by transformations of the form of (5), then one cannot prevent g_{44} from vanishing at $r = 2m$, but one can prevent g_{11} from being infinite. Thus, if one takes in place of r, a coordinate u given by (6)

$$r - 2m = u^2/8m, \tag{13}$$

one gets

$$ds^2 = -(1 + u^2/16m^2)\,du^2 - \frac{1}{64m^2}(u^2 + 16m^2)^2(d\theta^2 + \sin^2\theta\, d\phi^2) + \\ + \left[u^2/(u^2 + 16m^2)\right] dt^2. \tag{14}$$

We see that at $u = 0$ (corresponding to $r = 2m$), g_{11} is finite. However, since g_{44} vanishes for $u = 0$, the determinant g also vanishes and a singularity remains, for one needs a non-vanishing value of g in order to obtain finite values for the components of the contravariant metric tensor $g^{\mu\nu}$. In the present case, g^{44} will be infinite for $u = 0$.

It might be remarked that the form of solution given by (14), while useful for discussing the situation in the neighbourhood of $r = 2m$, does not go over into the form of Eq. (4) as u goes to infinity. However this can be easily remedied by taking, in place of (13),

$$r - 2m = \frac{w^2}{w + m/2}, \tag{15}$$

so that for $w \to \infty$, $r = w$, while for $|w| \ll m$, $w = u/4$. One then obtains

$$ds^2 = -\frac{(w + m)^4}{(w + \frac{m}{2})^4}\left[dw^2 + \left(w + \frac{m}{2}\right)^2(d\theta^2 + \sin^2\theta\, d\phi^2)\right] + \frac{w^2}{(w+m)^2}\,dt^2. \tag{16}$$

Incidentally, this is closely related to another, familiar form of the Schwarzschild solution: if one writes

$$w = r' - \frac{m}{2}, \tag{17}$$

one gets

$$ds^2 = -(1 + m/2r')^4\left[dr'^2 + r'^2(d\theta^2 + \sin^2\theta\, d\phi^2)\right] + \left[\frac{1 - m/2r'}{1 + m/2r'}\right]^2 dt^2, \tag{18}$$

the well-known isotropic form (7).

We see from the above that, if one restricts oneself to transformations that maintain the static form of the Schwarzschild solution, the singularity in the line element cannot be removed. However, it should be remarked that there is a more important point involved here than that of the singularity. For $r < 2m$, we see that, in (3), $g_{11} > 0$, $g_{44} < 0$, so that in this region r is time-like and t space-like. However, this is an impossible situation, for we have seen that r is defined in terms of the circumference of a circle, so that r is space-like, and we are therefore faced with a contradiction. We must conclude that the portion of space corresponding to $r < 2m$ is non-physical. This is a situation which a coordinate transformation, even one which removes the singularity, cannot change. What it means is that the surface $r = 2m$ represents the boundary of physical space and should be regarded as an impenetrable barrier for particles and light rays.

The standpoint presented here finds support in the recent work of Janis, Newman and Winicour (8). These authors showed that if one writes down a spherically symmetric solution of the Einstein field equations for the case in which a scalar field ψ is present and one then lets ψ tend to zero, one obtains in the limit a form of the solution which agrees with (3) for $r > 2m$ but which has a discontinuity at $r = 2m$, so that the Schwarzschild sphere $r = 2m$ becomes a singular point, i.e., $4\pi r^2 = 0$. This is another way of describing the fact that the interior of the Schwarzschild sphere has been excluded from the physical space.

There seem to be two ways of dealing with this non-physical space. One can simply exclude it from the physical space by requiring that the boundary given by $r = 2m$ be an impassable barrier for light or matter, or one can go over to a coordinate system which has been chosen so as to exclude the non-physical region without leaving any boundary.

As an example of such a coordinate system, one might take that in which the radial coordinate is the variable u given by Eq. (13). We see that for all real values of u, $r \geqslant 2m$, so that the non-physical region has been excluded. To each value of $r > 2m$, there correspond two values of u having equal magnitudes but opposite signs. These two sets of values can be regarded as describing a space of two sheets, the value $u = 0$ representing the bridge between them (6). Alternatively, one can regard them as providing a double valued labelling of the points in ordinary space.

If one takes, as the radial coordinate, the variable w given by Eq. (15), the situation is somewhat similar. As w goes from $-m/2$ to zero, r goes from ∞ to 2m, and as w goes from zero to ∞, r goes from 2m to ∞, so that here again to each value of $r > 2m$ there correspond two values of w. For $w < -m/2$, the corresponding value of r is negative, and one can therefore disregard such values of w.

In the case of the isotropic coordinates in the line element given by Eq. (18), we see from (17) that as r' goes from zero to $m/2$ and from $m/2$ to ∞, r goes from ∞ to 2m and then back to ∞. It is interesting to note that, in the case of the curvature tensor components given in Eq. (6), the singularity appearing there at r = 0 does not appear anywhere in the space described by the coordinate r'.

II. MOTION OF PARTICLES AND LIGHT RAYS

In order to obtain a better understanding of the nature of the Schwarzschild singularity, let us investigate the motion of particles and light rays in its vicinity. For this purpose, let us begin by taking as the radial coordinate

$$\xi = r - 2m. \tag{19}$$

Then the line element (3) takes the form

$$ds^2 = -(1 + 2m/\xi)\,d\xi^2 - (\xi + 2m)^2(d\theta^2 + \sin^2\theta\, d\phi^2) + (1 + 2m/\xi)^{-1} dt^2. \tag{20}$$

In view of the remarks of the previous chapter, we must now say that only for $\xi \geq 0$ do we get points of physical space. Let us restrict ourselves to radial motion (θ, ϕ = const.) and, in order to simplify the calculations, let us suppose that r is nearly equal to 2m, or that $|\xi| \ll m$. Then (20) goes over into

$$ds^2 = -(2m/\xi)\,d\xi^2 + (\xi/2m)\,dt^2. \tag{21}$$

On the other hand, if we take as radial coordinate the variable u given by (13), then for r nearly equal to 2m, or $|u| \ll m$, the line element (14) can be written in the present case

$$ds^2 = -du^2 + \alpha^2 u^2\, dt^2, \tag{22}$$

where $\alpha = 1/4m$. From (13) and (19) we have

$$\xi = u^2/8m, \tag{23}$$

and this can be used to go directly from Eq. (21) to (22).

Let us now consider the motion of a test particle in the space described by the line element (22). The geodesic equations can be obtained from the variational principle,

$$\delta \int_{s_1}^{s_2} (-\dot{u}^2 + \alpha^2 u^2 \dot{t}^2) ds = 0, \tag{24}$$

where a dot denotes a derivative with respect to s, and where u and t are given arbitrary infinitesimal variations. Varying t, we get the relation

$$\frac{d}{ds}(2\alpha^2 u^2 \dot{t}) = 0,$$

or

$$u^2 \dot{t} = \beta \quad (\beta = \text{const.}). \tag{25}$$

From (22) we can write

$$-\dot{u}^2 + \alpha^2 u^2 \dot{t}^2 = 1, \tag{26}$$

so that

$$\dot{u}^2 = \alpha^2 \beta^2 / u^2 - 1.$$

Let us suppose that the particle is released at $u = b$ with zero velocity. Then $\alpha\beta = b$, and we have

$$\dot{u}^2 = b^2/u^2 - 1, \tag{27}$$

which can be integrated to give

$$(b^2 - u^2)^{\frac{1}{2}} = \pm s + \text{const.} \tag{28}$$

If we take $s = 0$ for $u = b$, this can be written

$$u^2 + s^2 = b^2. \tag{29}$$

We have here the equation of a circle. However, in the course of the motion, s must increase monotonically. Hence the trajectory must be such that its graph is composed of semicircles joined together as in Fig. 1. The motion can be described in terms of an integer n ($\gtrless 0$) so that, in order to satisfy Eq. (28),

$$u = (-1)^n [b^2 - (s-2nb)^2]^{\frac{1}{2}}, \qquad (2n-1)b \leq s \leq (2n+1)b. \tag{30}$$

As can be seen from Eq. (27), whenever $u = 0$, $\dot{u}$ is infinite.

Let us now go back to Eq. (25) and make use of (29). Taking $t = 0$ for $s = 0$, one finds the relation

$$s = b \tanh \alpha t \qquad (0 \leq s \leq b, \quad 0 \leq t < \infty). \tag{31}$$

Substituting this into (29), one gets

$$u = b/\cosh \alpha t. \tag{32}$$

If, instead of a particle, we consider the motion of a light ray, then $ds = 0$. We can introduce, in place of s, a parameter p which varies in the course of the motion, so that instead of Eqs. (25) and (26), we have

$$u^2 \dot{t} = \beta \qquad (\beta = \text{const.}), \tag{33}$$

$$-\dot{u}^2 + \alpha^2 u^2 \dot{t}^2 = 0, \tag{34}$$

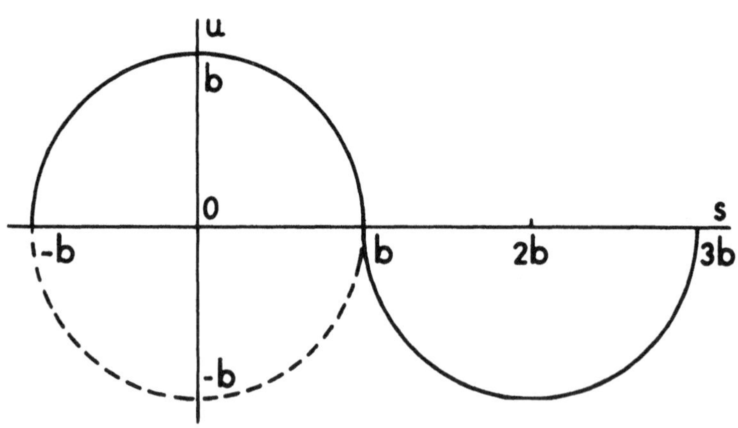

Fig. 1. Particle trajectory: coordinate u as function of proper time s.

where now a dot denotes a derivative with respect to p. Eliminating t between (33) and (34), one gets

$$u\dot{u} = \pm \alpha\beta ,$$

so that

$$u^2 = \pm 2\alpha\beta p + \text{const.} \tag{35}$$

If we take $u = b$ for $p = 0$ and let the ray move toward the origin, we have

$$u^2 = b^2 - 2\alpha\beta p \qquad (0 \leq 2\alpha\beta p \leq b^2, \quad u \geq 0). \tag{36}$$

Here β is an arbitrary scale factor associated with the parameter p. Substituting (36) into (33), and taking $t = 0$ for $p = 0$ we get

$$t = -\frac{1}{2\alpha} \ln\left(1 - \frac{2\alpha\beta p}{b^2}\right). \tag{37}$$

If we eliminate p between (36) and (37), we obtain

$$u = b e^{-\alpha t} . \tag{38}$$

We could use Eq. (36) to describe the continuation of the motion to negative values of u. However p would then not vary monotonically in the course of the motion. In order to have p vary monotonically, we must change the sign in (35) and take, in place of (36), the relation

$$u^2 = 2\alpha\beta p - b^2 \qquad (b^2 \leq 2\alpha\beta p \leq 2b^2, \quad u \leq 0). \tag{39}$$

The graph of the trajectory formed by joining the parabolas given by (36) and (39) is shown in Fig. 2.

From (32) and (38) we see that the coordinate time t required to reach the point $u = 0$ from an arbitrary point $u = b$ is infinite both for a particle and a light ray. One can readily verify that for a particle or a light ray to travel in the opposite direction, from $u = 0$ to a point having a finite value of u, also requires an infinite coordinate time. These facts form essentially the basis for the argument presented by Hilton (9) that the Schwarzschild singularity ($u = \xi = 0$) is an irremovable physical barrier.

One may argue against this by pointing out that, according to Eq. (29), for $u = 0$ $s = b$, so that the proper time of an observer moving with the particle, that elapses until the singularity is reached, is finite. However, it is difficult to accept the idea that the observer can accompany the particle to the point $u = 0$, for as the particle approaches this point, its velocity tends to that of light. This can be seen by noting that the metric velocity V, as measured with clocks and meter sticks, is given in our case by

$$V^2 = -\frac{g_{11}}{g_{44}}\left(\frac{dx^1}{dx^4}\right)^2 = \frac{1}{\alpha^2 u^2}\left(\frac{du}{dt}\right)^2. \tag{40}$$

Now, from Eqs. (25) and (26), one gets

$$\left(\frac{du}{dt}\right)^2 = \alpha^2 u^2 - u^4/\beta^2. \tag{41}$$

Hence, as $u \to 0$, $|V| \to 1$, the speed of light. It appears therefore that one should rather think of the observer as located (at rest or in motion) at some distance from the singularity and acquiring information

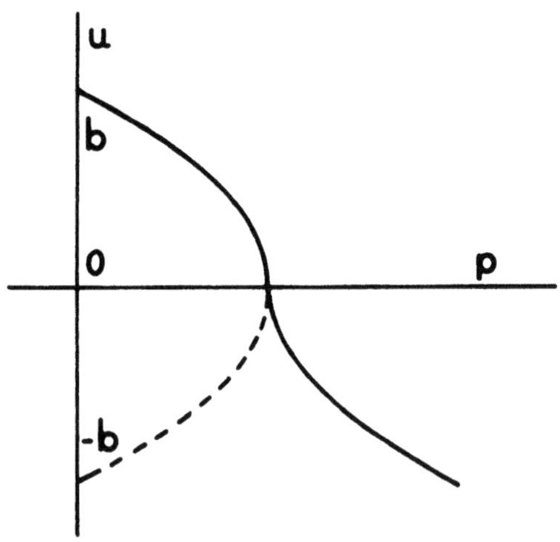

Fig. 2. Light-ray trajectory: coordinate u as function of parameter p.

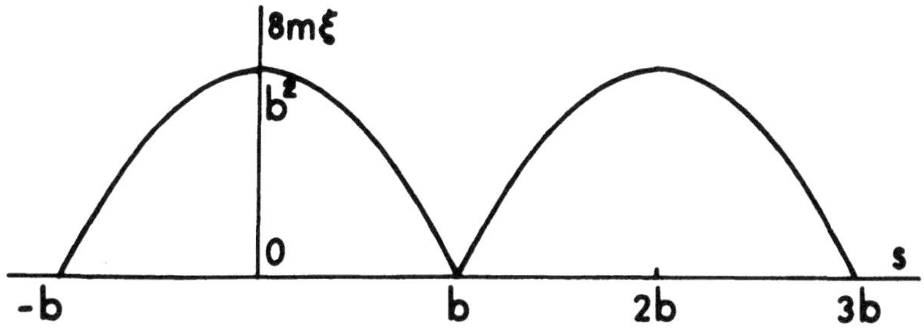

Fig. 3. Particle trajectory: coordinate ξ as function of proper time s.

by sending and receiving particles or light rays. In that case Hilton's argument seems convincing.

However, the present standpoint is that $u = 0$ corresponds to $\xi = 0$, or $r = 2m$, which represents an irremovable barrier because it separates the physical space $(r > 2m)$ from the non-physical space $(r < 2m)$.

Going back to Eqs. (23) and (29), we can write correspondingly for a test particle

$$8m\xi = b^2 - s^2, \tag{42}$$

so that

$$\dot{\xi} = -\alpha s. \tag{43}$$

We see that for $\xi = 0$, or $s = b$, $\dot{\xi}$ is finite. In the general case, corresponding to (30), we have

$$8m\xi = b^2 - (s - 2nb)^2, \qquad (2n-1)b \leq s \leq (2n+1)b. \tag{44}$$

Hence the graph of the trajectory consists of segments of parabolas, as shown in Fig. 3. We see that, when the particle strikes the boundary, it is reflected back elastically.

The light ray is also reflected at the boundary, $\xi = 0$. From (36) and (39), we get

$$8m\xi = b^2 - 2\alpha\beta p \qquad (0 \leq 2\alpha\beta p \leq b^2), \tag{45}$$

for the incident ray, and

$$8m\xi = 2\alpha\beta p - b^2 \qquad (b^2 \leq 2\alpha\beta p \leq 2b^2), \tag{46}$$

for the reflected ray. Hence, in terms of the parameter p, the trajectory consists of two straight lines, as shown in Fig. 4.

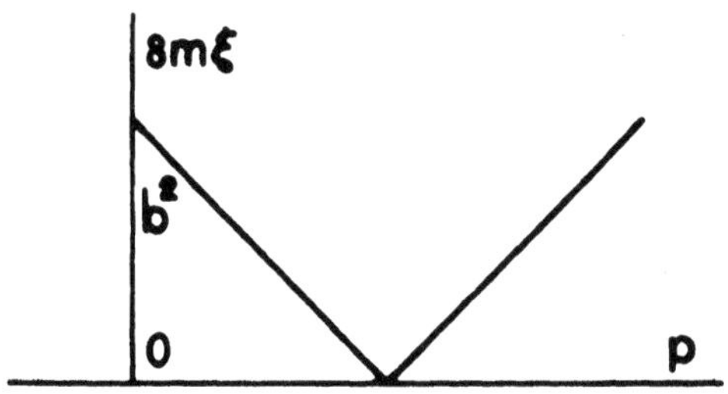

Fig. 4. Light-ray trajectory: coordinate ξ as function of parameter p.

We see then that we can deal in two ways with the Schwarzschild singularity and the non-physical region which it encloses. We can make use of a coordinate like r or ξ , which describes both the physical and non-physical regions, as well as the impassable boundary between them, or we can use a coordinate like u which corresponds only to points in the physical region and automatically excludes the non-physical one. In the first case, where there is a boundary separating the physical from the non-physical region, one must assume that in the physical space all the geodesics describing the motion of particles and light rays are reflected at this boundary.

III. EXTENDED SOURCE

In the previous chapters it has been argued that the Schwarzschild singularity represents an impassable boundary between the physical space outside and the non-physical space inside, and the motion of particles and light rays in the vicinity of this boundary has been investigated. However one can raise the question whether the Schwarzschild singularity exists in nature.

We know that the Schwarzschild solution is associated with the presence of a point-mass at the origin, for everywhere else the field equations for empty space are satisfied. The picture we get from this solution, then, is that every point-mass is enclosed in a Schwarzschild sphere, a kind of rigid impenetrable bubble. The existence of such a surface of discontinuity seems strange, particularly since we are accustomed in some other parts of physics to the description of phenomena by means of analytic functions. To be sure, one can see a certain analogy in the case of the light-cone at a point: the light-cone, in the space of four-velocities, provides a sharp boundary between that part of space accessible to material particles and that part which is inaccessible. Nevertheless, the idea that physical space has a boundary is somewhat repugnant.

The presence of the point-mass is a disturbing feature, and one cannot help wondering whether the difficulty may not be due to the assumption of such a mass singularity. Since it appears that the particles found in nature have finite extensions, let us examine the situation in which the source of the static, spherically symmetric field occupies a finite volume. This means that there is a finite region in which there exists a non-vanishing energy-momentum density tensor $T_\mu^{\ \nu}$ which also is static and has spherical symmetry. In this region the Einstein field equations take the form

$$G_\mu^{\ \nu} = -8\pi T_\mu^{\ \nu}, \tag{47}$$

instead of $G_\mu^{\ \nu} = 0$, as in the Schwarzschild case.

Let us now write the line element in the form

$$ds^2 = -\frac{1}{\sigma}dr^2 - r^2 d\theta^2 - r^2\sin^2\theta\, d\phi^2 + e^\nu dt^2, \tag{48}$$

with $\sigma = \sigma(r)$, $\nu = \nu(r)$. In the present coordinate system the only non-vanishing components of $T_\mu^{\ \nu}$ are $T_1^{\ 1}$, $T_2^{\ 2} = T_3^{\ 3}$, and $T_4^{\ 4}$, and they are all functions of r. The field equations (47) are found to have the form

$$\sigma\left(\frac{\nu'}{r} + \frac{1}{r^2}\right) - \frac{1}{r^2} = -8\pi T_1{}^1, \tag{49}$$

$$\sigma\left(\frac{\nu''}{2} + \frac{\nu'^2}{4} + \frac{\nu'}{2r}\right) + \sigma'\left(\frac{\nu'}{4} + \frac{1}{2r}\right) = -8\pi T_2{}^2, \tag{50}$$

$$\frac{\sigma'}{r} + \frac{\sigma}{r^2} - \frac{1}{r^2} = -8\pi T_4{}^4, \tag{51}$$

with $\sigma' = \dfrac{d\sigma}{dr}$, $\nu' = \dfrac{d\nu}{dr}$.

There exist the identities,

$$G_\mu{}^\nu{}_{;\nu} \equiv 0, \tag{52}$$

and hence, from the field equations (47), the corresponding equations of motion,

$$T_\mu{}^\nu{}_{;\nu} = 0. \tag{53}$$

In the present case, Eq. (53) gives only one non-trivial relation, namely, for $\mu = 1$:

$$\frac{d}{dr}(T_1{}^1) + T_1{}^1\left(\frac{2}{r} + \frac{1}{2}\nu'\right) - \frac{2}{r}T_2{}^2 - \frac{1}{2}\nu' T_4{}^4 = 0, \tag{54}$$

so that

$$T_2{}^2 = \frac{1}{2}r\frac{d}{dr}(T_1{}^1) + \left(1 + \frac{1}{4}r\nu'\right)T_1{}^1 - \frac{1}{4}r\nu' T_4{}^4. \tag{55}$$

This relation represents the condition for static equilibrium.

We see that $T_1{}^1$ and $T_4{}^4$ can be taken arbitrarily, and $T_2{}^2$ is then determined. It is sufficient to solve Eqs. (49) and (51); then Eq. (50) will be satisfied with $T_2{}^2$ given by (55).

For convenience, let us write

$$T_1{}^1 = S, \qquad T_4{}^4 = \rho . \tag{56}$$

It will be assumed that these are everywhere finite and that the energy (or mass) density $\rho \geqslant 0$. Let us suppose that we are dealing with a particle of radius a. Then ρ and S are substantially different from zero only for $r \lesssim a$.

Let us now integrate the field equations, beginning with (51). The general solution of this equation is given by

$$\sigma = 1 - \frac{8\pi}{r}\int_0^r \rho r^2 dr + \frac{C}{r}, \tag{57}$$

where C is an arbitrary constant. If there is not to be a singularity at $r = 0$, we must choose $C = 0$, so that our solution takes the form

$$\sigma = 1 - \frac{8\pi}{r}\int_0^r \rho r^2 dr. \tag{58}$$

We see that $\sigma(0) = 1$, $\sigma'(0) = 0$. For $r > a$, we can write, as in the Schwarzschild case,

$$\sigma = 1 - \frac{2m}{r}, \tag{59}$$

where the mass m is given by

$$m = 4\pi\int_0^a \rho r^2 dr. \tag{60}$$

From Eq. (49) one gets

$$\nu = \nu_0 + \int_0^r \frac{1}{r}\left(\frac{1}{\sigma} - 1\right) dr - 8\pi\int_0^r \frac{Sr}{\sigma} dr, \tag{61}$$

where $\nu_0 = \nu(0)$ is a constant of integration. If we require that, as $r \to \infty$, $\nu \to 0$, then (61) can be written

$$\nu = -\int_r^\infty \frac{1}{r}\left(\frac{1}{\sigma} - 1\right) dr + 8\pi\int_r^\infty \frac{Sr}{\sigma} dr. \tag{62}$$

For $r > a$, using (59) and taking $S = 0$, we get

$$\nu = \ln(1-2m/r), \tag{63}$$

as in the Schwarzschild solution, while for $r \lesssim a$ we can write

$$\nu = -\int_r^a \frac{1}{r}\left(\frac{1}{\sigma}-1\right)dr + 8\pi\int_r^a \frac{Sr}{\sigma}\,dr + \ln(1-2m/a). \tag{64}$$

The form of the solution for ν will depend on the nature of the radial stress S. Let us consider a few examples:

(a) Suppose that $S = 0$. Then Eq. (62) takes the form

$$\nu = -\int_r^\infty \frac{1}{r}\left(\frac{1}{\sigma}-1\right)dr, \tag{65}$$

and for $r \lesssim a$, Eq. (64) becomes

$$\nu = -\int_r^a \frac{1}{r}\left(\frac{1}{\sigma}-1\right)dr + \ln(1-2m/a). \tag{66}$$

In this case Eq. (55) gives

$$T_2{}^2 = -\frac{1}{4}r\nu'\rho = \frac{1}{4}\left(1-\frac{1}{\sigma}\right)\rho. \tag{67}$$

(b) One can choose S so that $\nu' = 0$. Eq. (49) then gives

$$S = \frac{1}{8\pi r^2}(1-\sigma), \tag{68}$$

or by (58)

$$S = \frac{1}{r^3}\int_0^r \rho r^2 dr. \tag{69}$$

From (55) one then gets (70)

$$T_2^{\ 2} = \tfrac{1}{2} r S' + S.$$

If one makes use of (69) and (58), one can write

$$\left.\begin{aligned} T_2^{\ 2} &= \tfrac{1}{2}\rho - \frac{1}{2r^3}\int_0^r \rho r^2 dr, \\ &= -\frac{\sigma'}{16\pi r}. \end{aligned}\right\} \qquad (71)$$

Since we are considering a particle of radius a, Eqs. (68) and (69), and hence the condition $\nu' = 0$, are valid only for $r \leqslant a$. At $r = a$ there will be, in general, a discontinuity in S, since $S = 0$ for $r > a$. According to (55), such a discontinuity will lead to an infinite value of $T_2^{\ 2}$. This can be interpreted as indicating the presence of a surface stress Q at $r = a$, given by

$$\begin{aligned} Q &= \int_{a-\varepsilon}^{a+\varepsilon} T_2^{\ 2} dr, \\ &= \tfrac{1}{2} a \Delta S, \end{aligned} \qquad (72)$$

where ε is a small (positive) quantity, and ΔS is the discontinuity in S at $r = a$.

We have ν = const. inside the particle. In order to have continuity at $r = a$, one takes

$$e^{\nu} = 1 - 2m/a \qquad (0 \leqslant r \leqslant a). \qquad (73)$$

(c) Another simple possibility is to take

$$S = \rho. \qquad (74)$$

This leads to the relation $e^{\nu} = \sigma$, so that, by (58)

$$e^{\nu} = 1 - \frac{8\pi}{r}\int_0^r \rho r^2 dr. \tag{75}$$

We now have, by (55) and (51),

$$\left.\begin{aligned} T_2^{\ 2} &= \tfrac{1}{2} r\rho' + \rho, \\ &= -\frac{(r\sigma)''}{16\pi r}. \end{aligned}\right\} \tag{76}$$

Here again a surface stress may be present at $r = a$.

(d) In the case of a fluid one has $T_2^{\ 2} = S = -p$, where p is the pressure. Eq. (54) takes the form

$$p' = -\tfrac{1}{2}\nu'(\rho + p), \tag{77}$$

and this is to be solved together with Eqs. (49) and (51) and the equation of state, which provides a relation between p and ρ. In general this set of equations is difficult to solve, except in the case of ρ = const., which corresponds to the well-known Schwarzschild interior solution (10). Because of the difficulties involved, the case of the fluid will not be considered here any further.

Let us now go back to the solutions for σ and ν as given, say, by Eqs. (58) and (62). If the density ρ is such that

$$8\pi\int_0^r \rho r^2 dr < r, \tag{78}$$

for all values of $r \leq a$, then we see from (58) that σ will be positive, and hence g_{11} will be finite and negative for all values of r. As for ν, we see from (62), or still better, from (63) and (64), that ν will be everywhere finite, so that g_{44} will be finite and positive everywhere. Hence in this case the metric will be free from singularities.

If the inequality (78) does not hold everywhere, then singularities will occur. For sufficiently small values of r it is obvious that (78) must hold. Let us suppose that, as r increases, we reach a value $r = R$ for which

$$8\pi \int_0^R \rho r^2 dr = R. \tag{79}$$

Then, by Eq. (58), σ vanishes and g_{11} is infinite. For $r > R$ we encounter a region in which $g_{11} > 0$, hence a non-physical region since, as we have seen, the coordinate r is a length ($1/2\pi$ times the circumference of a circle) and cannot be time-like. However, if we take r sufficiently large ($r > 2m$) we are again in a physical region where $g_{11} < 0$.

To see the situation more clearly, let us consider the simple case in which the density ρ is constant inside the particle. Then (58) gives

$$\left.\begin{aligned} \sigma &= 1 - r^2/R^2 \qquad (0 \leq r \leq a), \\ &= 1 - 2m/r \qquad (r \geq a), \end{aligned}\right\} \tag{80}$$

where we now take

$$R^2 = 3/8\pi\rho, \qquad m = 4\pi\rho a^3/3 = a^3/2R^2. \tag{81}$$

If $2m < a$, or $a < R$, then $\sigma > 0$, and g_{11} is positive and finite for all values of r. In case (a), with $S = 0$, one finds

$$\left.\begin{aligned} e^\nu &= (1 - a^2/R^2)^{\frac{3}{2}}/(1 - r^2/R^2)^{\frac{1}{2}} \qquad (0 \leq r \leq a), \\ &= 1 - 2m/r \qquad (r \geq a). \end{aligned}\right\} \tag{82}$$

Eq. (67) gives

$$T_2{}^2 = -\tfrac{1}{4} r^2 \rho/(R^2 - r^2) \qquad (0 \leq r < a). \tag{83}$$

In case (b), with $\nu' = 0$ for $r < a$, one finds that

$$S = \tfrac{1}{3}\rho \qquad (0 \leq r < a), \tag{84}$$

and, as we have seen

$$e^{\nu} = 1 - 2m/a \qquad (0 \leq r \leq a). \tag{73}$$

Eq. (70) gives in this case

$$T_2{}^2 = S \qquad (0 \leq r < a), \tag{85}$$

so that one has isotropic tension inside the particle. However, since there is a discontinuity in S, we see from (72) that there will be a surface stress at $r = a$,

$$Q = -\tfrac{1}{6} a\rho, \tag{86}$$

i.e., there will be a "negative surface tension", or surface compression on the spherical surface, $r = a$.

In case (c), with $S = \rho$, one has

$$e^{\nu} = 1 - r^2/R^2 \qquad (0 \leq r \leq a), \tag{87}$$

and by (76),

$$T_2{}^2 = \rho \qquad (0 \leq r < a). \tag{88}$$

Here again we have isotropic tension inside the particle. At $r = a$ there will be a surface stress

$$Q = -\tfrac{1}{2} a\rho. \tag{89}$$

Now let us suppose that we increase either the density ρ or the radius a, or both, until we reach the situation in which $R = a = 2m$. Then g_{11} becomes infinite at $r = a$. In case (c) above, the line element is given by

$$ds^2 = -(1-r^2/R^2)^{-1}dr^2 - r^2(d\theta^2+\sin^2\theta\, d\phi^2) + (1-r^2/R^2)dt^2$$

$$(0 \leq r < R), \tag{90}$$

and we see that this has the form corresponding to the de Sitter static model of the universe (11). This line element describes a space with a closed geometry, that is, a space from which no particle or light ray can escape. It follows that no particle or light ray can enter such a space from the outside. Hence, from the standpoint of the outer space, such a region must be regarded as inaccessible and therefore as not belonging to our physical space.

It should be remarked at this point that it is possible, by means of a coordinate transformation (12, 13), to put the de Sitter line element into a form corresponding to a vanishing spatial curvature, and hence to an infinite open universe. However, the required transformation involves the time, and the resulting line element is non-static. As was emphasized earlier, we are here concerned only with the static case.

In the cases (a) and (b) above, the solutions for ν are different from that in the case (c). However, for the present purpose, what matters is the form of the spatial line element, i.e., the expression for ds^2 with $dt = 0$, and this is the same in all three cases. Denoting it by $-dl^2$ and introducing, in place of r, an angle χ by the relation

$$r = R\sin\chi, \tag{91}$$

we get

$$dl^2 = R^2(d\chi^2 + \sin^2\chi\, d\theta^2 + \sin^2\chi \sin^2\theta\, d\phi^2), \tag{92}$$

which has the form of the line element on the three-dimensional surface of a four-dimensional sphere. Actually, the situation is a little more complicated. In the case of the sphere one has $0 \leq \chi \leq \pi$. According to (91), as χ increases from 0 to π, r increases from 0 to R and then decreases to 0. In our case, on the other hand, r goes only from 0 to R, so that the corresponding manifold represents only half of the sphere. Hence we have here not spherical, but elliptical, geometry, in which antipodal points of the sphere are regarded as identical (14). However, we are still dealing with a closed space.

Let us now go back to case (b). According to (73), $e^{\nu} = 0$, for $0 \leq r \leq R$, so that the time dimension seems to have disappeared. This has come about from the fact that $\nu' = 0$ and the requirement that ν be continuous for $r = a = R$. However, once we are aware that the space for $r \leq R$ is closed, then the continuity requirement becomes meaningless, and we can discard it. Let us then take $\nu = 0$, instead. This gives

$$ds^2 = -(1 - r^2/R^2)^{-1}dr^2 - r^2(d\theta^2 + \sin^2\theta d\phi^2) + dt^2 \qquad (0 \leq r < R), \tag{93}$$

which has the form corresponding to the Einstein static model of the universe (15).

In case (a) we have a similar situation. According to (82), $e^{\nu} = 0$ for $r \leq R$, but again this is a consequence of the continuity requirement, which no longer has any significance if the space is closed. We can discard the vanishing coefficient in the expression and write

$$ds^2 = -(1 - r^2/R^2)^{-1}dr^2 - r^2(d\theta^2 + \sin^2\theta d\phi^2) + (1 - r^2/R^2)^{-\frac{1}{2}}dt^2 \qquad (0 \leq r < R). \tag{94}$$

This can be regarded as the line element of a certain static, closed, anisotropic, inhomogeneous universe.

In the preceding examples, in which we have assumed that inside the particle the density ρ is constant, we have found that, if the particle radius a is equal to the Schwarzschild radius $2m$, the space occupied by the particle is closed. The closure of the space is associated with the fact that, for $r = a$, $\sigma = 0$ and g_{11} is infinite. One can expect that in the case of a non-uniform density the situation will be qualitatively similar, although the details will be more complicated. It is found that if $a = 2m$, so that $\sigma(a) = 0$, the space described by $0 \leq r \leq a$ will be closed, but it will be, in general, inhomogeneous and anisotropic. This is discussed in the Appendix.

However, if the space occupied by the particle is closed, then, as we have seen in the examples, the usual continuity conditions become meaningless for $r = a$. This means that there is no definite relation between the field outside the particle and the properties of this particle. Thus one could assume, for example, that for $r > a$ the space is flat. There is a certain plausibility in this assumption: if no test particle or light ray can get out of the closed space, then it is reasonable to suppose that no influence of the particle can reach the outside space.

One can go further along this line of thought. If we have a situation in which $\sigma(r) > 0$, for $r < R$ and $\sigma(R) = 0$, with $R < a$, where a is the particle radius, then a straight-forward integration of Eq. (51) leads to the result that, for r slightly greater than R, $\sigma(r) < 0$, and we have an unphysical region. However, if we know that the region corresponding to $r \leq R$ is a closed space, this result is questionable. We can write the solution of Eq. (51) for $r > R$ in the form

$$\sigma = 1 - \frac{8\pi}{r}\int_R^r \rho r^2 dr + \frac{C_1}{r}. \tag{95}$$

For this to agree with the form in Eq. (58), one must take $C_1 = -2m_1$, where

$$m_1 = 4\pi \int_0^R \rho r^2 dr, \tag{96}$$

and where $2m_1 = R$, since $\sigma(R) = 0$. From (95) it follows indeed that, in general, for r slightly greater than R, $\sigma(r) < 0$. However, if we take into account the fact that we are now outside of a closed space, we can replace (95) by the equation

$$\sigma = 1 - \frac{8\pi}{r}\int_R^r \rho r^2 dr, \tag{97}$$

by taking $C_1 = 0$. Then for r slightly greater than R, $\sigma(r)$ is now positive. If this is the case for all values of r such that $R \leq r \leq a$, then the field outside the particle is now given by the Schwarzschild solution corresponding to a particle mass $m - m_1$, and there is no Schwarzschild singularity. On the other hand, if there is a value of r in the range $R < r < a$ for which (97) gives $\sigma = 0$, then we have again a closed space, and we can essentially repeat the previous considerations. We thus end up with an exterior Schwarzschild solution (for $r > a$) corresponding to a mass smaller than m, say m_e, such that $2m_e < a$, and this solution will not have any singularity outside of the particle.

The conclusion that one is led to is that in nature there do not exist any Schwarzschild singularities. This means that, in all cases, $a \lesssim 2m$, where a is the particle radius and m is its observed mass, i.e., the mass corresponding to its gravitational field.

This conclusion is essentially in agreement with that reached by McCrea (16). On the basis of energy considerations, McCrea concluded that the smallest possible radius which a mass m could have is 2m, or that the greatest mass that could be compressed into a radius a is $\frac{1}{2}a$.

The cases $a = 2m$, regarded as a limiting case, requires some comment. Suppose we have a star which grows through the accretion of matter. Then it is conceivable that, in the course of time, it will reach a state for which $a = 2m$. However, the time required to reach this state will be infinite, from the standpoint of an outside observer, as McCrea (16) remarked. This can be seen from an order-of-magnitude calculation based on Eq. (32) or from a more accurate calculation based on the exact equations of motion:

Let us simplify the calculations by assuming that the star has a constant density ρ and a radius a so that, with $R = (3/8\pi\rho)^{\frac{1}{2}}$ and mass $m = a^3/2R^2$, we have $2m < a < R$. The line element (3) is then valid for $r \geqslant a$. Using it in the equations of the geodesic, we can calculate the motion of a test particle. Let us assume radial motion, the particle starting at $r = r_0$ at $t = 0$ with zero velocity. One finds that the time t at which the particle reaches a radius r ($< r_0$) is given by

$$t = \frac{r_0}{2}\left(\frac{r_0}{2m} - 1\right)^{\frac{1}{2}}\left[\sin\Phi + \left(1 + \frac{4m}{r_0}\right)\Phi\right] +$$

$$+ 2m \ln\left\{\frac{1 + \left(1 - \frac{4m}{r_0}\right)\cos\Phi + \left(\frac{8m}{r_0}\right)^{\frac{1}{2}}\left(1 - \frac{2m}{r_0}\right)^{\frac{1}{2}}\sin\Phi}{1 - \frac{4m}{r_0} + \cos\Phi}\right\}, \qquad (98)$$

where Φ is defined by

$$r = r_0 \cos^2\frac{\Phi}{2}. \qquad (99)$$

Let us express m in terms of R and a in the second term on the right-hand side of (98) and let us take $r = a$. The one finds

$$t = \frac{\pi r_0^{\frac{3}{2}}}{2(2m)^{\frac{1}{2}}} + \frac{a^3}{R^2} \ln \frac{R+a}{R-a} \quad (a \ll r_0). \tag{100}$$

We see that, as a approaches R (and therefore a approaches 2m), t tends to infinity.

One can expect that this conclusion will be valid also if the motion has an angular component, since the presence of angular momentum would, if anything, tend to slow down the radial motion for small values of r.

Let us consider next the case of a contracting star. If we assume that the pressure inside it vanishes, then each particle moves on a geodesic. One can see that the time required for the star radius to contract to the Schwarzschild radius will be infinite for an outside observer. (For a co-moving observer the time will be finite, but we have seen that his speed will approach that of light.) If pressure is present, its effect will be to slow down the contraction velocity. Hence we see again that the case $a = 2m$ cannot be reached in a finite time.

IV DISCUSSION

In the preceding chapters several different approaches were used, leading to different conclusions, and it is now necessary to try to arrive at a unified picture.

In Chapter I the Schwarzschild solution (often referred to as the exterior solution), associated with a point-mass source, was considered. It was concluded that for $r < 2m$ one has a non-physical region, because r is both space-like and time-like. Therefore $r = 2m$ represents an impassable barrier for particles and light rays, so that a real singularity exists. The motion of particles and light rays near this singularity was investigated in Chapter II. It was found that, when they reach this barrier, they are reflected back. However, the time required to reach the barrier, as judged by an outside observer, is infinite.

In Chapter III the point mass was replaced by a finite-sized particle. It was found, by a straight-forward integration of the field equations, that three kinds of regions may exist: if, in addition to the physical region, there is a non-physical region present, as in the case of the point mass, then near the center there will be another physical region which, however, will be closed and therefore inaccessible. However, the fact that this region is closed raises doubts about the method of integrating the field equations: the continuity condition used at the boundary between two regions becomes meaningless if one of them is closed. It is plausible to assume that the mass inside a closed space cannot produce a gravitational field outside of this region. If one assumes this, one finds that, for the particle mass corresponding to the observed field, there is no Schwarzschild singularity outside of the particle, i.e., that the particle radius is never smaller than the Schwarzschild radius. Hence, the picture of a point mass being the source of the Schwarzschild field, considered in Chaps. I and II, has to be discarded.

There remains the possibility of the particle radius being equal to the Schwarzschild radius, as a limiting case. However, it appears that, in the case of a star, for this state to be reached either as a result of contraction or through the mass growing by accretion of matter would require an infinite time. If one believes that the age of the universe is finite, then one would not expect to find this limiting case in nature, although cases close to this may perhaps occur.

REFERENCES

1. K. Schwarzschild, Sitzber, Preuss. Akad. Wiss. (Math.-Phys. Kl.) 1916, p. 189.
2. D. Finkelstein, Phys. Rev. **110**, 965 (1958).
3. A.S. Eddington, Nature **113**, 192 (1924).
4. M.D. Kruskal, Phys. Rev. **119**, 1743 (1960).
5. W. Israel, Phys. Rev. **143**, 1016 (1966).
6. A. Einstein and N. Rosen, Phys. Rev. **48**, 73 (1935).
7. A.S. Eddington, The Mathematical Theory of Relativity, 2nd ed., p. 93, Cambridge University Press, 1924.
8. A.I. Janis, E.T. Newman, and J. Winicour, Phys. Rev. Letters **20**, 878 (1968).
9. E. Hilton, Proc. Roy. Soc. (London) **A283**, 491 (1965).
10. K. Schwarzschild, l.c., p. 424.
11. W. de Sitter, Proc. Akad. Wetensch. Amsterdam **19**, 1217 (1917).
12. G.E. Lemaître, J. Math. and Phys. (M.I.T.) **4**, 188 (1925).
13. H.P. Robertson, Phil.Mag. **5**, 835 (1928).
14. R.C. Tolman, Relativity, Thermodynamics and Cosmology, p. 338, Clarendon Press, Oxford, 1934.
15. A. Einstein, Sitzber, Preuss. Akad. Wiss. (Math.-Phys. Kl.) 1916, p. 142.
16. W.H. McCrea, Astrophys. Norvegica **9**, 89 (1964).

APPENDIX

Let us consider a three-dimensional curved space imbedded in a four-dimensional Galilean space. Let the four-dimensional space be described by means of the Cartesian coordinates z_1, z_2, z_3, z_4, or the polar coordinates R, χ, θ, ϕ, given by

$$\left.\begin{aligned} z_1 &= R\cos\chi, \\ z_2 &= R\sin\chi\cos\theta, \\ z_3 &= R\sin\chi\sin\theta\cos\phi, \\ z_4 &= R\sin\chi\sin\theta\sin\phi, \end{aligned}\right\} \quad \text{(A1)}$$

so that

$$R^2 = z_1^2 + z_2^2 + z_3^2 + z_4^2. \quad \text{(A2)}$$

The line element in this space can be written

$$dl^2 = dz_1^2 + dz_2^2 + dz_3^2 + dz_4^2, \quad \text{(A3)}$$

or

$$dl^2 = dR^2 + R^2[d\chi^2 + \sin^2\chi(d\theta^2 + \sin^2\theta d\phi^2)]. \quad \text{(A4)}$$

If the three-dimensional surface is a sphere given by R = const., then (A4) goes over into Eq. (92). Let us now consider a more general surface given by

$$R = R(\chi) \qquad (0 \leq \chi \leq \pi). \quad \text{(A5)}$$

Then the line element on this surface is given by

$$dl^2 = \left[\left(\frac{dR}{d\chi}\right)^2 + R^2\right]d\chi^2 + R^2\sin^2\chi(d\theta^2 + \sin^2\theta d\phi^2). \quad \text{(A6)}$$

Let us assume that $R(\chi)$ is finite, single-valued and smooth, so that the surface is closed and has the topology of a three-sphere. Let us also assume that

$$R(\chi) = R(\pi - \chi), \quad \text{(A7)}$$

so that we can regard antipodal points as identical ("elliptical" geometry). Let us write

$$\left.\begin{array}{lll} R = R_0, & \dfrac{dR}{d\chi} = 0 & (\chi = 0), \\ R = R_1, & \dfrac{dR}{d\chi} = 0 & (\chi = \frac{\pi}{2}), \end{array}\right\} \tag{A8}$$

where the conditions on the derivatives describe the fact that there are no cusps.

Now, in the case of a static, spherically symmetric distribution of matter, we have from Eq. (48)

$$dl^2 = \frac{1}{\sigma} dr^2 + r^2(d\theta^2 + \sin^2\theta\, d\phi^2). \tag{A9}$$

We have seen that, for $r = 0$, $\sigma = 1$ and $\frac{d\sigma}{dr} = 0$. Let us consider the case where, as we increase r, we reach a value for which $\sigma = 0$. The question is whether the portion of space bounded by the surface $\sigma = 0$ is closed, i.e., whether, after carrying out a suitable coordinate transformation, we can identify it with the closed three-surface described by (A5).

If we compare (A6) and (A9) we can write

$$r = R \sin\chi, \tag{A10}$$

$$\frac{1}{\sigma} dr^2 = \left[\left(\frac{dR}{d\chi}\right)^2 + R^2\right] d\chi^2. \tag{A11}$$

We see that $r = 0$ for $\chi = 0$. Let us take the value of r, for which $\sigma = 0$, as corresponding to $\chi = \frac{\pi}{2}$. Then, in the present notation, this is given by $r = R_1$. If we think of R and χ as functions of r, then we have $R = R_0$ for $r = 0$ and $R = R_1$ for $r = R_1$. From Eq. (A10) one finds that

$$d\chi = \frac{1 - rR'/R}{(R^2 - r^2)^{\frac{1}{2}}} dr, \tag{A12}$$

where $R' = \frac{dR}{dr}$. Substituting this into (A11), one gets

$$\frac{1}{\sigma} = \frac{R^2(1+R'^2) - 2rRR'}{R^2 - r^2}. \tag{A13}$$

With σ a given function of r, this is a differential equation, the solution of which will give R as a function of r. If this is put into A(10) one then obtains a relation between r and χ, which is the required transformation. However, it is convenient to take new variables given by

$$z = R_1^2 - r^2, \qquad q = R^2. \tag{A14}$$

Then (A13) can be written

$$(R_1^2 - z)(q'^2 + 2q') + q = \frac{1}{\sigma}(q + z - R_1^2). \tag{A15}$$

Now let us consider the process of integrating this differential equation starting at $z = 0$ $(r = R_1)$, where we take $q = R_1^2$. Let us assume that for r nearly equal to R_1,

$$\sigma = \gamma(R_1 - r) \qquad (\gamma = \text{const.}). \tag{A16}$$

Then we can write

$$\sigma = \frac{\gamma z}{2R_1} \qquad (|z| \ll R_1^2), \tag{A17}$$

and from (A15) one finds that

$$q' = -1 + \frac{2}{\gamma R_1} \qquad (z = 0). \tag{A18}$$

If one writes for σ a power series in powers of $R_1 - r$, then one can solve the equation for q as a power series in powers of z. In principle, then, there is no difficulty in carrying out the integration. If, by integrating, one determines q for $z = R_1^2$ $(r = 0)$, then this gives R_0^2.

We see then that, in principle, one can determine $R(r)$ and, from A(10), r as a function of χ, so as to obtain the line element (A6) describing a closed three-dimensional space.

The closure of the space is associated with the vanishing of σ. One can see what happens by investigating the line element in the vicinity of $\chi = \pi/2$. Let us write

$$\chi = \psi + \frac{\pi}{2}, \tag{A19}$$

where ψ is small, and let us expand R as a Taylor series in ψ. Because of (A7) only even power will appear:

$$R = R_1 + \frac{1}{2} R_1'' \psi^2 + \cdots, \tag{A20}$$

where $R_1'' = \left(\frac{d^2R}{d\chi^2}\right)_{\chi = \frac{\pi}{2}}$.

Eq. (A10) takes the form

$$r = R_1 + \frac{1}{2}(R_1'' - R_1)\psi^2 + \cdots \tag{A21}$$

If one assumes again that σ is given by Eq. (A16), then one finds that in (A9)

$$\frac{1}{\sigma} dr^2 = \frac{2}{\gamma}(R_1 - R_1'') d\psi^2. \tag{A22}$$

On the other hand in (A6) one gets

$$\left[\left(\frac{dR}{d\chi}\right)^2 + R^2\right] d\chi^2 = R_1^2 d\psi^2. \tag{A23}$$

Comparing these expressions one obtains

$$R_1'' = R_1 - \frac{1}{2}\gamma R_1^2. \tag{A24}$$

This is the condition that has to be satisfied by $R(\chi)$ at $\chi = \frac{\pi}{2}$ in order for this value of χ to correspond to $r = R_1$ where σ vanishes.

SINGULARITIES

Robert Geroch

Department of Mathematics, Birkbeck College

London, England

The goals of this paper are (1) to outline in general terms the present status of work on singularities, and (2) to discuss some of the outstanding problems in the subject and to indicate possible lines of attack on these problems. Our point of view will be an optimistic one: we shall concentrate more on what one would like to have than on what one is likely to get in the near future.

We shall be particularly concerned with the long-range goal of obtaining a more unified approach to singularities. There are currently large numbers of assorted definitions, theorems, and ideas about singularities which must eventually be brought together into some comprehensible pattern. As a first step in such a program one would like to find a set of "standard" definitions. However, it appears to be an almost chronic problem with singularities that, whenever some intuitive idea is to be made precise, five or ten possibilities spring to mind. While this phenomenon is presumably a general feature of any new subject, perhaps we are beginning to reach the point where we may hope to be able to summarize the situation with a few standard definitions and reasonably simple theorems. Then will singularities take their place in elementary textbooks.

The three sections of this paper deal, respectively, with the definition, existence, and properties of singularities.

In Section 1 we discuss the question of formulating a suitable definition of a singular spacetime. We introduce a general framework in which various definitions can be constructed. The basic problem is that either one can ask for a definition with which to prove theorems (and invariably wind up with geodesic incompleteness), or one can ask for a definition which is intuitively more satisfying (in which case it is difficult to settle on any particular one). The hope is that we shall eventually find some definition satisfying both criteria.

In Section 2 we pose the question: "How are the singular solutions of Einstein's equations distributed among all solutions?" This facet of the problem is now comparatively well understood. We will be able to give a fairly complete, though not very rigorous, answer to our question. It would be particularly interesting to make this answer more precise, that is, to reformulate the theorems of Hawking and Penrose (15, 16, 17, 19, 30, 32) so as to give properties of the space of solutions of Einstein's equations rather than properties of individual solutions. The most serious difficulty in such a program turns out to be the lack of a suitable topology on the set of solutions.

In Section 3 we review attempts to classify and describe singularities. The idea is to define a collection of "singular points" which can be attached to the spacetime manifold. Properties of singularities are then to be described locally in terms of properties of these additional singular points. Of the three approaches considered here, the first gives, in principle, detailed information about the singular points but is rather too complicated to be practical, the second is fairly simple but is insensitive to finer details of the structure of the singularities, and the third is still some way from being properly formulated.

A number of more concrete - and more technical - ideas are introduced in the appendices. The first three appendices describe tools for working with singular spacetimes: definitions of "singular" based on the behavior of certain timelike curves, extensions of spacetimes, and topologies on collections of spacetimes. The fourth appendix contains a simplified (and less rigorous) proof of a theorem of Hawking (15, 17) on the existence of singularities. This proof is included here in order to give the reader a feeling for how the proofs go and a general idea of why various conditions are required for the theorems. Finally, the fifth appendix contains a list of problems, ranging from general formulations of what one would like to do to precise statements which require only proof or counterexample. Each appendix, with the exception of the last, may be read independently of the rest of the paper.

1. DEFINITION OF A SINGULAR SPACETIME

In my view we do not yet have a fully satisfactory definition of a singular spacetime*. Ideally, one would like to have some definition which is intuitively satisfying, simple to understand, and at the same time suitable for proving theorems on the existence of singularities. It is possible, of course, that we may be asking for too much: perhaps no definition exists which satisfies all our criteria. On the other hand, the problem is an important one, for the formulation of a suitable definition would represent a major step toward both unifying ideas and clarifying many other questions relating to singularities.

The first important point concerning singularities is that they cannot be represented as points of the spacetime manifold (8, 26, 35). The reason is essentially that, given only the "nonsingular" regions of spacetime, there is no prescription for how many and in what configuration additional singular points are to be included. This situation may be contrasted with that in, say, electrodynamics, in which the background Minkowskian metric allows us to define the entire manifold before the field of interest - the electromagnetic tensor - is set down. In general relativity, on the other hand, the "background metric" is the very field whose singularities we wish to describe. Strictly speaking, one should talk of "singular spacetimes" rather than of "singularities".

There are two distinct notions to be combined into the definition of a singular spacetime. Let us consider as an example the spacetime M consisting of a small open neighborhood of one of the homogeneous spacelike sections in a Friedmann model. Our M is not "singular" in the usual intuitive sense (e.g., no scalar invariants become infinite). We would, nontheless, like to rule out M as a model of the universe on the grounds that M represents only "part of the universe". It is because M is extendible that the singularities - which are certainly present in the Friedmann models - do not show up in M. The first step, then, is to recognize the extendible spacetimes. Suppose now that we have a spacetime M' which is inextendible (e.g., the full Friedmann model). We then require a second criterion to recognize that M' is "singular". The important point is that we do not expect at this stage to be able to recognize singularities in an extendible spacetime because singularities always appear "at the edge" of the spacetime manifold, and it is precisely this "edge" which may be missing from an extendible space. That is, we need only formulate a definition of "singular" which is

---

* By a <u>spacetime</u> we shall understand a (connected, Hausdorff) 4-dimensional manifold with a (C^∞) metric g_{ab} of signature $(+,-,-,-)$.

applicable to inextendible spacetimes. Finally, we may try to generalize our definition to extendible spaces. Thus, we envisage three steps in the definition: (1) define "extendible", (2) define "singular" for inextendible spacetimes, and (3) define "singular" for all spacetimes. (Only the second of these steps is difficult.)

It is easy to say what extendible means: a spacetime M is extendible if M is isometric to a (proper) subset of some other spacetime M', i.e., if M can be "enlarged" as a spacetime. We may regard inextendibility as a reasonable physical condition to be imposed on models of the universe. (Why, after all, would Nature stop building our universe at M when She could just as well have carried on to build M'?) Furthermore, an extendible spacetime provides a rather unpleasant environment for certain observers who, although they follow geodesics, nontheless experience only a finite proper time. A number of questions involving extensions are discussed in Appendix B.

The next step is to formulate a definition of "singular" which is applicable to inextendible spacetimes. One basic notion which can be used in finding such a definition is the fate of certain, sufficiently well-behaved, observers. We can imagine at least two dangers which might confront our observer: he may be on a world-line with only a finite total length, or he may be torn apart by unseasonably high tidal forces. The first danger leads to definitions in which an inextendible spacetime is singular if there exists a timelike world-line which is reasonable in some sense (e.g., has bounded acceleration) but which has finite total length (c.f., (8, 23), Appendix A). The second danger leads to definitions in terms of the behavior of the Riemann tensor or of certain scalar invariants along timelike curves. It is easy to formulate in this way numerous precise criteria for what a singular inextendible spacetime is: the problems are to decide just how serious the danger must become before we call the spacetime singular and to choose the relative weights to be assigned to the two dangers. This step - the definition of a singular inextendible spacetime - is the heart of the problem. Unfortunately, it is difficult to see in any case how a definition which differs essentially from geodesic incompleteness will lead to theorems on the existence of singularities.

Let us now assume that we have succeeded in formulating a definition of "singular" which gives reasonable answers for inextendible spacetimes. There is then a natural way to generalize this definition to all spacetimes. A spacetime M will be called singular if every extension M' of M which is itself inextendible is singular (as an inextendible spacetime). By this definition, every open subset of a nonsingular spacetime is nonsingular. For example, an arbitrary open subset of Minkowski space, while geodesically incomplete, would not be considered singular.

As an illustration of these remarks, let us see how geodesic incompleteness fits into this framework. For inextendible spacetimes, we would identify "singular" and "timelike geodesically incomplete" (i.e., the concern is entirely with the life-span of unaccelerated observers). The final definition would then read as follows: a spacetime M is G-singular if every extension of M is timelike geodesically incomplete. Thus, only geodesic incompleteness which cannot be eliminated by an extension would count as a singularity. Consider, for example, the Friedmann model (Fig. 1). The spacetime A is defined as a neighborhood of one of the homogeneous spacelike sections. While A is geodesically incomplete, it is not G-singular, for there certainly exists an extension of A (not the Friedmann model!) which is complete. The spacetime B, on the other hand, includes a region which "reaches the singularity at R = 0". Every extension of B is incomplete, and so B is G-singular. The full Friedmann model is inextendible and incomplete and, therefore, G-singular. Similarly, the (extended) Schwarzschild solution (22), the (extended) Kerr solution (3), and Taub space (38) (i.e., the region below the Misner boundary (26, 28)) are G-singular, while Minkowski space and the plane-wave solutions (20) are not.

But what does all this have to do with the theorems on the existence of singularities - theorems which do not even mention extensions? The typical statement is: "If a spacetime satisfies certain conditions, then it is geodesically incomplete". It is essentially because "geodesically incomplete" is closely related to both "extendible" and "G-singular" that incompleteness provides an acceptable substitute for the existence of a singularity. Suppose we have a

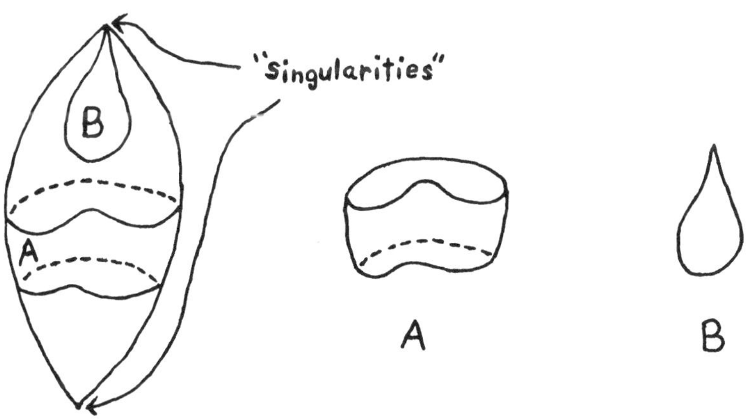

Fig. 1. The Friedmann model (two spatial dimensions suppressed: the spacelike sections (3-spheres) appear as circles in the figure). The open subsets A and B define spacetimes. While both A and B are geodesically incomplete, only B is G-singular.

spacetime M which satisfies the "certain conditions" of the theorem. Then M must be geodesically incomplete. It does not follow, however, that M is G-singular, for M may be incomplete only because it is extendible. (Every extendible spacetime is geodesically incomplete (26, 35).) That is, we have two possibilities: (1) M is extendible (in which case M is unsatisfactory as a model of our universe by the principle of inextendibility), or (2) M is inextendible (in which case M, being inextendible and geodesically incomplete, is G-singular). We could rephrase our theorem in the completely equivalent form: "If a spacetime M satisfies certain conditions and if M is inextendible, then M is G-singular". That G-singular reduces, effectively, to the much simpler notion of geodesic incompleteness is one of the fundamental reasons why it has been possible to obtain meaningful theorems about the existence of singularities.

2. THE EXISTENCE OF SINGULARITIES

Much of the work which has been done on singularities so far has been directed toward finding conditions under which a spacetime must be geodesically incomplete. Consequently, the situations under which (G-) singularities develop is one of the best understood aspects of the subject. The present collection of theorems (15, 16, 17, 19, 30, 32) together with certain physical arguments suggest a broad picture of how the singular solutions of Einstein's equations are distributed among all solutions - a picture which appears unlikely to change significantly with future results.

Let us represent the collection of all solutions of Einstein's equations diagramatically by a piece of paper. (By a "solution" we mean here a spacetime whose stress-energy tensor satisfies some reasonable inequality. See Appendix D.) Now blacken with a pencil all the points of this figure which represent "singular solutions" in some appropriate sense. The question is: What does the resulting figure look like? Is it mostly black with a few white points, a uniform shade of gray (how dark?), or covered with black and white patches? We now have a general idea of the answer to this question. We shall first discuss this answer and try to indicate why it is felt to be essentially correct. Later in this section we will be concerned with an approach to unifying our knowledge about the existence of singularities: to prove theorems not about individual solutions but about the appearance of this picture representing all solutions.

We should really have drawn two pictures - one for "closed universes" and one for "open universes" * - for the situation appears

* It is not yet clear what are the most appropriate definitions for closed and open universes. For example, there exist (continued)

to be quite different in the two cases.

Consider first the open universes. On the one hand, there are known to exist a number of nonsingular open universes: various plane-wave solutions (20), the electromagnetic solutions of Bertotti and Robinson (2, 34), the oscillating fluid ball solutions (37), the cylindrical dust-filled solution of Maitra (24), and many others. Furthermore, it is reasonable to suppose that, if the metric of certain of these solutions is perturbed slightly, the result will again be a nonsingular open universe. For example, we would certainly expect there to exist exact solutions which are "almost Minkowski space", but with a small amount of gravitational radiation coming in from infinity, scattering, and returning to infinity. Such families of solutions represent white patches in our space of open universes.

On the other hand, there are singularity theorems which are applicable to open universes (15, 16, 19, 30, 32). All of these theorems are, however, of the "threshold" variety: if some inequality happens to be satisfied, then a singularity will develop (e.g., the trapped surface condition of Penrose (30)). Such conditions represent not "reasonable physical conditions" on models of our universe, but rather signs of imminent collapse. We should certainly expect that, if a solution satisfies an inequality which signals the beginning of collapse, then any spacetime obtained by sufficiently small perturbations of that solution will also satisfy the inequality. Such families of singular solutions would be represented by black patches in our diagram.

That is, our picture of the space of open universes can be expected to contain black and white patches. It is at least possible that there would also be some gray areas in our picture, but this seems rather less likely.

The situation is quite different in the case of closed universes: the theorems are somewhat stronger. There are theorems (19) which assert that a closed universe must be singular if it satisfies conditions which might be expected to hold in the "generic" situation and if causality is not violated. (It is felt that these causality conditions will, in many cases, eventually be eliminated.) Theorems about closed universes do not characteristically have threshold

---

(continued) spacetimes which may be covered by a one-parameter family of compact spacelike sections and which may also be covered by a one-parameter family of noncompact spacelike sections. For present purposes, however, we shall take a <u>closed universe</u> to mean a spacetime which contains a compact, spacelike, 3-dimensional submanifold. (Unfortunately, there are compact spacetimes which are not even "closed universes" in this sense.)

conditions. (Although some do, e.g., (17).) Thus, we expect that the diagram for closed universes will be almost entirely black. There are, however, at least a few white points: there exist closed, geodesically complete flat spacetimes. (A class of such spacetimes can easily be constructed using the compact, flat 3-manifolds of Nowacki (29).) Perhaps there are a few other nonsingular closed universes, but these may be expected to appear either as isolated points or at least regions of lower dimensionality in an otherwise black diagram.

It was tacitly assumed in the discussion above that the cosmological constant Λ was zero. It may help in visualizing the situation if we indicate what happens when this assumption is dropped (Fig. 2). Roughly speaking, a cosmological constant of positive sign tends to create fewer singular solutions, while a negative Λ has the opposite effect. Thus, in the open case, an increase in Λ shrinks the black patches and expands the white ones. In the closed case, a positive cosmological constant tends to create white patches (where before there were only regions of lower dimensionality), while when $\Lambda < 0$, I would guess, there does not exist even one nonsingular closed universe. Thus, in the closed case, $\Lambda = 0$ is just

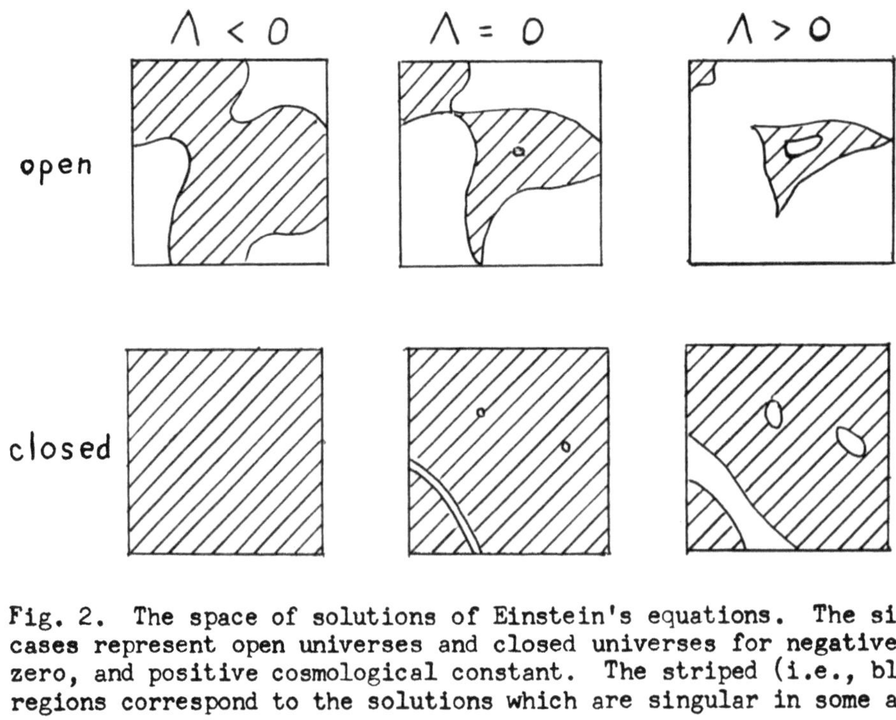

Fig. 2. The space of solutions of Einstein's equations. The six cases represent open universes and closed universes for negative, zero, and positive cosmological constant. The striped (i.e., black) regions correspond to the solutions which are singular in some appropriate sense.

on the borderline between the existence and nonexistence of nonsingular solutions*.

The remarks above suggest that we try to formulate theorems which assert not that individual solutions of Einstein's equations are singular, but rather that the picture (Fig. 2) representing all solutions has certain features. There are of course two obstacles to such a program: to state the theorems and to prove them. It appears, in fact, that saying what one wants to say is at least half the problem. We shall now indicate one possible approach.

Let M be a given (connected, Hausdorff) 4-dimensional manifold. We denote by $\mathfrak{D}(M)$ the collection of all symmetric, second rank, covariant tensor fields on M. We are interested only in elements of $\mathfrak{D}(M)$ which can be interpreted physically as the metrics for models of our universe. We therefore introduce the subset $\tilde{\mathfrak{D}}(M)$ of $\mathfrak{D}(M)$ consisting of metrics which have signature (+,-,-,-), are inextendible, and whose Ricci tensor satisfies the energy condition: $R_{ab}t^a t^b$ nonnegative for all timelike t^a (c.f., Appendix D). (Possibly, we should also include a causality condition - e.g., no closed timelike curves - for admission to $\tilde{\mathfrak{D}}(M)$.) Finally, we denote by $\tilde{\mathfrak{D}}_c(M)$ the subset of $\tilde{\mathfrak{D}}(M)$ consisting of closed universes - i.e., metrics which admit a compact spacelike 3-dimensional submanifold - and by $\tilde{\mathfrak{D}}_o(M)$ the open universes (metrics without such a submanifold)**. The sets $\tilde{\mathfrak{D}}_c(M)$ and $\tilde{\mathfrak{D}}_o(M)$ represent the pictures of all closed and all open universes discussed earlier (Fig. 2, in the case $\Lambda = 0$).

There is one further point that should be emphasized concerning our space $\tilde{\mathfrak{D}}(M)$. Let $g_{ab} \epsilon \tilde{\mathfrak{D}}(M)$ be a Lorentz metric on M, and let $g'_{ab} \neq g_{ab}$ be another metric which is obtained from g_{ab} by a diffeomorphism (smooth transformation) on M. The two metrics g_{ab} and g'_{ab} represent exactly the same physical universe, the only difference between them being a relabeling of the points of M. However, g_{ab} and g'_{ab} define distinct points of $\tilde{\mathfrak{D}}(M)$. It would be preferable, of course, if distinct points of $\tilde{\mathfrak{D}}(M)$ could represent physically distinct universes, but no suitable formalism for dealing with this situation has been developed. Fortunately, having our universes

* It is very difficult to see how these remarks concerning expanding and shrinking patches can be made precise because, when we change Λ, our space of solutions becomes an entirely different set. That is, we cannot point to a given white patch in the $\Lambda > 0$ picture and say that it is the "same" patch as another one in the $\Lambda = 0$ picture, only now somewhat larger.

** Note that whether a spacetime is closed or open depends in an important way on the metric. It is not simply a property of the underlying manifold.

over-represented in $\tilde{\mathfrak{D}}(M)$ would not detract from the intuitive content of theorems about $\mathfrak{D}(M)$.

The points of $\tilde{\mathfrak{D}}_c(M)$ and $\tilde{\mathfrak{D}}_o(M)$ consist of singular spaces and nonsingular spaces (for definiteness, G-singular). How would we say that $\tilde{\mathfrak{D}}_o(M)$ contains singular patches and nonsingular patches? One possibility would be in terms of a topology: one could interpret a "patch" to mean the existence of a set with an interior. Thus, we might ask whether the singular and nonsingular elements of $\tilde{\mathfrak{D}}_o(M)$ form sets with nonempty interior. That there are no gray regions in $\tilde{\mathfrak{D}}_o(M)$ could be interpreted to mean that every point of $\tilde{\mathfrak{D}}_o(M)$ lies either in one of the patches or on the edge of one of the patches. That is, we might try to prove that every point of $\tilde{\mathfrak{D}}_o(M)$ is either in the closure of the interior of the set of singular elements, or else in the closure of the interior of the set of nonsingular elements. In the case of closed universes, we would like to prove that the "generic" point of $\tilde{\mathfrak{D}}_c(M)$ is singular. In topology, we may interpret a property as holding "generically" when it holds on an open and dense set. That the set of singular solutions be open means that arbitrary sufficiently small variations in a singular solution again produces a singular solution, while that this set be dense means that, no matter how small the permitted perturbations, one can always get from any nonsingular solution to a singular solution.

Thus, we would at least be in a position to formulate a number of conjectures if we could find a suitable topology on $\tilde{\mathfrak{D}}(M)$. In Appendix C we discuss the properties of two such topologies, a coarse topology C^p and a fine topology F^p. In which topology should we assert, for example, that the singular spacetimes are open and dense in $\tilde{\mathfrak{D}}_c(M)$? The problem is that neither topology is suitable. The C^p topology is so coarse (so few open sets) that it is too easy for a set to be dense, while the F^p topology is so fine that it is too easy for a set to be open. One answer might be to prove that the singular spaces are open in C^p and dense in F^p*. There is another way to interpret this suggestion. Let us imagine that there is some "correct" topology in which to establish that the singular closed universes are open and dense. We do not know precisely what this topology is, but we certainly expect that it will be finer than C^p and coarser than F^p. To prove that a set is open in the C^p topology and dense in the F^p topology is completely equivalent to proving that it is open and dense in <u>every</u> topology finer than C^p

---

* This statement is probably false in the precise form given here. It may be more reasonable to prove that some subset of the set of singular solutions is open (in C^p) and dense (in F^p). For example, the subset of the plane consisting of points (x,y) such that either $x \neq 0$ or $x = y = 0$ is not open and dense, but contains an open and dense subset.

and coarser than F^p. Thus, one would establish that the singular solutions define a set which is open and dense in the "correct" topology, even though we do not know precisely what this topology is!

Alternatively, one might try to find a more suitable topology on $\tilde{\mathfrak{S}}(M)$. A general method for constructing topologies is given in Appendix C. It is important, I feel, that one settles on one (or possibly two) topologies in which to work rather than discovering a new topology for each new theorem.

3. PROPERTIES OF SINGULARITIES

Having now established that large classes of solutions of Einstein's equations are singular, one would next like to obtain a better understanding of the physics of the singularities themselves. It would be most interesting, for example, to have a classification of singularities into three or four main types. One could then contemplate more refined versions of the theorems: certain conditions imply the existence of singularities of a certain type. The current outstanding difficulty in classifying singularities according to their properties is that we have no fully satisfactory way of taking a singular spacetime (G-singular, say) and producing a collection of "singular points" whose local properties can be studied in detail. In this section we shall briefly outline three different approaches to this problem of finding and classifying singular points.

In the first approach (9, 10), one begins by associating with each incomplete timelike geodesic an ideal "endpoint". Certain of these endpoints are then identified to give a collection of singular points. Roughly speaking, the endpoints associated with geodesics Γ_1 and Γ_2 are identified provided that Γ_1 enters and remains in the region consisting of all those points which are reached by geodesics resulting from "small variations" in Γ_2 (Fig. 3). These identifications are then made into an equivalence relation. The resulting set of equivalence classes of endpoints is called the g-boundary. The g-boundary, it turns out, has a very rich structure: one can discuss the topological, causal, and in certain cases even the metric structure of this set of singular points. In fact, the richness of the structure is one of the main problems with this approach. The number of conceivable definitions is so large that it is difficult to decide which considerations will be most fruitful. Furthermore, the entire construction is rather complicated: it is difficult to see how one can prove nice theorems with such cumbersome machinery. On the other hand, an advantage of the g-boundary approach is that it is tied directly to the definition of a singular spacetime in terms of incomplete geodesics. Thus, there is at least a possibility that certain theorems about the existence of singularities might be reinterpreted as theorems about the structure of the g-boundary.

In the second approach (13), one begins by associating with each inextendible timelike curve (not necessarily a geodesic) an ideal "endpoint". The endpoints associated with curves γ_1 and γ_2 are identified provided that the past* of γ_1 is identical to the past of γ_2 (Fig. 3). The corresponding equivalence classes are called the <u>ideal points</u>. These ideal points must then be divided into two classes, those which represent "true singular points" and those which represent "points at infinity". (In Minkowski space, for example, the ideal points define Penrose's null surface at future infinity (31).) The two types of ideal points enter the discussion on an equal footing because the entire construction of the ideal points is conformally invariant. The distinction between points at infinity and singular points, on the other hand, will not be conformally invariant. No precise formulation of this distinction has been found. Aside from this last problem, the above char-

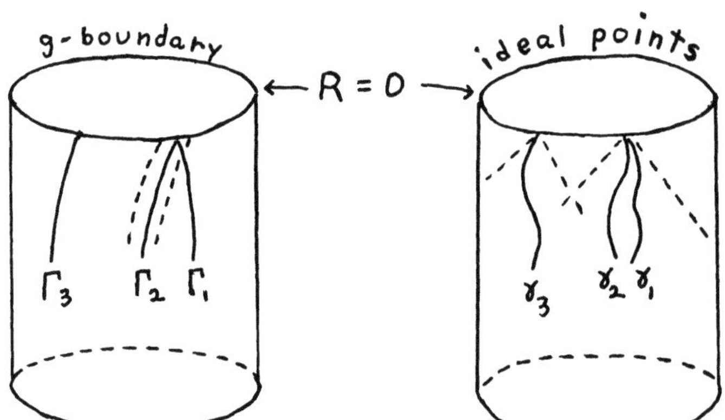

Fig. 3. Two analyses of the singularity in the Friedmann universe (two spatial dimensions suppressed). In the first figure we use the g-boundary approach. The "endpoint" of Γ_2 is identified with that of Γ_1 but not with that of Γ_3. The region between the dashed lines represents the set of points reached by geodesics resulting from small variations of Γ_2. The g-boundary is a 3-sphere. In the second figure we use the ideal points construction. The endpoint associated with curve γ_1 is identified with that of γ_2 but not with that of γ_3. The dashed lines indicate the pasts of the curves. The set of ideal points is a 3-sphere.

* The <u>past</u> of a set is defined as the collection of all points which can be reached from that set by a past-directed timelike curve. (The spacetime is assumed to be time-oriented and to satisfy a causality condition).

acterization of singular points is attractively simple. Unfortunately, the set of ideal points has a rather weak structure. (Not even a reasonable topology has been found on this set.) Furthermore, the ideal points do not, in certain cases, distinguish what one might consider to be different singular points. Misner (27) has discussed some spacetimes which contain but a single ideal point! (Compare with the Friedmann model, Fig. 3.)

As an example of these two approaches, we consider the closed Friedmann model (Fig. 3). In each case the set of singular points consists of a 3-dimensional sphere (not just a point!) attached to the spacetime at R = 0. ("R" represents the radius of the spherical spacelike sections.) In the case of the g-boundary construction, one defines (with some effort) a topology, a differentiable structure, and a metric on the set of singular points. With the ideal points construction, the set of singular points can be found by inspection, but, although one certainly knows what topology he would like to have on this set, there is no prescription which is reasonable in the general case and which gives the desired topology in Fig. 3.

A third approach to the classification of singularities might be based on embedding theorems. It has recently been shown by Clarke (5) that any spacetime may be isometrically embedded in a flat space of signature (+,+,+,-, . . . ,-) (with 87 minuses!). The idea is to associate with the spacetime M a space with "singular points attached" by taking the closure of M in its embedding space. Of course, there is no hope that the embedding will be unique with such a high dimension for the embedding space. However, it might be possible to prove that the closure of M is essentially unique, or that certain properties of the closure are independent of the choice of embedding.

All three approaches outlined above have the common feature that they attempt to associate singular points with an arbitrary spacetime. It seems quite likely that the singular points of a spacetime satisfying an energy condition (c.f., Appendix D) might be far simpler to classify than those in the general case. It would be interesting to look for a method of defining singular points which uses, in some essential way, the fact that we need only consider spacetimes which satisfy an energy condition.

Finally, it should not be too difficult to refine the above characterizations of singular points along the following lines. (For simplicity, we discuss only the g-boundary.) A spacetime consisting of a small open set in Minkowski space is G-nonsingular, as we have seen in Section 1, yet it certainly has a nontrivial g-boundary. It would be more pleasant, however, if we could redefine the g-boundary so that it would be empty whenever the spacetime is G-nonsingular. We might therefore proceed as follows. Let M be any spacetime, and let M' be an inextendible extension of M (such exist, c.f., Appendix B). Points of the g-boundary of M' which are not "near" M should not count as singular points of M. Let us consider, therefore, the set $\bar{M}$ consisting of the union of M with the set of points of the g-boundary of M' which are also in the closure of M. This $\bar{M}$ will not in general be independent of the extension M' of M. (It is an instructive exercise to find an example. Choose for M any open proper subset of Minkowski space.) However, it might be possible to find some sort of "minimal" $\bar{M}$ by taking the "intersection" (in a suitable sense) of all the $\bar{M}$'s obtained in this way. In Fig. 1, for example, the "refined g-boundary" of the spacetime A should be empty, while that of B should consist of one or more points all located at the tip of the teardrop.

ACKNOWLEDGEMENT

I wish to thank M. Walker and R. Penrose for reading the manuscript and suggesting numerous improvements. This paper was written while the author was under a National Science Foundation Postdoctoral Fellowship.

APPENDIX A. T-COMPLETENESS

T-completeness was introduced by Ehresmann (7, see also 1), and has since been used, especially by Avez (1), as a hypothesis in the proof of various global results about spacetimes. T-completeness is concerned entirely with the fate of certain well-behaved observers. We shall present in this appendix a slightly different formulation of T-completeness with a view toward applications of these considerations to a more satisfactory definition of a singular spacetime.

Let M be a spacetime, and let $\gamma(s)$ be a timelike curve parameterized by its proper length s from a past endpoint p. Let ξ^a denote the unit tangent vector to γ. Choose three unit vectors

η^a_α (α = 1, 2, 3) on γ which are orthogonal to each other and to ξ^a, and which are Fermi propagated along γ:

$$\xi^b \nabla_b \eta^a_\alpha = - \xi^a (A_b \eta^b_\alpha),$$

where $A_b = \xi^a \nabla_a \xi_b$ is the acceleration of γ. The components of the acceleration with respect to the η^a_α,

$$A_\alpha(s) = A_b \eta^b_\alpha$$

will be called the <u>acceleration functions</u> of γ. (Note that, since $A_b \xi^b = 0$, the $A_\alpha(s)$ and η^a_α determine A_b uniquely.) These functions represent the information required by the pilot if he wishes to steer his rocket ship so as to traverse the world-line γ.

It is clear that an orthonormal tetrad (ξ^a, η^a_α) at a point of M along with a set $A_\alpha(s)$ of acceleration functions defines a unique timelike curve in M. This curve need not, however, be defined for the full range of s-values for which the $A_\alpha(s)$ were originally given: the curve may, at some time, "fall off the edge" of M.

A spacetime M is said to be <u>T-complete</u>* if, for every set of acceleration functions $A_\alpha(s)$ which represents a timelike curve in Minkowski space, and for every orthonormal tetrad at a point of M, the corresponding curve in M assumes s-values in the same range as its counterpart in Minkowski space. (Intuitively, T-completeness means that a pilot in the spacetime M who steers his rocket as though he were in Minkowski space, and so that he would not "fall off the edge" of Minkowski space**, will not fall off the edge of the spacetime M either.) The above definition has two unpleasant

---

* In fact, in the usual definition of t-completeness one admits also null curves. We have not done this here because it would destroy our simple characterization of curves in terms of acceleration functions and because it is perhaps more reasonable from the physical point of view to concentrate on world-lines of observers rather than of photons.

** To construct a timelike curve which "falls off the edge" of Minkowski space, draw a curve which goes off to null infinity (31) in such a way that its tangent vector becomes lightlike so quickly near infinity that the curve has finite total length.

features: (1) it appears to be a rather complicated task - involving all acceleration functions which arise from curves in Minkowski space - to check in practice whether or not a given spacetime is T-complete, and (2) the definition appears to depend critically on Minkowski space as the model of a complete spacetime. Why not choose some other spacetime, such as De-Sitter space, for the model? In fact, both of these objections will soon be overcome by a reformulation of the definition.

Since we are concerned with what happens to observers after each finite time interval, it is natural to concentrate on acceleration functions defined only for a finite range of s-values. Given a set $A_\alpha(s)$ of (C^∞) acceleration functions, defined for $s \in [0, s_0)$, we say that the spacetime M <u>accommodates</u> the $A_\alpha(s)$ if, for every orthonormal tetrad at a point of M, the corresponding curve in M can be extended beyond $s = s_0$. (That is, we may safely give to our pilot in the spacetime M any set of acceleration functions which M accommodates.) The crucial factor determining whether or not M will accommodate a given set of $A_\alpha(s)$ is the behavior of these functions near $s = s_0$. In fact, we could easily formulate a great variety of definitions of a singular spacetime by asking whether or not the spacetime accommodates all acceleration functions having certain properties near $s = s_0$. For example, a spacetime is bounded acceleration complete (23) if it accommodates acceleration functions which are bounded in a neighborhood of s_0. Similarly, a spacetime is (timelike) geodesically complete if it accommodates zero acceleration functions.

We assert that a spacetime is T-complete if and only if it accommodates every set of acceleration functions which assume a limiting value at $s = s_0$, and which are C^∞ in a neighborhood of s_0. (The proof is easy.) Note that this formulation does not even mention Minkowski space. We conclude that exactly the same notion of T-completeness would result if "Minkowski space" in the original definition were replaced by any other spacetime (even an incomplete spacetime)!

The above remarks suggest the following definition: a spacetime is <u>T_n-complete</u> ($n = 0, 1, \ldots, \infty$) if it accommodates every set of acceleration functions which assume a limiting value at $s = s_0$ and which are C^n in a neighborhood of s_0. Thus, T_∞-complete-

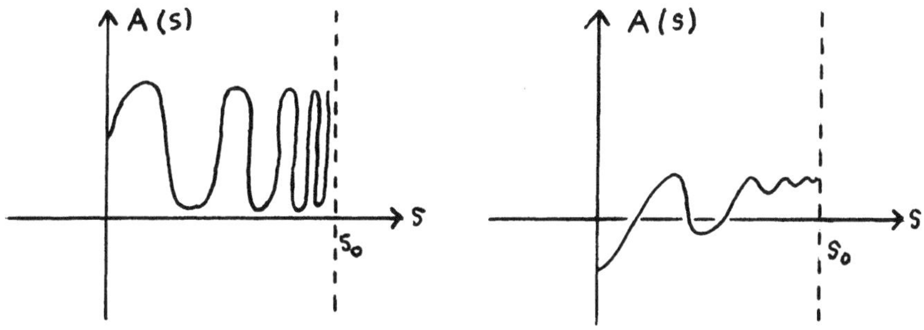

Fig. A1. Typical acceleration functions for testing whether or not a spacetime is BA-complete (first graph) and T_0-complete (second graph).

---

ness is equivalent to T-completeness. The definitions are related by the following string of implications

$$BA \Rightarrow T_0 \Rightarrow T_1 \Rightarrow \ldots \Rightarrow T_\infty \Rightarrow G, \qquad (A.1)$$

where "BA" stands for bounded acceleration complete and "G" stands for (timelike) geodesically complete. (Misner completeness (26, 35), on the other hand, would appear to the right of geodesic completeness on the list.)

To illustrate the physical meaning of T_n-completeness, let us compare BA and T_0. There is an example (8) of a spacetime which is geodesically complete, but not BA-complete. The acceleration functions of the incomplete curve in that example is shown in Fig. A1. We see that BA-completeness is, physically, a rather strong requirement on a spacetime, for no rocket pilot could possibly adjust his throttle according to the first graph in Fig. A1. T_0- completeness, on the other hand, deals with the fate of pilots whose acceleration functions are as in the second graph in Fig. A1. This clearly represents a more reasonable task for our rocket pilot. T_0 is, perhaps, physically more interesting than BA.

Finally, we may ask which of the implications in (A.1) can be reversed. The only known result is that G does not imply BA. It is probably true that not a single implication is reversible.

APPENDIX B. EXTENSIONS

In this appendix we shall present a formalism for dealing with extensions of spacetimes. The formalism leads not only to very concise statements of the two fundamental constructions involving extensions, but also to a number of important and unsolved problems.

Let M and M' be two (connected) spacetimes, and let us write $M \leq M'$ to mean that M' is an extension of M (i.e., that there exists an isometry of M onto a subset of M'.) The relation defined in this way is not a very useful one, however, for it does not even define a partial ordering (i.e., a relation such that $M \leq M'$, $M' \leq M \Leftrightarrow M = M'$ and $M \leq M' \leq M'' \Rightarrow M \leq M''$) on the collection of all spacetimes*. The point is that the collection of all spacetimes is not the most convenient set from the point of view of extensions. Define a <u>framed spacetime</u> as a connected spacetime M along with an orthonormal tetrad at a single point of M. Let $\mathcal{F}$ denote the collection of all framed spacetimes**. We write $M \leq M'$, where now $M, M' \in \mathcal{F}$ if there exists an isometry of M onto a subset of M' which takes the preferred frame in M into the preferred frame in M'. The reason for introducing frames is that now, if there exists one such isometry, it is unique (12): $\leq$ thus defines a partial ordering on $\mathcal{F}$. Although $\mathcal{F}$ could certainly be endowed with more structure than merely this partial ordering, it turns out that most of the interesting questions involving extensions can be formulated as properties of $\leq$.

A (nonempty) subset $\mathcal{P} \subset \mathcal{F}$ will be called <u>directed</u> (36, pp3) if, whenever $M, M' \in \mathcal{P}$, there exists an $M'' \in \mathcal{P}$ such that $M \leq M''$ and $M' \leq M''$. For example, given any $M \in \mathcal{F}$, the set

$$I^-(M) = \{M' \mid M' \leq M\}$$

is directed. (This $I^-(M)$ can be interpreted as the collection of all open connected subsets of M which contain the preferred tetrad.)

Two very useful results concerning extensions may now be expressed in the following form:

---

* For example, let M denote the subset $\{(t,x,y,z) \mid t < 0\}$ and M' the subset $\{(t,x,y,z) \mid t < 0,\ t < x^2 + y^2 + z^2 - 1\}$ of Minkowski space. Then $M \leq M'$ and $M' \leq M$, but M and M' are not equal (isometric).

** It is in the set $\mathcal{F}$ that limits are defined (12). In fact, these limits induce a topology on $\mathcal{F}$ very similar to the C^∞ topology (Appendix C).

<u>Theorem B1</u>*. Let $\mathcal{P}$ be directed. Then there exists a unique framed spacetime M such that (1) $\mathcal{P} \subset I^{-}(M)$, and (2) M is minimum with respect to property (1), i.e., if $\mathcal{P} \subset I^{-}(N)$, then $M \leq N$.

<u>Theorem B2</u>*. Let M, $M' \in \mathfrak{F}$, and suppose that some element of preceeds both M and M'. Then there exists a unique framed spacetime M'' such that (1) M'' preceeds both M and M', and (2) M'' is maximum with respect to property (1), i.e., if $N \leq M$ and $N \leq M'$, then $N \leq M''$.

The first theorem describes the operation of "patching together" a collection of spacetimes to form a sort of "union", but in which certain isometric regions of the spacetimes are identified. The second theorem asserts that, if two spacetimes agree in some neighborhoods of their preferred frames, then there exists unique maximal neighborhoods in which they agree. The essential ideas of the proofs of both theorems were given in (4) (although not in quite the generality stated here). In particular, it follows immediately from Theorem B1 and Zorn's Lemma (21) that every framed spacetime is contained in a (not in general unique) inextendible spacetime.

With any property P of spacetimes, for example, "is geodesically complete" or "has two Killing vectors", we associate the subset $\mathfrak{F}(P)$ of $\mathfrak{F}$ consisting of all framed spacetimes which have the property P. In fact, "a subset of $\mathfrak{F}$" means exactly the same thing as "a property of spacetimes". It is natural to ask which features of the property P show up in the partial ordering on $\mathfrak{F}(P)$. There are three particularly interesting conditions on $\mathfrak{F}(P)$:

1. If $M' \in \mathfrak{F}(P)$ and $M \leq M'$, then $M \in \mathfrak{F}(P)$.
2. Given any directed subset of $\mathfrak{F}(P)$, the unique element M defined by Theorem B1 is also contained in $\mathfrak{F}(P)$.
3. If M is maximal in $\mathfrak{F}(P)$ (i.e., if $M \leq M' \in \mathfrak{F}(P)$ implies $M = M'$), then M is maximal in $\mathfrak{F}$ (i.e., $M \leq N \in \mathfrak{F}$ implies $M = N$).

Condition (1) states that every connected open subset of a spacetime with property P also has property P, (2) states that the operation of "patching together", applied to spacetimes with property P, again yields a spacetime with property P, and (3) states that a spacetime with property P which cannot be extended to a larger spacetime with property P cannot be extended to <u>any</u> larger spacetime.

It is normally quite easy to decide whether or not a given

---

* A construction analagous to that of Theorem B1 is well-known in many areas of mathematics, e.g., in groups or in topological spaces. The collection of framed spacetimes forms a category. (The morphisms are isometries into which preserve the preferred frame.) Theorem B1 asserts that direct limits (36, pp 18) exist in this category. Inverse limits, on the other hand, do not exist. Instead, we have Theorem B2.

property satisfies conditions (1) and (2). For example, "is a source-free solution of Einstein's equations", "has at least n Killing vectors", and "has a Weyl tensor everywhere type . . ." all satisfy the first two conditions. (In fact, the statement that a property satisfy conditions (1) and (2) provides a good definition of what we mean by a "local" property of spacetimes.) On the other hand, condition (1) fails for "is asymptotically simple" and "has a Cauchy surface", while (2) fails for "is geodesically incomplete" (global properties).

A more difficult - and more interesting - problem is to decide which properties satisfy condition (3). For example, to determine whether "is a source-free solution of Einstein's equations" satisfies condition (3), we must decide whether or not the following is true: If a source-free solution cannot be further extended as a source-free solution, then it cannot be further extended as a spacetime. While this statement is probably true, no proof is known. Suppose that the above statement were false, e.g., for some framed spacetime M. In Section 1 we discussed the possibility of introducing inextendibility as a new condition to be imposed on spacetime models. Let us give our spacetime M to Nature and ask Her to make it into a model of the universe. By the principle of inextendibility, She must extend M. But M cannot be extended without violating Einstein's equations. In order to aviod severely restricting the class of available models in this way, one would be tempted to revise the principle of inextendibility - to require of models of the universe only that they cannot be further extended as solutions of Einstein's equations. Such unpleasant modifications can be entirely avoided if "is a solution of Einstein's equations" satisfies our condition (3). Similar remarks apply to the question of whether "has no closed timelike curves" satisfies condition (3). In fact, the status of closed timelike curves with respect to condition (3) would have an important bearing on whether or not the program for refining the notion of a "singular point" by using extensions can be carried out for the ideal points construction (c.f., the end of Section 3).

Finally, we remark on the role of our conditions in finding a suitable definition of a singular spacetime. The property "is geodesically complete" satisfies conditions (2) and (3), but not (1). That it does not satisfy condition (1) - a subset of a complete spacetime need not be complete - is, in a sense, what is wrong with completeness as a definition of a nonsingular spacetime. On the other hand, by introducing extensions directly into the definition, we obtained in Section 1 a notion of nonsingularity which satisfies conditions (1) and (3). (It does not satisfy (2). However, we cannot hope for a definition which satisfies both (1) and (2) because "nonsingular" should be a global property of spacetimes.) This reshuffling of conditions represents part of the motivation for Section 1.

APPENDIX C. TOPOLOGIES ON COLLECTIONS OF SPACETIMES

For much of the discussion of Section 2 we needed a topology on certain collections of spacetimes. It turns out that there exists a fairly general method for constructing such topologies. In this appendix we shall briefly outline this method and show how the two most common topologies are derived from it. We must emphasize at the beginning that none of the topologies discussed here has been found to be entirely suitable for general relativity. It is useful, nontheless, to have a general idea of the advantages and disadvantages of the various topologies which are available.

Let M be a fixed (connected, Hausdorff) 4-dimensional manifold, and let $\mathscr{D}(M)$ denote the collection of all C^∞, symmetric, second rank, covariant tensors on M*. It is not necessary at this stage to restrict consideration to metrics of signature (+,-,-,-) - the set of Lorentz metrics is a subset of $\mathscr{D}(M)$, and so will inherit each topology from $\mathscr{D}(M)$. In fact, it is more convenient to consider the general case, for the ideas can then be taken over immediately to other types of geometrical objects on M.

Let h_{ab} be any positive-definite metric on M (with associated covariant derivative operator ∇_a), C any (nonempty) closed subset of M, and p any nonnegative integer (possibly infinity). With this data we define a distance function on pairs of elements of $\mathscr{D}(M)$ as follows:

$$\rho(g_{ab}, g'_{ab}) = \underset{C}{\mathrm{LUB}} \sum_{n=0}^{p} 2^{-n} \frac{|g-g'|_n}{1+|g-g'|_n} \qquad \text{(C.1)}$$

where

$$|g-g'|_n = \left\{[\nabla_{a_1} \cdots \nabla_{a_n}(g_{rs}-g'_{rs})][\nabla_{b_1} \cdots \nabla_{b_n}(g_{uv}-g'_{uv})]\, h^{a_1 b_1} \cdots h^{sv}\right\}^{1/2}$$

That is, two metrics g_{ab}, $g'_{ab} \in \mathscr{D}(M)$ are "close" if their values and first p derivatives are close (with respect to h_{ab}) on the set C. (The complicated form of (C.1) was necessary for two reasons: (1) to ensure that the least upper bound exists even though some $|g - g'|_n$ may be unbounded on C, and (2) so that the case $p = \infty$ will not have to be treated separately. The actual topology one obtains from ρ is independent of the details of Eqn. (C.1).)

Fix the integer p. Then each arbitrary choice of h_{ab} and C defines, via the distance function (C.1), a topology on $\mathscr{D}(M)$. That is, a neighborhood of a metric $g_{ab} \in \mathscr{D}(M)$ consists of all $g'_{ab} \in \mathscr{D}(M)$ such that $\rho(g_{ab}, g'_{ab}) < \epsilon$, where $\epsilon > 0$. However, we are

---

* It is easy to generalize to metrics of lower differentiability class, but this generalization offers nothing new and would somewhat obscure the discussion.

interested in topologies which do not depend on such arbitrary choices. To obtain "invariant" topologies, we proceed according to the following plan. Specify in some invariant way an appropriate collection of pairs (h_{ab}, C), where each h_{ab} is positive-definite and each C closed. Each pair defines a distance function (C.1) and hence a family of open sets on $\mathcal{D}(M)$. The aggregate of all finite intersections and arbitrary unions of all open sets obtained from the pairs (h_{ab}, C) in our collection defines a topology on $\mathcal{D}(M)$. Hence, for a fixed value of p, the problem of constructing certain topologies reduces to that of selecting a suitable collection of pairs (h_{ab}, C).

The $\underline{C^p}$ <u>topology</u> on $\mathcal{D}(M)$ is defined by the collection of all pairs (h_{ab}, C) for which C is compact*. A neighborhood of a metric $g_{ab} \in \mathcal{D}(M)$ consists of all metrics whose value and first p derivatives are close to those of g_{ab} in some compact set, and whose behavior outside that compact set is unrestricted. The C^p topology is rather coarse, i.e., there are very few open sets. For example, in the C^p topology the subset of $\mathcal{D}(M)$ consisting of all metrics of Lorentz signature is not even open (for M noncompact). If $g_{ab} \in \mathcal{D}(M)$, the one-parameter family of metrics λg_{ab} defines a continuous curve in $\mathcal{D}(M)$. Thus, $\mathcal{D}(M)$ is connected. (However, the Lorentz metrics will not in general form a connected subset of $\mathcal{D}(M)$. There are normally only a few components, and these are closely related to the "homotopy classes of Lorentz metrics".) The sequence of metrics diag $(\mu_m, -1, -1, -1)$, where $\mu_m = 1 + m[1 + (x-m)^2]^{-1}$ $(m = 1, 2, \ldots)$, approach Minkowski space in the C^p topology. (The "bump" in the metrics becomes larger as it receeds to infinity.)

The $\underline{F^p}$ <u>topology</u> (18) on $\mathcal{D}(M)$ is defined by the collection of all pairs (h_{ab}, C) for which C = M**. Thus, a neighborhood of a metric $g_{ab} \in \mathcal{D}(M)$ consists of metrics whose value and first p derivatives are within a certain range of those of g_{ab} at each point, such that this "range" varies continuously but otherwise arbitrarily over M. The F^p topology is rather fine, i.e., there are very many open sets. For example, the sequence of metrics diag $(\nu_m, -1, -1, -1)$, where $\nu_m = 1 + (m^2 + x^2 + y^2 + z^2)^{-1}$ $(m = 1, 2, \ldots)$, do not approach Minkowski space in the F^p topology. (The "bump" in the metrics remains centered at the origin and decreases in amplitude.) The one-parameter family of metrics λg_{ab} is not continuous in the F^p topology (provided g_{ab} has noncompact support, e.g., if g_{ab} has Lorentz signature). Finally (again assuming M noncompact), $\mathcal{D}(M)$ has many different connected components in the F^p topology.

---

* Fix a given positive-definite metric $\bar{h}_{ab}$. Then the C^p topology is identical to that obtained from all pairs $(\bar{h}_{ab}, C)$ with C compact.

** The F^p topology is identical to that obtained from <u>all</u> pairs (h_{ab}, C).

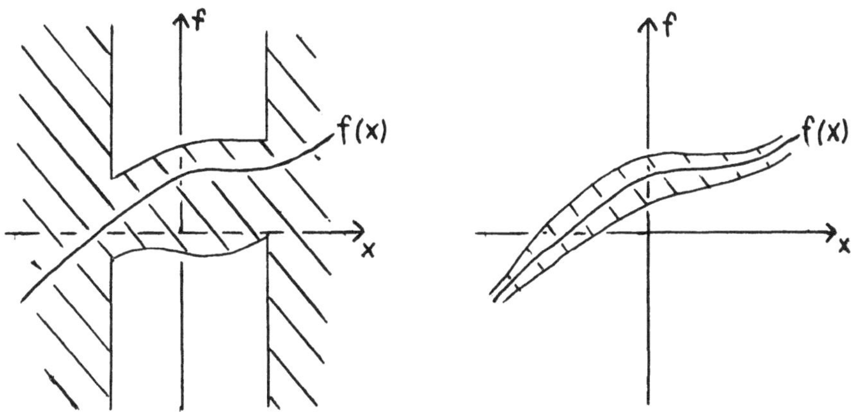

Fig. C1. Neighborhoods of a function f(x) in the C^0 and F^0 topologies, respectively. Typical neighborhoods of f(x) consist of all functions whose graphs lie in the shaded regions.

For example, the set of all g_{ab} for which M has finite volume is both open and closed.

As a specific example, we describe analogous topologies for functions on the real line. Each function is represented by its graph (Fig. C1). The graphs corresponding to functions in a neighborhood of a given function f(x) are indicated in the figure.

The C^{p+1} topology is finer than the C^p topology, the F^{p+1} finer than the F^p, and the F^p finer than C^p. That is, we have

$$\begin{array}{ccccccccc} C^0 & \longrightarrow & C^1 & \longrightarrow & C^2 & \longrightarrow & \dots & \longrightarrow & C^\infty \\ \downarrow & & \downarrow & & \downarrow & & & & \downarrow \\ F^0 & \longrightarrow & F^1 & \longrightarrow & F^2 & \longrightarrow & \dots & \longrightarrow & F^\infty \end{array}$$

where arrows denote increasing fineness. For a noncompact manfiold M, none of the above topologies are equivalent, while for a compact manifold C^p and F^p coincide. (The resulting topology is quite "reasonable" for compact M.)

We conclude this appendix with a few technical remarks about the topologies. Write $\mathcal{T}$ for the product of the set of all positive-definite metrics on M and the set of all nonempty closed subsets of M. Then, as we have seen, each subset of $\mathcal{T}$ defines a topology on $\mathcal{D}(M)$. The group of diffeomorphisms on M acts in an obvious way as a transformation group on $\mathcal{T}$. A subset of $\mathcal{T}$ which is invariant under the action of this group defines what we have called an invariant topology. (More generally, the group of diffeomorphisms acts as a transformation group on $\mathcal{D}(M)$, and an <u>invariant</u>

topology is one for which these transformations are homeomorphisms.) A topology is Hausdorff if and only if the union of the C's is dense in M. In particular, every invariant topology (except the indiscrete one) is Hausdorff.

The coarsest topology which can be constructed in this way is that in which each closed set C is finite. This topology is strictly coarser than C^p. The finest topology is F^p. It is certainly possible to construct invariant topologies intermediate between C^p and F^p. For example, we could consider the collection of all pairs (h_{ab}, C) for which the closure of the interior of each C is compact. Finally, there do exist invariant topologies on $\mathcal{D}(M)$ which cannot be constructed in this way from pairs (h_{ab}, C). For example, in Eqn. (C.1) for the C^p topology, we could add a term $V(1 + V)^{-1}$, where V is the volume of M with respect to $(g_{ab} - g'_{ab})$.

APPENDIX D. A THEOREM ON THE EXISTENCE OF SINGULARITIES

Many of the same arguments are used over and over in the various existence theorems. Therefore, by going through the proof of just one theorem, one can get a good general idea of how the proofs go and why certain conditions enter into the theorems. In this appendix we shall give a simplified discussion of a result due to Hawking (15, 17). In order not to obscure the idea of the proof, certain technical details have been replaced by qualitative arguments. For other discussions of the theorems, see, for example, (32, 39).

There are essentially two steps in the proof: (1) to show that, under certain circumstances, there exists no longest timelike curve from a point to a spacelike 3-surface, and (2) to show that this absence of longest curves leads to incomplete geodesics.

Let M be a spacetime with metric g_{ab}, and let S be any spacelike 3-surface in M. Given a point p to the future of S, we ask whether or not, among all timelike curves from p to S, there exists a curve of maximal length. (In the indefinite case, timelike geodesics tend to be the longest curves between points. Curves which "bend" more than geodesics lie nearer the light cones, and so have shorter length.) In order to have an example before us to illustrate the argument, let us consider the hyperboloid S consisting of points unit distance from the origin in Minkowski space (Fig. D1). As we shall see shortly, the past null cone of the origin is the boundary between those points which can be joined to S by a curve of maximal length and those points which cannot.

Returning now to the general case, let us assume for the moment that we have obtained some timelike curve γ from p to S whose length

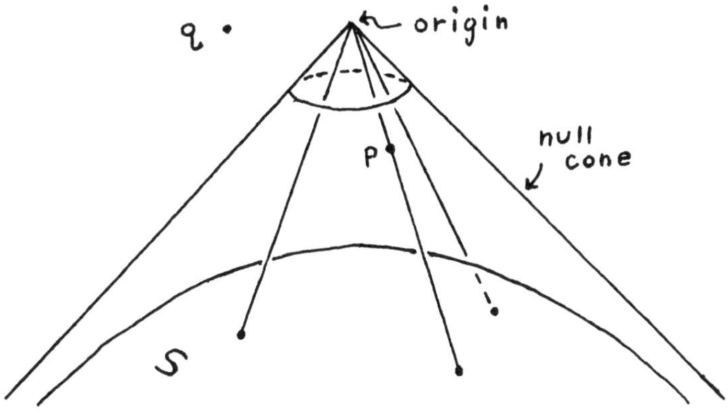

Fig. D1. A hyperboloid S in Minkowski space (one spatial dimension suppressed). Through any point p between the past null cone of the origin and S there passes a timelike curve to S of maximal length (the straight lines in the figure). Through any point q above this light cone, one can find a timelike curve to S of arbitrarily long length.

---

is maximal. Then γ must be a geodesic (for otherwise we could lenghten γ by straightening it out), and γ must intersect S orthogonally (for otherwise we could lenghten γ by moving slightly its point of intersection with S). We are thus led to study the normal geodesics from S.

Through each point of S draw the (unique) timelike geodesic beginning orthogonally to S (e.g., the straight lines in Fig. D1). The field ξ^a of unit tangent vectors to this geodesic congruence is regular in some neighborhood of S (i.e., before the geodesics begin to cross in caustics). There is an important equation (33) - which underlies all of the theorems - governing the behavior of such a congruence. We define the <u>convergence</u>, $c = -\nabla_a \xi^a$, of our congruence. Then

$$\begin{aligned} \xi^b \nabla_b c &= -\xi^b \nabla_b \nabla_a \xi^a \\ &= -\nabla_a(\xi^b \nabla_b \xi^a) + (\nabla_a \xi_b)(\nabla^b \xi^a) - R_{ba}{}^a{}_d \xi^b \xi^d \\ &= (\nabla_a \xi_b)(\nabla^b \xi^a) + R_{ab} \xi^a \xi^b \end{aligned} \quad \text{(D.1)}$$

where we have used the fact that ξ^a is geodetic. We next modify the two terms on the right in this equation. For the first term, observe that, by construction, $\nabla_a \xi_b = \nabla_{(a}\xi_{b)}$ and $\xi^a \nabla_a \xi_b = 0$: that is, $\nabla_a \xi_b$ may be considered as a symmetric tensor in the 3-space orthogonal to ξ^a. Therefore, $(\nabla_a \xi_b)(\nabla^b \xi^a) \geq \frac{1}{3} c^2$. (This is a well-known matrix inequality: for any symmetric 3x3 matrix A, trace $A^2 \geq 1/3$ (trace $A)^2$.) We next assume that the second term on the right in Eqn. (D.1) is nonnegative. Expressed in terms of the stress-energy tensor via Einstein's equations, this assumption requires $(T_{ab} - \frac{1}{2}Tg_{ab})t^a t^b \geq 0$ for all timelike t^a: the famous energy condition. (For a perfect fluid, for example, the condition becomes $\rho + p \geq 0$, $\rho + 3p \geq 0$.) Eqn. (D.1) now takes the final form

$$\xi^b \nabla_b c \geq \frac{1}{3} c^2. \qquad \text{(D.2)}$$

Eqn. (D.2) has a simple physical interpretation: when the energy condition is satisfied, there is an irreversible tendency for c to increase along the congruence, i.e., for the timelike geodesics to converge.

The idea is to use Eqn. (D.2) to cause a catastrophe for the normal geodesics from S. One way of doing this is to assume - as is the case in Fig. D1 - that the geodesics are already converging as they leave S. That is, we now assume that $c \geq c_0 > 0$ on S, where c_0 is a constant. It then follows from Eqn. (D.2) that c becomes infinite within a distance $s_0 = 3/c_0$ from S along each (future-directed) normal geodesic. An infinite c is the signal that our geodesics have begun to cross. The fact that each geodesic runs into a caustic within a distance s_0 from S does not yet imply any singular behavior on the part of the spacetime itself: caustics can form in a regular region of a spacetime (e.g., at the origin in Fig. D1).

The reason why we are interested in caustics is slightly more subtle than this. Let γ be one of the timelike geodesics normal to S, and choose a point p beyond $s = s_0$ on γ (Fig. D2). This γ has passed through a caustic before reaching p, and so some neighboring geodesic γ' (which also began normal to S) intersects γ at a point r before p. The angle between these two geodesics at r is of the order of ϵ while the difference in distance along the two geodesics from r to S is of the order of ϵ^3 *. It now follows that, by "rounding off the corner" (the dashed line in Fig. D2) of the broken geodesic prb' we obtain a timelike curve from p to S whose length is greater than that of γ.

To summarize, (1) if there is a longest timelike curve from p to S, it must be a normal geodesic, and (2) if any normal geodesic has length greater than $3/c_0$, then it does not represent the longest timelike curve from p to S. In other words, we have shown that,

---

* Expand the difference in distance in powers of ϵ. The ϵ term vanishes because γ meets S orthogonally. The ϵ^2 term vanishes because the curvature of S at b is just such that γ' meets γ at r. Hence, the first term is order ϵ^3.

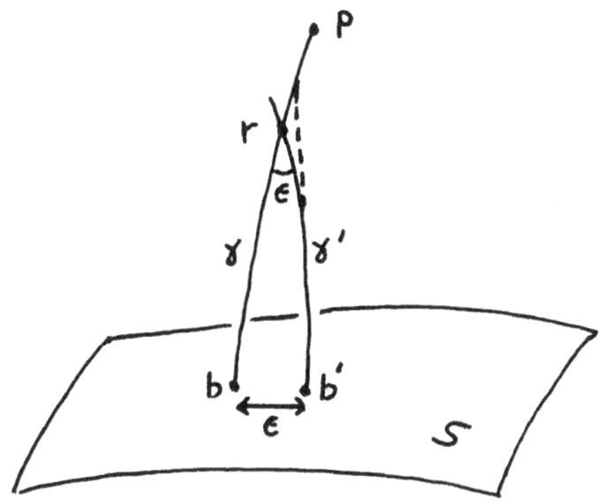

Fig. D2. Geodesic γ' intersects geodesic γ before γ reaches p. The difference in length, prb - prb', is of the order of ϵ^3. Therefore, the route from p to S via the dashed line (ending at b') is longer than the route prb along γ.

---

if we choose any point p a distance greater than $3/c_0$ from S (along any timelike curve), then there exists no timelike curve from p to S of maximal length. (Note that this is precisely the situation in Fig. D1.)

In order to make this difficulty in finding maximal curves to S into a contradiction, we now make the final assumption, that S is a Cauchy surface (11, 15, 30), i.e., that every timelike curve (without endpoints) intersects S exactly once. The hyperboloid S in Fig. D1 is not a Cauchy surface (certain timelike curves, e.g., from q, fail to reach S), while a timelike plane in Minkowski space is a Cauchy surface. To say that S is a Cauchy surface for M means, intuitively, that the entire evolution of M is completely and uniquely determined by initial data on S. Let p be any point to the future of S. Since every timelike curve from p intersects S, there must be a curve from p to S of maximal length. (This statement, which seems reasonable intuitively, is proven in (11, 15).) But the underlined statement above tells us how to find a point p from which there is no timelike curve to S of maximal length. The only way to escape a contradiction is to conclude that we cannot find such a point p, i.e., that no future-directed timelike curve from S has length greater than $3/c_0$. We have proven:

Theorem D1 (Hawking). Let M be a spacetime whose Ricci tensor satisfies the energy condition: $R_{ab}t^a t^b \geq 0$ for all timelike t^a. Suppose that M contains a Cauchy surface S whose convergence c satisfies $c \geq c_0 > 0$. Then every future-directed timelike curve from S has length no greater than $3/c_0$. (In particular, every timelike geodesic in M is incomplete.)

Theorem D1 is rather weak. Suppose in a spacetime M we manage to find a spacelike 3-surface S whose convergence is everywhere greater than some c_0 (e.g., the surface in Fig. D1). As the spacetime evolves from S we begin looking for the appearance of a singularity. However, we may well be disappointed: because of some later development in the spacetime, it may turn out that S was not a Cauchy surface after all. (For example, the hyperboloid in Fig. D1 is not a Cauchy surface for Minkowski space.) The problem is that, by examining only a neighborhood of S, it is impossible to determine whether or not S will ultimately be a Cauchy surface. Quite generally, theorems about the existence of singularities which require a Cauchy surface are considerably weaker than those which do not. We now briefly indicate how Theorem D1 can be modified to eliminate the assumed existence of a Cauchy surface.

If S is not a Cauchy surface, then there exists some region, called the domain of dependence (or Cauchy development) (6, 11, 15, 32) of S, which consists of those points from which all past-directed timelike curves do intersect S. (In Fig. D1, the domain of dependence of S is the region between S and the past null cone of the origin.) Our previous arguments are valid within this domain of dependence. The future boundary of the domain of dependence is called the Cauchy horizon (11, 17, 32). It is always a null surface (possibly with "corners": the Cauchy horizon of S in Fig. D1 is the past null cone of the origin). The idea is to see what happens on this Cauchy horizon. Let us assume that S is compact (i.e., "finite": a sphere and a torus are compact, a plane and a cylinder are not). The central question is: "Is the Cauchy horizon H of S also compact?" Either answer leads to a difficulty.

Suppose first that H is compact. But H is a null surface. A null curve drawn in the compact surface H must therefore wander about until it eventually comes back arbitrarily near to itself. This situation represents a type of causality violation - almost-closed null curves - practically as serious as the existence of closed timelike curves. (This first possibility is what happens in Taub-NUT space.)

Suppose that H is not compact. Then we may find a sequence of points p_1, p_2, . . . as in Fig. D3 such that the p_i have no point of accumulation. For each of the p_i there is a timelike geodesic γ_i from p_i to S whose length is maximal and (therefore) less than s_0. Each of these geodesics is uniquely determined by its unit tangent vector at the point of intersection with S. Thus, we have a sequence of vectors normal to S. But S was assumed compact, and so this sequence accumulates at some unit vector η^a normal to S. If the geodesic corresponding to this vector could be extended to length greater than s_0, we would be able to construct an accumulation point for the p_i (c.f., Fig. D3). Hence, this geodesic γ is incomplete.

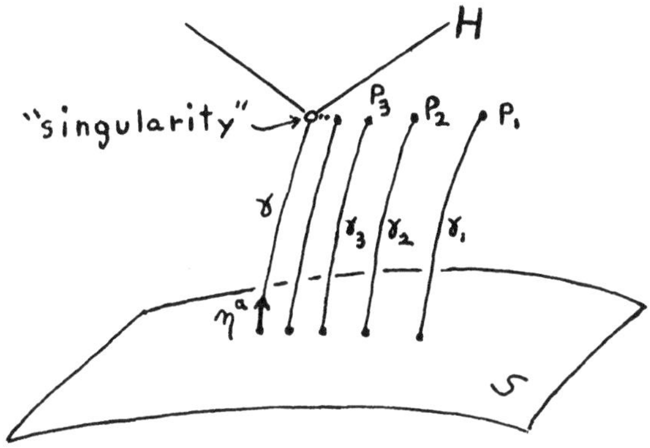

Fig. D3. If H is not compact, choose a sequence of points p_i between H and S such that the p_i do not have an accumulation point. The limit γ of the corresponding sequence of maximal geodesics from p to S defines our incomplete geodesic.

We have proven:

Theorem D2 (Hawking). Let M be a spacetime whose Ricci tensor satisfies the energy condition, and suppose that M contains a compact spacelike surface S whose convergence $c \geq c_0 > 0$. Then, assuming that M contains no almost-closed null curves, M must have an incomplete timelike geodesic.

Note that Theorem D2 differs from Theorem D1 in three essential ways: (1) we need not assume a Cauchy surface, (2) we must assume that S is compact, and (3) we obtain only a single incomplete geodesic rather than the incompleteness of all geodesics. In fact, Theorem D2 is true even if we do not exclude almost-closed null curves (17).

APPENDIX E. PROBLEMS

The problems listed below are for the most part fairly difficult. As far as I know, all are unsolved. (For easier problems, there are numerous statements, especially in the appendices, which are given without proof.) Most of the problems have been mentioned at one point or another in the text, but are repeated here in order to give a general picture of certain lines of attack on singularities. Note that the essence of many of the problems is simply to find a suitable precise definition for some intuitive notion.

Definition of a Singular Spacetime

1. Investigate definitions of a singular spacetime based on the tidal forces experienced by certain observers. Is it true that,

if tidal forces are assumed not to grow too large, then certain types of completeness (e.g., in Appendix A) become equivalent?

2. Find counterexamples to show that the implications in (A.1) are not reversible. (It is almost certainly possible to modify the example of (8) to show that G does not imply T_n for any n.)

3. Find an example of a spacetime which is null incomplete but spacelike and timelike complete. (The above represents the only case of this type for which no counterexample is known. See (8, 23).)

4. Investigate definitions of a singular spacetime based on Misner incompleteness (26, 35).

5. Find a definition of a singular spacetime which is intuitively satisfying, fairly simple, and such that the existing theorems on singularities can be reinterpreted to obtain theorems in terms of the new definition.

Existence of Singularities

6. Prove or find a counterexample: Every inextendible spacetime containing a compact spacelike 3-surface and whose stress-energy satisfies a suitable energy condition is either G-singular or flat. (The energy condition must be strong enough to exclude the Einstein universe. Perhaps one should first prove that the 3-surface cannot be, topologically, a 3-sphere.)

7. Prove that when $\Lambda < 0$ there does not exist even one nonsingular closed universe.

8. Prove, in a suitable topology, that there can never be gray regions in Fig. 2. (It might be possible to prove such a result without using any energy conditions or other theorems, just the definition of a singular spacetime.)

9. Find an acceptable topology on $\tilde{\mathcal{D}}(M)$. Prove, in that topology, some theorems reflecting our intuitive discussion of the space of closed and open universes. (In the open case, perhaps one can prove something about asymptotically flat spacetimes (31).)

10. Perhaps the attempt to interpret Fig. 2 in terms of a topology is the wrong approach. Think of some other structure on the set $\tilde{\mathcal{D}}(M)$ for which precise statements can be made. (For example, one might try to define a measure on $\tilde{\mathcal{D}}_c(M)$ and prove that the nonsingular metrics have measure zero.)

11. Characterize in some simple way all invariant topologies on $\mathcal{D}(M)$ which may be constructed using our technique of pairs (h_{ab}, C). More generally, find all invariant topologies on $\mathcal{D}(M)$. Are any of these topologies suitable for a discussion of Fig. 2?

12. The difficulty with the F^p topology is that, since h_{ab} may become very small at infinity, we tend to require that metrics in a neighborhood of g_{ab} be "too close" to g_{ab} near infinity. This remark suggests that we consider pairs (h_{ab}, C) such that $C = M$ and h_{ab} is complete. Do we thus obtain a topology different from F^p? Is the topology interesting?

13. Find a precise way to formulate the following statement: "A cosmological constant of positive sign tends to create fewer

singular solutions." Prove it.

Properties of Singularities

14. Find some way of defining singular points which combines the simplicity of the ideal point construction with the richness of the structure of the g-boundary. (A definition which directly involves an energy condition would, perhaps, be most likely to succeed.) Divide the resulting singular points into a small number of types.

15. We outlined, at the end of Section 3, a program for refining existing notions of singular points. Carry out such a program with the g-boundary and/or with the ideal points.

16. Find a definition of singular points based on Clarke's embedding theorems (5).

17. To what extent is a singular spacetime characterized by the acceleration functions it accommodates? Can one classify singularities according to the acceleration functions which are or are not accommodated?

Miscellaneous

18. In $\mathcal{D}(M)$, physically identical metrics are represented by many distinct points. Study the set of physically distinct metrics. For example, does this set have any interesting topologies?

19. The definition of a "closed universe" given in Section 2 was chosen because the existence of a compact spacelike 3-surface is involved in certain of the singularity theorems. Find a more appropriate definition which, however, does not change the qualitative features of Fig. 2. (In particular, one would like compact universes to be "closed".)

20. Determine whether or not "is a source-free solution of Einstein's equations" and "has no closed timelike curves" satisfy condition 3 of Appendix B. (Similarly for "satisfies an energy condition, "has a Killing vector", and many other properties of spacetimes.)

REFERENCES

1. Avez, A., Essais de Geometrie Riemannienne Hyperbolique Globale, Ann. Inst. Fourier, 13, 105 (1963).

2. Bertotti, B., Uniform Electromagnetic Fields in the Theory of General Relativity, Phys. Rev., 116, 1331 (1959).

3. Boyer, R.H., Lindquist, R.W., Maximal Analytic Extension of the Kerr Metric, J. Math. Phys., 8, 265 (1967).

4. Choquet-Bruhat, Y., Geroch, R., Global Aspects of the Cauchy Problem in General Relativity, Comm. Math. Phys., (submitted for publication).

5. Clarke, C.J.S., On the Global Isometric Embedding of Pseudo-Riemannian Manifolds, (preprint, Camb. Univ., 1969).

6. Courant, R., Hilbert, D., Methods of Mathematical Physics II, (Interscience, New York, 1965).

7. Ehresmann, C., Colloque de Topologie Algebrique (espaces fibres), Bruxelles, 1951.

8. Geroch, R., What is a Singularity in General Relativity?, Ann. Phys., 48, 526 (1968).

9. Geroch, R., Local Characterization of Singularities in General Relativity, J. Math. Phys., 9, 450 (1968).

10. Geroch, R., The Structure of Singularities, article in Battelle Recontres, C. DeWitt, J. Wheeler, ed., (Benjamin, New York, 1968).

11. Geroch, R., The Domain of Dependence, J. Math. Phys., (to be published).

12. Geroch, R., Limits of Spacetimes, Comm. Math. Phys., (to be published).

13. Geroch, R., Kronheimer, E.H., Penrose, R., Ideal Points in Spacetime, Proc. Roy. Soc., (submitted for publication).

14. Graves, J.C., Brill, D.R., Oscillatory Character of the Reissner-Nordström Metric for an Ideal Charged Wormhole, Phys. Rev., 120, 1507 (1960).

15. Hawking, S.W., The Occurrence of Singularities in Cosmology I, Proc. Roy. Soc., 294A, 511 (1966).

16. Hawking, S.W., The Occurrence of Singularities in Cosmology II, Proc. Roy. Soc., 295A, 490 (1966).

17. Hawking, S.W., The Occurrence of Singularities in Cosmology III, Proc. Roy. Soc., 300A, 187 (1967).

18. Hawking, S.W., The Existence of Cosmic Time Functions, Proc. Roy. Soc., 308A, 433 (1969).

19. Hawking, S.W., Penrose, R., The Singularities of Gravitational Collapse and Cosmology, Proc. Roy. Soc., (submitted for publication).

20. Jordan, P., Ehlers, J., Kundt, W., Strenge Lösungen der Feldgleichungen der Allgemeinen Relativitätstheorie, Akad. Wiss. der Lit. Mainz (1960).

21. Kelly, J.L., General Topology, (Van Nostrand, New York, 1955)

22. Kruskal, M.D., Maximal Extension of the Schwarzschild Metric, Phys. Rev., 119, 1743 (1960).

23. Kundt, W., Note on the Completeness of Spacetimes, Zeit. für Physik, 172, 488 (1963).

24. Maitra, S.C., A Stationary Dust-Filled Cosmological Solution with $\Lambda = 0$ and Without Closed Timelike Lines, (preprint, Univ. of Maryland, 1965).

25. Melvin, M.A., Pure Magnetic and Electric Geons, Phys. Lett., 8, 65 (1964).

26. Misner, C.W., The Flatter Regions of Newman Unti and Tamburino's Generalized Schwarzschild Space, J. Math. Phys., 4, 924 (1963).

27. Misner, C.W., The Mix-Master Universe, (preprint, Univ. of Maryland, 1969).

28. Newman, E.T., Tamburino, L., Unti, T., Empty Space Generalization of the Schwarzschild Metric, J. Math. Phys., 4, 915 (1963).

29. Nowacki, V.W., Die Euklidischen, dreidimensionalen geschlossenen und offenen Raumformen, Comm. Math. Helv., 7, 81 (1934).

30. Penrose, R., Gravitational Collapse and Space-Time Singularities, Phys. Rev. Lett., 14, 57 (1965).

31. Penrose, R., Zero Rest Mass Fields Including Gravitation, Asymptotic Behaviour, Proc. Roy. Soc., 284A, 159 (1965).

32. Penrose, R., Structure of Space-Time, article in Battelle Recontres, C. DeWitt, J. Wheeler, ed., (Benjamin, New York, 1968).

33. Raychaudhuri, A., Relativistic Cosmology, Phys. Rev., 98, 1123 (1955).

34. Robinson, I., A Solution of the Maxwell-Einstein Equations, Bull. Acad. Polon. Sci., 7, 351 (1959).

35. Shepley, L., SO(3,R) Homogeneous Cosmologies, (PhD thesis, Dept of Phys., Princeton, 1965).

36. Spanier, E.H., Algebraic Topology, (McGraw Hill, New York, 1966).

37. Synge, J.L., Relativity: The General Theory, (North-Holland, Amsterdam, 1960).

38. Taub, A.H., Empty Spacetimes Admitting a Three-Parameter Group of Motions, Ann. Math., 53, 472 (1951).

39. Thorne, K.S., Gravitational Collapse, a Review-Tutorial Article (1968, to be published).

ENERGY-MOMENTUM OF RADIATING SYSTEMS

J. Winicour

Aerospace Research Laboratories

In this paper I shall discuss the null hypersurface treatment of asymptotically flat systems, with particular emphasis on the concept of energy. Before I get deeply involved in that subject, let me first recall the developments which made this approach attractive. At least in the far field region, where gravitational effects are weak, one would expect the linearized theory to give satisfactory results. These can be readily obtained through the standard introduction of harmonic coordinates and the consequent reduction of Einstein's equations into flat space wave equations. However, one is also interested in the next approximation to the linearized theory, in which the gravitational field now serves as its own source, so that effects due to the non-linearity of general relativity can also be investigated. Fock,[1,2] has carried out this next approximation in detail. When outgoing radiation is present in the linear approximation, he found that, aside from the Minkowski metric limit, an asymptotic log r/r behavior dominates the harmonic metric at large distances along the outward null cones. Although Fock was able to extract from this first approximation scheme finite values for physically relevant quantities such as energy flux, the appearance of these log r/r terms obscured the validity of the convergence of the approximation expansion and the consistency of any outgoing radiation conditions of the Sommerfeld type.[3,4]

Sachs[5,6], Bondi, Metzner, and van der Burg[7], and Newman and Penrose[8] subsequently investigated the appropriate far field behavior for a radiating but asymptotically flat system from a different point of view within the context of the exact Einstein equations. Their treatments were very closely related and had in common the critical idea of describing gravitational fields in terms of the properties of the exact null hypersurfaces determined by the geometry. The limit of proceeding to null infinity along these hypersurfaces gave a geometric prescription for passing to the radiation zone. These investigations showed that if asymptotic flatness conditions were properly posed at null infinity, then no logarithmic dependence on affine parameter or luminosity distance would occur to leading orders asymptotically. In fact, the existence of null coordinate systems tailored upon the parameters picked out by a family of null hypersurfaces was established in which the metric had a well defined asymptotic series expansion to the first few orders of $1/r$.

How does this resolve with Fock's results? It turns out[9] that the asymptotic connection between the retarded time u, defined by an outgoing null hypersurface, and harmonic time t is given by

$$u \sim t - r - 2m \log (r-2m) - \frac{\log r}{r} \int_{-\infty}^{u} \Delta E(u', \Theta, \psi)\, du',$$

where m is the rest mass measured at spatial infinity and $\Delta E(u,\Theta,\psi)$ is the amount of energy which has been radiated in the (Θ,ψ) direction up to retarded time u. In carrying out the transformation from null to harmonic coordinates, the $m \log(r-2m)$ term, which is present in the Schwarzschild case, does not introduce any logarithmic behavior because m is a constant. But the $\log r/r$ term, which is present only for systems with radiative energy flux, does introduce logarithmic behavior in the harmonic metric in exactly the manner found by Fock.

This logarithmic behavior is clearly a harmonic coordinate effect. What is of basic physical significance is the existence of null coordinate systems in which such logarithmic terms do not appear. This feature formed the basis for Penrose's[10,11] geometrization of the concept of asymptotic flatness. Penrose's criterion for asymptotic flatness entails the existence of a geometry, conformal to the physical geometry, which has a certain finite and differentiable structure at its boundary, null infinity. Consider a function with asymptotic behavior

$$f \sim \log r/r.$$

Introducing $s = 1/r$ to give a finite description of the limit gives

$$f \sim -s \log s.$$

Although f vanishes at $s = 0$, its derivative is clearly infinite. By such considerations, Penrose's conformal requirements rule out the invariant significance of any asymptotic logarithmic or similar r-dependence.

Moreover, this technique provides a neat geometrical picture of asymptotically flat space-times. From the point of view of its conformal geometry, the physical manifold can be visualized as the interior region between two null cones (see Fig.1). The upper null cone J^+ represents points at future null infinity; and the lower null cone J^-, points at past null infinity. Their vertices I^+ and I^- represent future and past time infinity, and their intersection I^0 represents spatial infinity. The regularity of the conformal geometry at null infinity, coupled with Einstein's equations, leads to the peeling-off property of the Weyl tensor which Sachs[5,6] and Newman and Penrose[9] had previously discovered was the key requirement in a definition of asymptotically flat fields. The peeling-off property describes how successive terms in an asymptotic expansion of the Weyl tensor assume more specialized Petrov types as we approach null infinity. In Maxwell's theory, the analogous property is the relationship between E, H, and the propagation direction in the radiation zone. The uncanny success of the conformal technique is exemplified by the basic simplicity of the geometric assumptions which lead to this property. I say "uncanny" because so far attempts to extend this technique to study asymptotic behavior for time-like approaches to I^+ or I^- or for space-like approaches to I^0 have completely failed, except in the flat-space case.

The conformal picture provides a covariant means of defining the asymptotic symmetries of asymptotically flat systems.[12,13,14] Consider future null infinity J^+. We define the symmetry generators of J^+ by imposing Killing's equations on the boundary of the conformal manifold,

$$[\xi^{(\mu;\nu)}]_{J^+} = 0.$$

Explicit analysis of the solutions to these equations leads to the Bondi-Metzner-Sachs (BMS) group [15], which was first obtained from a more coordinate dependent point of view. The BMS group is essentially the Lorentz group plus a group of supertranslations. The supertranslation group is isomorphic to the additive group of functions on the 2-sphere. These functions describe the manner in which a sphere-like slice of J^+ is dragged along J^+ by the action of the asymptotic Killing vectors. The 4-parameter subgroup built out of spherical harmonics with $\ell \leq 1$ corresponds to the Poincare translations.

For any physical reasonable space-time, it is always possible to place additional restrictions on the BMS group at I^0 to eliminate the remaining supertranslations with $\ell \geq 2$, and thus reduce the BMS group to the Poincare group.[9,16] However, all such restrictions which have been proposed to date are coordinate dependent and not in the same geometric spirit in which the BMS group can be interpreted as an asymptotic symmetry group. These restrictions are based upon asymptotic conditions at spatial infinity obtained by first going out to J^+ and then sliding back along null infinity to I^0. On the other hand, Bergmann[17] has shown that the Poincare group cannot be extracted as a legitimate asymptotic symmetry group by means of a space-like approach to spatial infinity. Consequently, there is some confusion at present as to the exact sense in which the BMS group can be reduced to the Poincare group.

It turns out, however, that the BMS group is in any case just as good as the Poincare group for the purpose of defining energy and momentum.[12] Let me show you how this can be done. Give any global Killing vector ξ^μ, Komar[18] showed that the functional of a closed two-surface Σ,

$$K_\xi(\Sigma) = \oint_\Sigma \xi^{[\mu;\nu]}\, dS_{\mu\nu},$$

is completely independent of deformations of the surface in any simply connected vacuum region. For, say, the Kerr metric the time-translational and rotational Killing vectors lead in this way to the correct mass and angular momentum of the source.[12] For radiating systems, however, this technique can never be used to find the mass because there cannot exist a global time-translational symmetry.

For such systems, the idea is to utilize the asymptotic symmetries. Consider some spherical slice Σ^+ of J^+ (see Fig. 1). The obvious generalization which comes to mind is to calculate Komar's functional on Σ^+ by inserting one of the asymptotic Killing vectors. To do this necessitates knowledge of $\xi^{[\mu;\nu]}$ to order $1/r^2$, since the surface element is of order r^2. However, radiative metrics do not in general admit asymptotic solutions of Killing's equations to this required order in $1/r$. To circumvent this difficulty, consider the unique outgoing null hypersurface N which intersects J^+ at the retarded time determined by Σ^+. The following projection of Killing's equation into N can always be satisfied in any geometry:

$$[\xi^{(\mu;\nu)}k_\nu - \frac{1}{2}\,\xi^\nu{}_{;\nu}k^\mu]_N = 0,$$

where k^μ is the null direction in N. Furthermore, these projected Killing's equations uniquely propagate ξ^μ along N in terms of its boundary values at Σ^+. In this way any generator of the asymptotic symmetry group on J^+ determines a unique vector field on N. It is clear that when the asymptotic symmetry is also a global symmetry, this propagation equation is automatically satisfied. In fact, this property makes the propagation equation itself unique, if we restrict it to be linear and of first differential order.

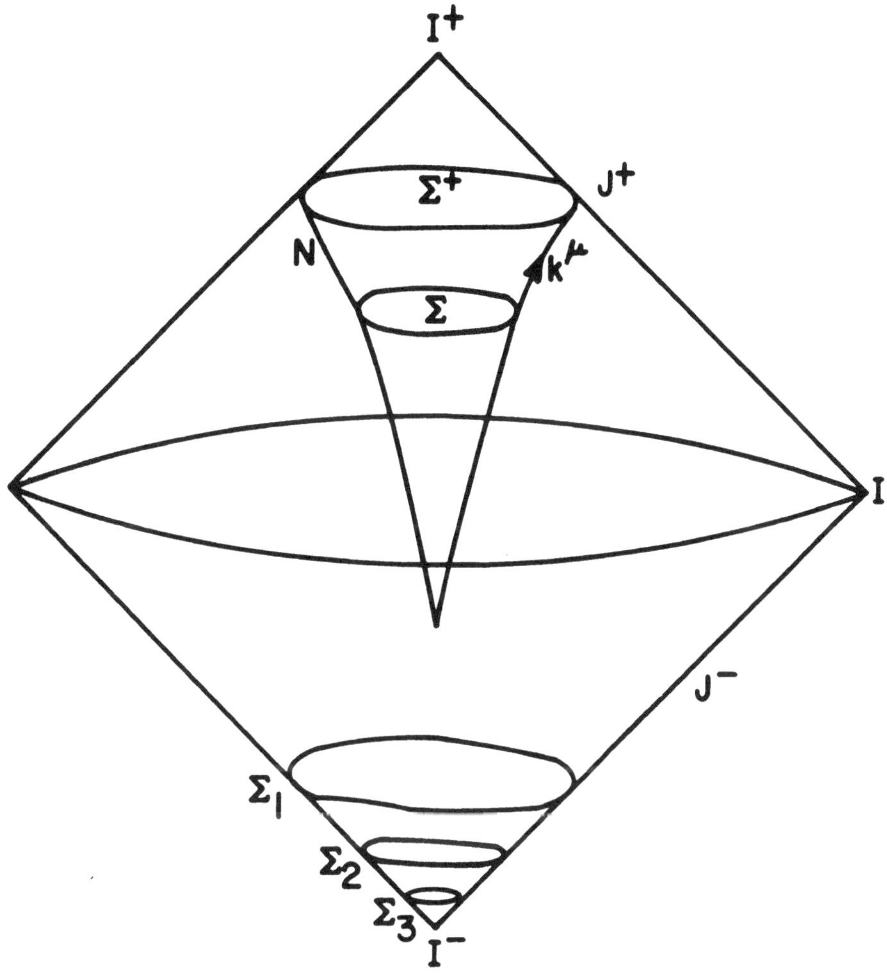

Fig. 1. The Picture of Null Infinity. Each slice Σ^+ of J^+ determines an outgoing null hypersurface N. Here N is depicted as a null cone. A sequence of 2-spheres converging to I^- along J^- is indicated.

Can we now plug the propagated Killing vectors into Komar's functional? The answer is still no, because the propagation equation determines ξ^μ only on the null hypersurface N and Komar's functional involves derivatives of ξ^μ in directions leading out of N. The idea now is to subtract from the integrand terms which vanish modulo Killing's equations so as to make the functional depend only on the knowledge of ξ^μ on N. Again by restricting the modification to be linear in ξ^μ and of first differential order, we are uniquely led to the functional

$$L_\xi(\Sigma) = \oint_\Sigma (\xi^{[\mu;\nu]} - \xi^\rho{}_{;\rho} k^{[\mu} m^{\nu]}) dS_{\mu\nu},$$

where m^ν is any vector leading out of N and normalized by $k^\mu m_\mu = -1$, so that the integrand is independent of the extensions of k^μ and m^μ.

The final result, which has been called the asymptotic symmetry linkage through Σ, works for any closed two-surface Σ whose null rays emanate to null infinity without developing caustics. By construction, the linkage functional is identical to Komar's functional in the case of a global Killing vector. An even more important feature of this construction is the linear dependence of this functional on the generators of the BMS group at null infinity. It is through this property that the asymptotic symmetry linkages provide a representation of the asymptotic symmetry group. The idea here is exactly analogous to a general covariant method of treating energy, momentum, and angular momentum in special relativity. Let $\xi_Q{}^\mu$ (Q running through the parameters of the Poincare group) label in some conventional way the Killing vectors of Minkowski space and let $T_{\mu\nu}$ be the energy-momentum tensor of some system. Then the linear functionals on a space-like surface σ.

$$L_Q(\sigma) = \int_\sigma \xi_Q{}^\mu T_\mu{}^\nu dS_\nu,$$

describe the total energy, and so forth, of the system. If observers using a slightly different, but isometric, coordinate system make the same conventional labelling of Killing vectors, the difference is given by the Lie commutator,

$$\xi_{Q'}{}^\mu = \xi_Q{}^\mu + c_{PQ}{}^R \, \xi_R{}^\mu \, \varepsilon^P,$$

where the $c_{PQ}{}^R$ are the structure constants of the Poincare group and the parameters ε^P describe the infinitesimal motion connecting the two coordinate systems. The functionals, being linear in the Killing vectors, undergo the same transformation

$$L_{Q'} = L_Q + c_{PQ}{}^R \, L_R \, \varepsilon^P.$$

This represents in the usual way the mixing of energy, momentum, and angular momentum under the action of the Poincare group. If we now let Q range over the parameters of the BMS group, as assigned by observers at null infinity, and if we put $L_Q(\Sigma) = L_{\xi_Q}(\Sigma)$, then the same transformation formulae apply to the asymptotic symmetry linkages, except that the structure constants must be taken to be those of the BMS group. However, because the Poincare translations commute with the supertranslations, the energy-momentum linkages $P_a(\Sigma)$ still transform as a Lorentz four-vector,

$$P_{a'}(\Sigma) = L_a{}^b \, P_b\,(\Sigma).$$

In this way, by passing to the limit $\Sigma \to \Sigma^+$ we can define, up to a Lorentz transformation, the total energy-momentum at the retarded time determined by Σ^+. Explicit analysis shows that the total energy obtained this way is precisely the energy functional first identified by Bondi, Metzner, and van der Burg[17] through less formal considerations. What I have tried to emphasize here is that Bondi's result can be given a coordinate-independent geometrical meaning.

Let me now return to more physical considerations. The important physical property which led Bondi to identify this energy expression was the monotonically decreasing nature of the energy as we slide Σ^+ into the future along J^+ due to the presence of gravitational radiation. In fact, Sachs[15] has shown that the energy-momentum flux carried off by the radiation is a time-like four-vector which vanishes if and only if there is no radiation.

What about the positive-definite properties of the energy-momentum itself? First, consider the energy-momentum measured along past null infinity at advanced times, such as indicated by Σ_1, Σ_2, and Σ_3 in Fig. 1. The time reversal of Sach's result implies that the energy is a monotonically increasing functional of these slices as we move up toward I^0. We can characterize a system which has no incoming non-zero rest mass fields by the condition $P_a(I^-) = 0$, in the limiting sense of a sequence of slices along J^-. In such a case, it follows that the energy-momentum measured at spatial infinity $P_a(I^0)$ must be time-like and have positive energy. The generalization to include non-gravitational incoming fields which are positive-definite is fairly obvious, but for simplicity I will restrict my remarks to the pure gravitational case.

This argument for the time-like nature of $P_a(I^0)$ is not completely convincing because it somewhat artificially circumvents the region of the physical manifold where gravitational interactions take place. The deficiency here is that we have precluded the existence of a strongly bound gravitational system with negative rest mass entering through I^-. However, the positive definiteness of the energy measured at spatial infinity can also be based upon the properties of a space-like hypersurface, provided assumptions limiting the generality of the gravitational field are made.[19-22] However, none of these treatments involve very satisfactory assumptions. Perhaps the most physically appealing argument is that due to Brill, Deser, and Faddeev.[23,24,25] They assume

(i) the existence of a simply connected space-like Cauchy hypersurface,
(ii) the existence of a coordinate system based upon this hypersurface for which the metric approaches the Minkowski metric with $g_{\mu\nu,\alpha} = 0(r^{-2})$, and
(iii) that the geometry of this Cauchy hypersurface can be reached from flat space by a continuous path in function space.

Assumption (i) limits the strength of the gravitational field, but only in a topological sense. Assumption (ii) imposes physically reasonable limitations on the strength of the radiation tail in the neighborhood of spatial infinity, although it should be pointed out that these limitations are more severe than simply demanding a finite energy flux in the radiation tail. Assumption (iii) is necessary because the argument relies upon variational techniques in function space. They first show, by examining the properties of weak fields, that the energy as a functional of the Cauchy data has a strict local minimum at flat space, for which the energy is zero. They then establish that the only critical points of the energy functional, that is values of the Cauchy data where the first variation of the energy vanishes, correspond to zero energy. From these results Brill, Deser, and Faddeev conclude that the energy measured at spatial infinity is a positive-definite functional of the Cauchy data provided that the following one-dimensional variational theorem can be extended to the energy functional. In one dimension, given a function $f(x)$ such that $f(0)$ is a strict local minimum and such that grad $f(x) = 0$ only when $f(x) = 0$, it follows that $f(x) > 0$ except at $x = 0$. In two-dimensions, however, Geroch (private communication) has found a simple construction for generating counter examples to this statement. It is thus clear that this variational conjecture cannot be easily extended to the function space of the energy functional. Nevertheless, the work of Brill and Deser[23] does indicate that for sufficiently weak fields the energy measured at spatial infinity must be positive.

After we move past I° along J^+ the energy monotonically decreases due to radiation. Will the energy continue to remain positive at later retarded times? We have investigated this question for retarded times which have the "good cone" property.[26] By this is meant retarded times determined by the null cone N of a non-singular point which emanates out to a slice Σ^+ of future null infinity without developing focal points. In this case, the constraint free initial data consists solely of the shear σ of N. The remaining initial data which occur in the truncated or sumptotic versions of the characteristic initial value problem, in Bondi's terminology[7] the mass aspect, the angular momentum-dipole moment aspects, and the news function, are all determined in terms of σ by virtue of the regularity of the vertex of N. To investigate the positive-definiteness we first note that the total energy-momentum $P_a(\Sigma^+)$ is zero when $\sigma = 0$. In fact, all the asymptotic symmetry linkages $L_Q(\Sigma)$ are independent of deformations of Σ lying in N when N is shear-free.[27] Thus $P_a(\Sigma^+)$ can be collapsed to vanishing surface integrals about the vertex of N when $\sigma = 0$. Next, it is not too difficult to show that $P_a(\Sigma^+)$ also vanishes to first order in σ, and that to second order in σ it is time-like with $E(\Sigma^+)$ strictly positive. This establishes the positive- definite and time-like nature of the energy-momentum for sufficiently weak systems for which we can neglect the $O(\sigma^3)$ terms.

We have also found it possible to demonstrate that the only value of the total energy for which the first variation with respect to σ vanishes is $E\ (\Sigma^+) = 0$. If the Brill-Deser variational conjecture were correct, this result would imply a much larger region in function space in which positivity could be established. However, I do not believe anything can be concluded on the basis of present theorems.

Summing up, the argument for positive-definitenes applies only for those values of the shear which are close enough in function space to $\sigma = 0$ to allow the existence of a "good cone" and the neglecting of $O(\sigma^3)$ terms. The need of a "good cone" is probably only a mathematical convenience, since positive energy densities can lead to the focussing effects which destroy the "good cone" property without affecting the positive energy property. At least for sufficiently shear-free fields such as represented by small perturbation from spherical symmetry, these results indicate that a system cannot radiate away more energy than it started with.

References

1. V. Fock, Rev. Mod. Phys. 29, 325 (1957).
2. V. Fock, "The Theory of Space Time and Gravitation" (Pergamon Press, Inc., New York, 1959).
3. A. Trautman, Lectures on General Relativity, King's College.

London, 1958 (unpublished).
4. A. Trautman, in Proceedings of the International Conference on Relativistic Theories of Gravitation, London, 1965 (unpublished).
5. R. K. Sachs, Proc. Roy. Soc. (London) A264, 309 (1961).
6. R. K. Sachs, Proc. Roy. Soc. (London) A270, 103 (1962).
7. H. Bondi, M.G.J. van der Burg, and A.W.K. Metzner, Proc. Roy. Soc. (London) A270, 103 (1962).
8. E. Newman and R. Penrose, J. Math. Phys. 3, 566 (1962).
9. R. A. Isaacson and J. Winicour, Phys. Rev. 168, 1451 (1968).
10. R. Penrose, Phys. Rev. Letters 10, 66 (1963).
11. R. Penrose, Proc. Roy. Soc. (London) A284, 159 (1965).
12. J. Winicour and L. Tamburino, Phys. Rev. Letters 15, 601 (1965).
13. L. Tamburino and J. Winicour, Phys. Rev. 150, 1039 (1966).
14. J. Winicour, J. Math. Phys. 9, 861 (1968).
15. R. K. Sachs, Phys. Rev. 128, 2851 (1962).
16. E. Newman and R. Penrose, J. Math. Phys. 7, 863 (1966).
17. P. G. Bergmann, Phys. Rev. 124, 274 (1961).
18. A. Komar, Phys. Rev. 113, 934 (1959).
19. D. Brill, Ann. Phys. (N.Y.) 7, 466 (1959).
20. R. Arnowitt, S. Deser, and C. W. Misner, Phys. Rev. 116, 1322 (1959).
21. A. Komar, Phys. Rev. 129, 1873 (1963).
22. A. Peres, see Ref. 21.
23. D. Brill and S. Deser, Phys. Rev. Letters 20, 75 (1968).
24. D. Brill and S. Deser, Ann. Phys. 50, 543 (1968).
25. D. Brill, S. Deser, and L. Faddeev, Phys. Letters 26A, 538 (1968).
26. D. Robinson and J. Winicour, to be published.
27. L. Derry, R. A. Isaacson, and J. Winicour, "Shear-Free Gravitational Radiation", submitted to Phys. Rev.

THE THEORY OF SUPERSPACE

Arthur Elliot Fischer

Palmer Physical Laboratory, Princeton University

Department of Mathematics, University of California
Berkeley

ABSTRACT

In this work a theory of superspace is introduced. The superspace, or space of all geometries, of a fixed closed (compact without boundary) 3-dimensional manifold M is defined as the orbit space $\mathcal{S}(M) = \frac{\text{Riem }(M)}{\text{Diff }(M)}$ of the group of diffeomorphisms, Diff (M), acting by "coordinate-transformation" on the space of Riemannian metrics, Riem (M). A geometry is then a point in $\mathcal{S}(M)$, i.e. an equivalence class of isometric Riemannian metrics. Superspace, as the space of physically distinguishable states, is the proper configuration space for a dynamical theory of relativity. It is the space in which the momentary geometries of space itself evolve. Our first result states that this space is in fact a metric space.

Since some Riemannian metrics have non-trivial symmetry groups, the associated symmetric geometries do not have neighborhoods homeomorphic to neighborhoods of geometries which have no (non-trivial) symmetries. Since these symmetric geometries are singular points in superspace, $\mathcal{S}(M)$ itself does not have a manifold structure. This Structure Problem is resolved by the Decomposition and Stratification Theorems. Superspace is shown to be partitioned into manifolds of geometries, the strata, such that the geometries of high symmetry are completely contained in the boundary of geometries of lower symmetry. A limit of geometries of the same type must therefore be either of the same symmetry type or more symmetrical. In this way a "generalized" dynamical analysis on superspace is possible. Together, these theorems begin the resolution of the fundamental question of both the classical and quantum theory of gravity; to wit, What is the structure of the domain manifold for the quantum-mechanical state functional [7, p.1115]?

The second part of this work is concerned with the implications of a symmetric Riemannian structure on the underlying topology. A complete determination of all topologies compatible with geometries which exhibit continuous groups of symmetries is made in the Superspatial Topology Theorem. This theorem also gives the possible groups that can occur as symmetry groups. The strata over these "symmetric" topologies are examined. The Stratum Theorem proves that the homogeneous geometries of a given symmetry type are parameterized by a finite-dimensional manifold. The Superspatial Stratum Theorem then applies these results to determine explicitly all finite-dimensional and contractible strata for the topologies on which such symmetric geometries can occur. Lastly, we prove that superspace always has an extension to a proper manifold.

I. INTRODUCTION

I.1. Superspace: Structure and Strata

Just as the geometry of space-time is the proper configuration space in which to carry out an analysis of the behavior of matter and energy, superspace is the proper configuration space in which the momentary spatial geometries themselves evolve. It is the space of all possible such geometries; a point of superspace is an instantaneous configuration of space itself.

Geometry does more than witness the action and reactions of matter and energy. From Einstein's field equations we know that geometry acts on matter and energy, telling them how to move. But these same equations tell us that matter and energy react on geometry; matter and energy curve space. In relativity, these actions are not separated, as they are in electrodynamics. They are different sides of the same coin. Similarly, the geometry of superspace itself acts on the momentary configuration of space, telling the geometry of space how to move. Could these geometries in turn react on the geometry of superspace? Is the basic object the geometry of space-time, or is the basic object the geometry of superspace? Is physics geometry, as Riemann and Clifford reasoned, or is physics the geometry of superspace? With these ideas in mind, we turn our attention to an investigation of the structure of superspace itself.

In this work, a theory of superspace is introduced. The Main Problem is that superspace, although a topological space, does not have a manifold structure. Some geometries in superspace are more symmetric than others. These geometries do not have neighborhoods homeomorphic to neighborhoods of geometries which are less symmetric (or not symmetric at all). Consequently, the symmetric geometries are singular points in superspace. Because of their presence superspace cannot be the configuration space of a dynamical analysis, as it is classically understood. This Structure Problem is resolved by the Stratification Theorem of Section IV. Superspace, although itself not a manifold, is shown to have a partition into manifolds of geometries, the strata, which fit together in an extremely regular way. Geometries of high symmetry are shown to be completely contained in the boundary of geometries of lesser symmetry. A limit of geometries of a particular type must therefore either be of the same type or more symmetrical. By knowing "where" in superspace the symmetric geometries occur, it is possible to pass smoothly from one manifold of geometries to another, with "care" being exercised as the more symmetric geometries are approached (i.e., at the boundary of the strata). In this way a "generalized" dynamical analysis on superspace (now stratified into manifolds) is possible.

Section I.2 is a brief introduction to Global Physics. Some properties peculiar to 3-manifolds are collected, not as evidence but as innuendo of the importance of the underlying topology for physical laws. Section I.3 gives the coordinate notion of superspace in terms of laboratory metric fields in a language which can be globalized to a manifold. A discussion of classical symmetries and their relation to singularities in this local superspace is included.

Section II, after some preliminary comments on the complete superspace of all 3-dimensional topologies, gives a global mathematical model for superspace as the orbit space of a group acting on a manifold. The model is justified from a global point of view. A preliminary connection of the model with equivalence of metrics under coordinate transformation is presented. The importance of the existence of a slice for a group acting on a manifold is discussed, and the important result of Ebin and Palais, the existense of a slice for superspace, is given. Concluding statements on the connection of superspace with coordinate-invariance are made. Section III gives the natural identification space topology to superspace and establishes some properties of the orbit projection map needed in the Stratification Theorem. In this topology superspace is shown to be a metric space, thereby answering an important open question of Stern. Section IV, after a preliminary discussion of stratifications in the infinite-dimensional setting, presents the Decomposition and Stratification Theorems for Superspace. The Decomposition Theorem partitions superspace into manifolds of geometries, indexed by the symmetry type of the geometry. Locally each of these manifolds parameterizes geometries of the given type. The Stratification Theorem tells us that these manifolds do not just pile up but fit together so as to stratify the ambient space, superspace.

In Sections V and VI we shift ground in order to make specific statements about the superspaces of some particular topologies and about the strata in the superspaces of these topologies. In Section V we are interested in the global restrictions required by the presence of a symmetric geometry. It is important to determine these topologies from two points of view. From the mathematical point of view the superspaces of these topologies have the most varying strata. From the physical point of view, these topologies are the rarest and therefore are of the most physical interest. Using recent results in the theory of compact transformation groups, a complete determination of all topologies compatible with any geometry which exhibits a continuous group of symmetries can be made. These topologies are classified in the Superspatial Topology Theorem. From them, some new and interesting topologies for highly symmetric cosmological models can be chosen. Section VI turns toward an investigation of the strata that occur in superspace. Some of the strata are finite-dimensional (as manifolds) and the Stratum Theorem tells us that these strata correspond precisely to the homogeneous geometries. Moreover, these strata are just open balls in a Euclidean space. A condition is given for a stratum to be contractible. The Superspatial Strata Theorem combines these results with Section V to give a complete determination of all the finite-dimensional and contractible strata.

Throughout the theory of superspace, there is a mixture of infinite-dimensional analysis and the theory of compact transformation groups. This interaction comes about because we have two actions, the action of Diff (M) on Riem (M), whose orbit space is superspace, and the action of the isometry group $I_g(M)$ on M. Sections II, III, and IV emphasize the first action whereas Sections V and VI the second. Nonetheless it is the coupling of these actions which gives to superspace a rich physical and mathematical content.

I.2 Superspace and Global Physics

Our approach to superspace will be global; the whole underlying space will come into play, rather than just a small portion. What could be more natural than bringing the whole topology into play? We do live on a manifold, and not just an open set of a manifold. Should we therefore not expect the total space to play a part in our observations? Nonetheless, the classical physicist is completely unaware of the role of space in the large. Spurred on by his need for calculations he works only over open sets of a manifold. As a result, the topological structure is completely lost.

The proper setting is recovered in the theory of superspace by defining geometries over the whole space. Physical laws can therefore be sought out topologically, as well as over a coordinate domain. We refer to the physics in which a topological approach is employed (at the expense of coordinate charts and calculations) as Global Physics. The central idea of global physics is that the topology of the underlying space is fundamental in all physical interactions. Using the techniques of global physics topological statements about the underlying space can be made (see Section V).

As prime evidence of the importance of the underlying space as a total rather than fractured entity, I would like to turn around an old problem. I feel that we should view the difficulties experienced in the classification of closed 3-dimensional manifolds [45] not as a failure but as a measure of the variety inherent in this class of objects. (Note that there is no hope of classifying compact trianguable 4-manifolds by algebraic invariants [24], whereas the classification of compact 2-dimensional surfaces has been known essentially since Poincare. For a proof, see [47].) Could, for example, this variety "explain" the elementary particles as different topological types, whose symmetries reflect some algebraic invariants? Moreover, 3-manifolds have some very singular properties which do not appear in higher dimensions. Every topological 3-manifold has a triangulation [27], which then admits a smoothing to a C^1- and hence C^∞- differentiable structure, unique up to diffeomorphism [35]. The homeomorphism classes therefore coincide with the diffeomorphism classes. For 3-manifolds, all the homology groups can be determined from the fundamental group alone. (By Abelianization, $H_1(M)$ is determined, which, by Poincare duality, determines $H_2(M)$. $H_o(M)$ and $H_3(M)$ are determined by the number of components and the orientability of M respectively). It was for this reason that Poincare called $\pi_1(M)$ the fundamental group. Lastly, only for 3-dimensional spaces does the dimension of the orthogonal group equal the dimension of the space; rotational degrees of freedom therefore equal the tranlational degrees of freedom. Is the fact that the space in which we live has these properties a coincidence, or do the varieties that appear before us have a topological basis?

I.3 Superspace and the Coordinate-Gauge Problem of General Relativity

In this section we rephrase the coordinate gauge problem of general relativity in a language which we hope will make its globalization in the next section appear natural. We consider only the classical (local) setting to describe the physicist's concept of superspace and its connection with the coordinate gauge problem of general relativity.

Let $g_{ij}(x^k)$ be any spatial 3-metric whatsoever. $g_{ij}(x^k)$ is the actual result of a physical experiment relative to the coordinate system (x^k), and will therefore be referred to as a laboratory metric relative to (x^k). Relative to another set of coordinates $x^\ell = x^\ell(\bar{x}^k)$, with the same physical experiment yields the result

$$\bar{g}_{mn}(\bar{x}^k) = \frac{\partial x^i}{\partial \bar{x}^m}(\bar{x}^k)\,\frac{\partial x^j}{\partial \bar{x}^n}(\bar{x}^k)\; g_{ij}(x^\ell(\bar{x}^k)).$$

The new coefficients of the metric field are now different functions of the new variables, but the fundamental principle of general relativity states that these new functions are describing the same physical (metrical) property of the space. If we consider all possible coordinate transformations of (x^k) applied to $g_{ij}(x^k)$, we will generate a chain of laboratory metrics

$$\{\bar{g}_{mn}(\bar{x}^k) \,|\, x^\ell = x^\ell(\bar{x}^k) \text{ and } \bar{g}_{mn}(\bar{x}^k) = \frac{\partial x^i}{\partial \bar{x}^m}(\bar{x}^k)\,\frac{\partial x^j}{\partial \bar{x}^n}(\bar{x}^k)\; g_{ij}(x^\ell(\bar{x}^k))\}.$$

Since each laboratory metric in the chain describes the same intrinsic properties of the space (i.e., laboratory metrics in the same chain cannot be distinguished by physical means), the chain is identified with one physical state, a geometry of the laboratory. If we consider all possible chains generated by all possible laboratory metrics, an arbitrary metric $\tilde{g}_{ab}(\tilde{x}^c)$ must lie in at least one chain (the chain generated by it) and cannot lie in more than one. (If $\tilde{g}_{ab}(\tilde{x}^c)$ were related by the coordinate transformations $\tilde{x}^c = \tilde{x}^c(x^k)$ and $\tilde{x}^c = \tilde{x}^c(\bar{x}^\ell)$ to two laboratory metrics $g_{ij}(x^k)$ and $\bar{g}_{mn}(\bar{x}^\ell)$ respectively, then $\bar{g}_{mn}(\bar{x}^\ell)$ lies in the chain generated by $g_{ij}(x^k)$ by the coordinate transformation $x^k(\tilde{x}^c(\bar{x}^\ell))$.) Now suppose a physicist goes into his laboratory and does an experiment relative to a coordinate system and determines a laboratory metric. He wishes to know what his laboratory (intrinsic) geometry is. By examining all possible chains, he finds his laboratory metric in one and only one chain, his laboratory geometry. For a concrete example, think of a physicist determining a spatial Schwarzschild metric

$$ds^2 = \frac{dr^2}{1 - \frac{2m}{r}} + r^2(d\theta^2 + \sin^2 d\phi^2)$$

with respect to (r,θ,ϕ). The chain which containes this metric is composed of those metrics related to the Schwarzschild metric by a coordinate transformation (e.g., the Schwarzschild metric in Kruskal and isotropic coordinates will appear in this chain). Since each chain describes a single physical state, we identify all the laboratory metrics in a single chain to a single point in some space of geometries. The resulting space is the set of distinguishable physical states, or laboratory geometries, the superspace $\mathscr{S}(L)$ of the laboratory L (a local superspace). Because the coordinates have already been factored out, doing physics in $\mathscr{S}(L)$ is automatically coordinate-invariant. Since doing coordinate-invariant physics on the set of laboratory metrics $\{g_{ij}(x^k)\}$ is difficult, it might appear that introducing $\mathscr{S}(L)$ as the configuration space would be a considerable advantage. Unfortunately, $\mathcal{S}(L)$ is more complicated than the space of laboratory metrics. The fields $\{g_{ij}(x^k)\}$ form an open set in a Banach space (in the C^r-topology) and therefore have a C^∞-differentiable structure, making them "ideal" for a dynamical analysis. On the other hand, $\mathcal{S}(L)$, as an identification space, shares few of the properties of the manifold $\{g_{ij}(x^k)\}$. Its greatest shortcoming is that it fails to be a manifold so that physics (which is differential equations) cannot be done on $\mathcal{S}(L)$. In short, the problems of coordinate-invariance become questions about the structure of $\mathcal{S}(L)$.

In this local setting, we discuss why $\mathcal{S}(L)$ has singular points. Suppose a laboratory metric $g_{ij}(x^k)$, transformed to $\bar{g}_{mn}(\bar{x}^\ell)$ in another system of coordinates, is the same function of the variables (x^k) as the transformed metric $\bar{g}_{mn}(\bar{x}^\ell)$ is of the variables $(\bar{x}^\ell)$. The $g_{ij}(x^k)$ is <u>left fixed</u> by this coordinate transformation and the laboratory metrics $g_{ij}(x^k)$ and $\bar{g}_{mn}(\bar{x}^\ell)$ are in <u>isometric correspondence</u>. In this situation the corresponding geometry (chains of laboratory metrics) <u>exhibits a symmetry</u>. It may happen that the coordinate transformations which leave $g_{ij}(x^k)$ fixed depend on one or more parameters, i.e., all coordinate transformations related to (x^k) by $x^k = x^k(\bar{x}^\ell;\lambda_1,\ldots,\lambda_i,\ldots,\lambda_r)$, $|\lambda_i|$ small, also leave $g_{ij}(x^k)$ fixed. Such coordinate transformations define a <u>continuous motion</u> and the corresponding geometry exhibits a <u>continuous</u> symmetry. The laboratory metrics generated by a continuous motion are in isometric correspondence, and the coordinate transformations define a <u>symmetry</u> for any of these isometrically related metrics. If a laboratory metric has a symmetry we do not consider a generated isometric metric (which is the same functional form) as anything new. In the associated chain of laboratory metrics we do not allow repetition, i.e., we identify metrics which are the same functions of their variables. (Otherwise, any chain could be extended indefinitely by duplicating each metric by writing primes on the metric coefficients and their variables.) If a laboratory metric is symmetric, its associated chain will be "shorter" than the chain of a laboratory metric which exhibits no symmetries. When this shorter chain is identified to a point in $\mathcal{S}(L)$, a neighborhood of this point is different (not homeomorphic) from the neighborhoods of longer chains. These symmetric geometries are the singular points in $\mathcal{S}(L)$.

II. FOUNDATIONS

II.1 The Complete Superspace

The goal of Foundations is to present a global construction of the space of geometries for a fixed 3-manifold. This space of geometries is the superspace, $\mathcal{S}(M)$, of the manifold. The construction will be a globalization of the coordinate-equivalence of laboratory metrics known to the local physicist. All of his problems of coordinate-invariance become geometrized into questions about the structure of superspace itself. In particular, the Central Question, Do the equivalence classes of laboratory metrics have a manifold structure?, and the Main Problem, If they do not, what can be done about it?, have a formulation in terms of superspace. The existence of singularities in superspace answers the first question negatively. The second problem is the subject of Section IV.

An analysis of the space of geometries of a single 3-topology is proper within the framework of classical geometrodynamics, where changes of compact topology cannot occur [16]. Nevertheless, it is hoped that the complete superspace, $\mathcal{S}$, of all possible topologies can be pieced together from the individual superspaces, $\mathcal{S}(M_i)$, so that quantum fluctuations in the topology can be described [53,54]. This I believe will be possible only after a great deal is known about the individual $\mathcal{S}(M_i)$, for the following reasons. To put a topology on $\mathcal{S}$, it will be necessary to make identifications at the boundaries (i.e., boundaries as subspaces of the yet to be constructed $\mathcal{S}$) of the $\mathcal{S}(M_i)$. The problem is to determine when two geometries of different topologies are "close". Consider for example the torus pinching off to a banana (elongated sphere) and of a sphere elongating itself into a banana, passing through a pinched torus stage, and becoming a torus. The geometries of the pinched torus and of the elongated sphere are closed. If we follow the geometric development of each motion, i.e., its geometric motion, in $\mathcal{S}(T^3)$ and $\mathcal{S}(S^3)$; the limit points of each evolution provide a natural identification point of $\mathcal{S}(T^3)$ and $\mathcal{S}(S^3)$. By finding the limits of all deformations of all geometries of all closed topologies, it should be possible to make the proper identifications among the $\mathcal{S}(M_i)$. This is obviously a long program. In such a program, pinched manifolds will play the crucial intermediary role in the topology change. As a set, $\mathcal{S} = \cup_i \mathcal{S}(M_i)$, where the union is over all 3-manifolds. If we restrict to compact manifolds, the union is countable (because every 3-manifold has a triangulation). On the other hand, there are an uncountable number of open (contractible) 3-manifolds [25]. Together with the pinched manifolds, these manifolds could provide intermediary transition states between changes in compact topology. In such a picture, they would play the same role the irrational numbers do in filling the spaces between the rationals.

II.2 The Mathematical Model for Superspace and its Justification

We now construct superspace. The general idea is perhaps best illustrated by the following example. Consider a single topological 2-sphere with two Riemannian metrics on it. With respect to the first, S^2 is perfectly round except for a small bump on the north pole. With respect to the second, S^2 is perfectly round except for the exact same bump on the south pole. These two physical situations however are not distinct. A physicist working in the vicinity of the first bump does not detect any change in his environment in any transformation of the underlying space. He and all his equipment are carried along by this transformation into the second physical situation, unaware of the transformation. As these two situations can never be distinguished by physical means, they are identified (together with all intermediary states) to the same physical state.

Superspace is constructed as an orbit (or identification) space. The points in superspace are equivalence classes of isometric Riemannian metrics. The construction provides a natural means of identifying inequivalent physical situations to physically distinguishable states. Superspace coincides with the set of these states. The connection of superspace with the coordinate-gauge problem of general relativity will be deferred until Section II.3, but we conclude this section with a preliminary discussion.

Let M be a smooth (C^∞) compact connected orientable (with a chosen orientation) Hausdorff 3-dimensional manifold without boundary. M is then said to be superspatial. If g is a C^∞-Riemannian metric on a superspatial M, then (M,g) is also said to be superspatial. (The dimension of M will not be crucial to the arguments except for Sections V.2 and VI.3. Always dim M>1, see Section VI.2). The closed and orientable conditions for M are at present a matter of faith (or philosophy, or worse, religion), but the physical hope is clear and the mathematical necessity (at least for compactness) will be apparent. The condition of C^∞-metrics is quite crucial as far as the mathematics goes. If one believes in an empty world, following the topological dreams of Clifford and Riemann, then all apparent discontinuities disappear in the very small and everything becomes C^∞.

If M is superspatial, then Riem (M) will denote the space of C^∞-Riemannian metrics on M and Diff (M) the C^∞-group of C^∞-orientation preserving diffeomorphisms of M. Diff (M) acts as a transformation group on Riem (M) by pulling back metrics on Riem (M),

$$\mathrm{Diff}(M) \times \mathrm{Riem}(M) \to \mathrm{Riem}(M)$$

where the action sends $(f,g) \to f^*g$, $f \in \mathrm{Diff}(M)$, $g \in \mathrm{Riem}(M)$ (see [21, p. 99] for a description of the natural action of a diffeomorphism on a metric g).

Diff(M) will be the analog of the group of coordinate transformations, Riem(M) the space of laboratory metrics, and the action will correspond to the transformation of laboratory metrics under coordinate-transformation. For a fixed $g \in$ Riem(M), the action embeds Diff(M) in Riem(M) via the orbit map $_g$: Diff(M) $\to$ Riem(M), $_g(f) = f^*g$, and the image of Diff(M) by $\mathcal{O}_g$ is <u>the orbit through g</u>, $\mathcal{O}(g) = \{f^*g \mid f \in \text{Diff}(M)\}$. If two metrics g and $\bar{g}$ are on the same orbit, there exists a diffeomorphism f of M onto itself such that $f^*g = \bar{g}$, and conversely. But this is precisely the condition for g and $\bar{g}$ to be isometric (by an orientation-preserving map) [17, p. 60], so that two metrics are isometric if and only if they lie on the same orbit. The orbits of Diff(M) in Riem(M) therefore partition Riem(M) into isometry classes of Riemannian metrics (an equivalence relation). Furthermore, for each $g \in$ Riem(M), there may be some diffeomorphisms that leave g fixed; i.e., $f^*g = g$. These diffeomorphisms form a closed subgroup of Diff(M), the <u>isotropy group</u> of the action, $I_g(M) = \{f \mid f \in \text{Diff}(M),\ f^*g = g\}$. More importantly, $I_g(M)$ is the <u>isometry</u> or <u>symmetry group</u> of the Riemannian manifold (M,g), and is therefore a compact Lie group of diffeomorphisms whose Lie algebra $\mathscr{I}_g(M)$ is isomorphic to the Killing vector fields of (M,g) [20, p. 239]. The orbit space of the action, denoted

$$\mathscr{S}(M) = \frac{\text{Riem}(M)}{\text{Diff}(M)} ,$$

is obtained by identifying (or collapsing) each orbit in Riem(M) to a single point in $\mathscr{S}(M)$. The <u>orbit projection map Π</u> : Riem(M) $\to \mathscr{S}(M)$ identifies all isometric Riemannian metrics to a single equivalence class. The image of $g \in$ Riem(M) under Π, $g^* = \Pi(g)$, is the <u>geometry of M associated with g</u>, and <u>g represents the geometry g^*</u>. $\mathscr{S}(M)$ is the set of geometries of M, or the <u>superspace of M</u>. As a physicist can only determine metrical properties of his space, the isometric metrics must be identified to the same physical state. The points of superspace are these physically distinguishable states, the configuration space for general relativity.

To connect $\mathscr{S}(M)$ with the coordinate-invariance of general relativity, we identify an open set U of M (preferably a coordinate neighborhood) with the points of a laboratory L (walls excluded). Let ϕ be a chart for U, $\phi : U \to U' \subset R^n$, and f a C^∞-diffeomorphism of U onto itself. f is the physicist's "point map" or "active transformation of the laboratory". In local representation, $f_{\phi\phi} = \phi \circ f \circ \phi^{-1} : U' \to U'$ maps an open set of R^n onto itself and therefore "looks like" a coordinate transformation. In fact, by considering $\bar{\phi} = \phi \circ f$ as a new chart on U, the "passive" coordinate transformation $\phi \to \bar{\phi}$ given by $\bar{\phi} \circ \phi^{-1}$ is indistinguishable from f in local coordinates. Conversely, if we are given a diffeomorphism f_{12} of U' onto U' (i.e., a coordinate transformation) and a chart ϕ for U, $\phi : U \to U'$, then there exists a unique diffeomorphism $f = \phi^{-1} \circ f_{12} \circ \phi : U \to U$ whose local representation relative to ϕ is f_{12}. By choosing a chart ϕ for an open set U, the diffeomorphisms of U onto itself are in 1-1 correspondence with the diffeomorphisms of $U' = \phi(U)$ onto itself. This trivial observation is sometimes summarized by saying that "passive and active transformations are equivalent".

It may happen that for $g \in \text{Riem}(M)$, $g|U$ is invariant under a map $f|U: U \to U$ (i.e., $f|U$ is a local isometry, $(f|U)^* g = g$), but g is not invariant under any extension of $f|U$ to a map $f \in \text{Diff}(M)$. In this case g is said to <u>exhibit a local symmetry</u>. Local symmetries are left to the domain of classical differential geometry. As far as global physics goes, g exhibits no symmetries whatever.

II.3. Superspace and the Slice Theorem

In the theory of compact transformation groups (G-spaces), perhaps the single most important tool in the analysis of the orbit space is the existence of a slice for the group action. As soon as non-compact groups are considered (in which case a slice need not exist), there are essentially no corresponding theorems (see [43]). On the other hand, if a slice exists, the action begins to resemble the case of the compact group and many similar theorems begin to appear.

As superspace is the orbit space of a group acting on a manifold, it is natural to ask if a slice exists for this action. The <u>Slice Theorem for Superspace</u> has a long history, dating back to an unpublished letter of Palais to Lang (1961). More recently, Ebin began an extensive investigation of this question. The bulk of the theorem was proved in [8], although it was not clear whether or not $\frac{\text{Diff}(M)}{I_g(M)}$ was a homeomorphism onto its orbit $\mathcal{O}(g)$ (although it was shown to be an injective immersion). This annoying feature was removed by an idea of Palais. Anyway you slice it, the situation in the C^∞-case is now satisfactory [9,10].

When a slice exists, the action locally is completely determined. Roughly speaking, a slice at m is a subspace "orthogonal" to the orbit through m, which, together with a small neighborhood of the orbit, fills out an open neighborhood of m. The action of the group in a neighborhood of m is thus factored into a transition action along the orbit and an action of the isotopy group on the slice. If the action is differentiable, the slice can be chosen to be an open ball of some vector space on which the isotopy group acts as a group of linear operators. If we regard G-spaces as.

generalizations of principal fiber bundles, then a slice is the analog of a local cross-section and is the best that can be hoped for (because G does not act freely). The fact that such a slice exists for Diff(M) is an extremely strong result. Our Structure Theorems will be based on its existence.

(Slice Theorem for Superspace, Ebin, Palais [3, 8, 9, 10]):

For each $g \in \mathrm{Riem}(M)$, there exists a contractible submanifold S of Riem(M) containing g such that

(1) $f \in I_g(M) \Rightarrow f^*S = S$ (2) $f \notin I_g(M) \Rightarrow f^*S \cap S = \emptyset$

(3) there exists an open set Q of the orbit through g, $\mathcal{O}(g)$, containing g, and a local cross-section $\Gamma : Q \to \mathrm{Diff}(M)$ such that $F(q,s) = (\Gamma(q))^*s$ is a diffeomorphism of $Q \times S$ onto an open neighborhood U_g of g.

(3) is equivalent to

(3') there exists a local cross-section

$$\Gamma : Q \subset \frac{\mathrm{Diff}(M)}{I_g(M)} \to \mathrm{Diff}(M) \text{ (Q an open neighborhood}$$

of the identity $I_g(M)$ in the coset space $\frac{\mathrm{Diff}(M)}{I_g(M)}$) such that $F(q,s) = \Gamma(qI_g(M))^*s$ is a diffeomorphism of $Q \times S$ onto an open neighborhood U_g of g. ■

Remark: Diffeomorphism and submanifold are in the sense of C^∞ -Fréchet manifolds [3, 37].

U_g is a splitting neighborhood of g and (2) is the splitting neighborhood property. An immediate consequence of the Slice Theorem is that $I_g(M)$ cannot increase locally, which we refer to as the local decreasing property of the isometry group. This property plays a central role in the Stratification Theorem. From the local decreasing property, it follows that those metrics which have no symmetries i.e., $I_g(M) = \{id_M\}$, are an open set in Riem(M). They are also dense [8] and are therefore generic. Although few, the non-generic metrics lead to singularities in the orbit space, for their associated geometries fail to have neighborhoods homeomorphic to neighborhoods of generic geometries. Consequently, superspace cannot support a manifold structure. It is to this problem that we address ourselves shortly.

We first establish our notation and collect some standard results of infinite-dimensional analysis regarding Diff(M) and Riem(M). At various points we give a statement in coordinates to clarify the connection of superspace with coordinate-invariance of laboratory metrics (discussed in I.2), and to add weight to the interpretations of Diff(M), Riem(M), and the action of Diff(M) on Riem(M) as globalizations of coordinate-transformations of laboratory metrics.

Let M be a superspatial manifold, TM its tangent bundle, $L_s^2(TM) = L_s^2(M)$ the tensor bundle of continuous symmetric bilinear forms (bundle of 2-covariant tensors) and $L_s^{2+}(TM) = L_s^{2+}(M)$ the subspace of positive definite forms ($g_m(X_m, X_m) > 0$ if $X_m \neq 0, m \in M$). Each fiber of $L_s^{2+}(M)$ is the set of positive definite, symmetric, bilinear maps of $T_m M$ (i.e., all inner products for $T_m M$). $L_s^2(M)$ is a C^∞-tensor bundle and $L_s^{2+}(M)$ as a subspace inherits its C^∞-manifold structure. $\Gamma^r(L_s^2(M))$, $0 \leq r \leq \infty$, are the C^r-cross-sections of $L_s^2(M)$ and Riem(M) $= \mathcal{M}$ the C^∞-cross-sections of $L_s^{2+}(M)$. Each $\Gamma^r(L_s^2(M))$, $r < \infty$, is a second countable (and hence separable) Banach space in the C^r-topology (uniform convergence in r derivatives) [2, p. 31]. $\Gamma^\infty(L_s^2(M)) = \bigcap_{k=o}^{\infty} \Gamma^r(L_s^2(M))$ is a Fréchet space (locally convex topological vector space which admits a complete metric) and is dense in each $\Gamma^r(L_s^2(M))$. Riem(M) (not a linear space) is an open positive convex cone ($P + P \subset P$, $\lambda P \subset P$ for $\lambda > 0$) in $\Gamma^\infty(L_s^2(M))$ and is therefore a C^∞-(local) Fréchet manifold, modeled on its ambient space $\Gamma^\infty(L_s^2(M))$. As a local manifold, its tangent bundle is trivial, $T(\text{Riem}(M)) = \text{Riem}(M) \times \Gamma^\infty(L_s^2(M))$.

Diff(M) is a C^∞-manifold modeled on the Fréchet space $\Gamma^\infty(TM)$ (the space of vector fields on M). Its group structure defined by composition has C^∞-group operations (a Fréchet Lie group) [22]. Diff(M) acts naturally on all tensor bundles over M by differentiation and therefore on the cross-sections of these bundles. If $f \in$ Diff(M), f acts on TM by its tangent map, $T_m f : T_m M \to T_{f(m)} M$. If (U_1, ϕ_1) and (U_2, ϕ_2) are charts at m and f(m) respectively, then $f_{12} = \phi_2 \circ f \circ \phi_1^{-1}$ can be written as n-functions $x_1(\bar{x}_1, \ldots, \bar{x}_n), \ldots, x_n(\bar{x}_1, \ldots, \bar{x}_n)$. Relative to these coordinates, $T_m f$ is represented by the Jacobian matrix of f_{12}, i.e.,

$$\left(\frac{\partial \pi_j (f_{12})}{\partial \bar{x}_i}\right)_{\bar{x}=\phi_1(m)} = \left(\frac{\partial x_j(\bar{x}_1, \ldots, \bar{x}_n)}{\partial \bar{x}_i}\right)_{\bar{x}=\phi_1(m)},$$

$1 \leq i, j \leq n$. In charts, f looks exactly like a "passive" coordinate transformation, although it is an "active" (point) map of the manifold. By pulling back a tensor at f(m) to m, f induces a map on the cross-sections of covariant tensor bundles. For $\Gamma^\infty(L_s^2(M))$ we have:

$$(f^*g)_m(X_m,Y_m) = g_{f(m)}(T_mf(X_m),T_mf(Y))$$

with $f \in \mathrm{Diff}(M)$, $g \in \Gamma^\infty(L_s^2(TM))$ and $X_m, Y_m \in T_mM$. If $g \in \mathrm{Riem}(M)$, f^*g is also, so that Diff(M) restricts to Riem(M). The action is linear on Riem(M), i.e., for fixed f, $f^*(\lambda_1 g_1 + \lambda_2 g_2) = \lambda_1 f^* g_1 + \lambda_2 f^* g_2$. If in the local coordinates (ϕ_2, U_2), g is given as $g_{ij}dx^i \otimes dx^j$, the action is given by

$$\bar{g}_{mn}(\bar{x})d\bar{x}^m \otimes d\bar{x}^n = g_{ij}(x(\bar{x}))\frac{\partial x^i}{\partial \bar{x}^k}(\bar{x})\frac{\partial x^j}{\partial \bar{x}^\ell}(x)dx^k \otimes dx^\ell$$

with $\bar{g} = f^*g$, $\bar{x} = \phi_1(m)$ and $x = \phi_2 \circ f(m) = f_{12}(\bar{x})$. In charts, the action corresponds to the transformation of the laboratory metric under coordinate transformation.

For fixed $g \in \mathrm{Riem}(M)$, the action embeds Diff(M) as a differentiable submanifold via the orbit map, i.e.,

$$\mathcal{O}_g : \mathrm{Diff}(M) \to \mathrm{Riem}(M)$$

by $\mathcal{O}_g(f) = f^*g$. The tangent of this map at the identity in Diff(M) is

$$T_{id}\mathcal{O}_g : T_{id}\,\mathrm{Diff}(M) \to T_g\,\mathrm{Riem}(M)$$

i.e.,

$$T_{id}\mathcal{O}_g : \Gamma^\infty(TM) \to \Gamma^\infty(L_s^2(M)).$$

This map is defined as follows. Let $X \in \Gamma^\infty(TM)$ be a vector field on M, or equivalently, the tangent to a curve c in Diff(M), $c: I \to \mathrm{Diff}(M)$ with $c(t) = f_t \in \mathrm{Diff}(M)$, $I = (-1,1)$, and $f_o = id_M$. X is mapped by $T_{id}\mathcal{O}_g$ onto the tangent of the associated curve in Riem(M), i.e., $T_{id}\mathcal{O}_g(X(m)) = \frac{d}{dt}(\mathcal{O}_g \circ c(t))\Big|_{t=o} = \frac{d}{dt}(f_t^*g)\Big|_{t=o} = L_Xg$ where L_Xg is the Lie derivative of g with respect to the vector field X. Therefore $T_{id}\mathcal{O}_g(X) = L_X g$. Moreover $I_g(M)$ is an embedded submanifold of Diff(M), whose Lie algebra, $T_{id}I_g(M) = \mathcal{I}_g(M)$, (a finite-dimensional subspace of T_{id} Diff(M) = $\Gamma^\infty(TM)$ and hence a splitting subspace), coincides with the Killing vector fields of (M,g). Since $L_X g = 0$ defines a Killing vector field, kernel $(T_{id}\mathcal{O}_g) = \mathcal{I}_g(M)$. Since $\mathcal{O}_g$ is a differentiable embedding, $T_{id}\mathcal{O}_g$ maps $\frac{\Gamma^\infty(TM)}{\mathcal{I}_g(M)}$ injectively onto a splitting subspace of $\Gamma^\infty(L_s^2(M))$, which will be identified with $\frac{\Gamma^\infty(TM)}{\mathcal{I}_g(M)}$.

III. THE TOPOLOGY OF SUPERSPACE

III.1. The Orbit Projection Map and the Quotient Topology

As the space of Riem(M) (or of laboratory metrics) is a metric space (an open ball in a Fréchet space), it is natural to ask what topological properties the orbit space $\mathcal{S}(M)$ inherits? Unfortunately, quotient spaces have little or nothing to do with their covering spaces. Even in the case of orbit spaces, the existence of a slice, compact isotropy groups, and closed orbits (as we have for $\mathcal{S}(M)$) are not enough to insure that the orbit space is Hausdorff (see [43] for an example). The topological structure of superspace was first investigated by Stern in [48], where $\mathcal{S}(M)$ was shown to be Hausdorff. In the next section $\mathcal{S}(M)$ is shown to be metrizeable and an explicit metric is given. From this metric a quantitative description of the "closeness" of the two geometries can be computed. In this section we give the natural quotient or identification space topology to superspace and then describe some properties of the orbit projection map which will be needed in the Stratification Theorem.

$\mathcal{S}(M)$ is given the strongest topology in which the orbit projection map Π is continuous (the usual quotient space topology). In this topology a subset F^* of $\mathcal{S}(M)$ is open (respectively closed) if and only if $F = \Pi^{-1}(F^*)$ is open (respectively closed). If F is open in Riem(M), its saturation $\mathcal{D}(F) = \bigcup_{f\in\mathcal{D}} f^*F$ is the union of open sets (f^* is a homeomorphism of Riem(M) onto itself) and hence open. Since $\Pi(F)$ is open if and only if $\Pi^{-1}(\Pi(F)) = \mathcal{D}(F)$ is open, $\Pi(F)$ is open, so Π is an open map. Since there is at most one topology (if there is any) such that Π is both continuous and open, the topology of $\mathcal{S}(M)$ is characterized by the requirement that Π be both continuous and open.

If F is an arbitrary subset of $\mathcal{M}$, $\Pi^{-1}\overline{(\Pi(F))}$ is a closed set containing F and therefore $\bar{F}$. Hence $\Pi(\bar{F}) \subset \overline{\Pi(F)}$. In the theory of compact transformation groups, a simple argument shows that Π is also a closed map [41, p.2]. If Π were closed, $\Pi(\bar{F}) \supset \overline{\Pi(F)}$ so that $\Pi(\bar{F}) = \overline{\Pi(F)}$. For non-compact groups the argument does not go through but the equation $\Pi(\bar{F}) = \overline{\Pi(F)}$ can be recovered if we restrict to invariant sets of $\mathscr{M}$. (If F is invariant, so is $\bar{F}$, so that $\Pi^{-1}(\Pi(\bar{F})) = \mathscr{D}(\bar{F}) = \bar{F}$ is closed. $\Pi(\bar{F})$ is then closed and so contains the closed set $\overline{\Pi(F)}$.) This relation will be used in the Stratification Theorem to shuttle back and forth between the closed sets of $\mathscr{S}(M)$ and the closed invariant sets of Riem(M).

III.2. THE METRIZEABILITY OF SUPERSPACE

We collect in the following theorem the topological properties of superspace that will be needed in the structure theorems. These properties are the strongest separation and countability properties that a topological space can exhibit. From these properties we conclude that $\mathscr{S}(M)$ is Hausdorff, regular, completely regular, a Tynchoff space, normal, and paracompact. It is also separable, a Lindelöf space, and homeomorphic to a subset of the Hilbert cube in R^∞.

Theorem 1 (The Metrization Theorem for Superspace): Superspace is a connected second countable metrizeable space.

Proof: As the continuous image of the contractible (and hence connected) space Riem(M), $\mathscr{S}(M)$ is connected.

Since the C^∞-sections are dense in the C^r-sections, Riem(M) is a subspace of a second countable space and is therefore second countable. Since Π is both continuous and open, the image under Π of a countable base for the topology of Riem(M) is also a countable basis for the topology of $\mathscr{S}(M)$, so that $\mathscr{S}(M)$ is also second countable.

Since Riem(M) is modeled on a topological vector space whose topology cannot come from an inner product, nor from a norm (i.e, $\Gamma^\infty(L_s^2(M))$ is not Hilbertable), it is not possible to introduce a Riemannian metric on Riem(M) whose associated (topological) metric d induces the original manifold topology of Riem(M). Nonetheless it is possible to introduce a metric d which induces the manifold topology (since Riem(M) is a complete metric space) and such that Diff(M) acts as a group of isometries relative to d. This construction is carried out in [8] by introducing the spaces of Sobolev Riemannian metrics on M, denoted $\text{Riem}^s(M) = \mathscr{M}^s$, i.e. those metrics which have square integrable partial derivatives up to order s, s an integer, $s > \frac{n}{2} = \frac{\dim M}{2}$. These spaces of metrics are modeled on Hilbert Space so that it is possible to define a Riemannian metric G^s on $\mathscr{M}^s$ whose associated distance function d^s induces the original manifold topology. Moreover, G^s is invariant under the action of Diff(M), i.e. for $g \in \text{Riem}(M)$, $\pi^1, \pi^2 \in T_g\,\text{Riem}(M) = \Gamma^\infty(L_s^2(M))$,

$$G^s_g(\pi^1,\pi^2) = G^s_{f^*g}(T_g f^*(\pi^1),\ T_g f^*(\pi^2)) = G^s_{f^*g}(f^*\pi^1, f^*\pi^2)$$

(since $T_g f^* = f^*$).

G^s is given in local coordinates by

$$G^s_g(\pi^1,\pi^2)=\int_M g^{km}(x)g^{ln}(x)\pi^1_{kl}(x)\pi^2_{mn}(x)\sqrt{\det(g_{ij}(x))}\,dx^1\wedge\ldots\wedge dx^n$$

$$+\int_M g^{km}(x)g^{ln}(x)g^{qr}(x)\pi^1_{kl;q}(x)\pi^2_{mn;r}(x)\sqrt{\det g_{ij}(x)}\,dx^1\wedge\ldots\wedge dx^n$$

$$+\ldots+\int_M g^{km}(x)g^{ln}(x)\underbrace{g^{qr}(x)\ldots g^{uv}(x)}_{s}\pi^1_{kl;\underbrace{q;\ldots;u}_{s}}(x)\pi^2_{kl;\underbrace{r;\ldots;v}_{s}}(x)\cdot\sqrt{\det g_{ij}(x)}\,dx^1\wedge\ldots\wedge dx^n$$

The use of the covariant derivative makes G^s invariant under the action of Diff(M) since its expression in local coordinates is invariant with respect to coordinate transformations. The induced metric d^s is therefore also invariant under the action of Diff(M), i.e., $d^s(g_1,g_2) = d^s(f^*g_1,f^*g_2)$. By using the sequence $\{d^s\}$, an invariant metric d is defined on Riem(M) by

$$d(g_1,g_2) = \sum_{s>n/2}\frac{1}{2^s}\frac{d^s(g_1,g_2)}{1+d^s(g_1,g_2)}$$

which induces the original manifold topology on Riem(M).

Using this invariant metric we show that $\mathscr{S}(M)$ is metrizeable with metric defined by

$$d_{\mathscr{S}}(\Pi(g_1),\Pi(g_2)) = d_{\mathscr{S}}(\mathscr{O}(g_1),\mathscr{O}(g_2)) = \inf_{f,h\in\mathscr{D}}\{d(f^*g_1,f^*g_2)\}.$$

Note that this is the usual set distance on the subsets of a metric space, restricted to the sets which are orbits, which by [8] is a subset of the set of closed subsets. So restricted, $d_{\mathscr{S}}$ is actually a metric for $\mathscr{S}(M)$. Symmetry is clear and by the invariance of d we have

$$d_{\mathscr{S}}(\mathscr{O}(g_1),\mathscr{O}(g_2)) = \inf_{f,h\in\mathscr{D}}\{d(g_1,(h\circ f^{-1})^*(g_2))\}$$

$$= \inf_{f\in\mathscr{D}}\{d(g_1,f^*g_2)\}.$$

$d_{\mathscr{S}}$ is therefore positive definite because if two orbits intersect they are not distinct, and because in a metric space the set distance between a compact subset and a disjoint closed subset is greater than zero. In fact this distance is actually assumed, i.e.

there exists a $g_2' \in \mathcal{O}(g_2)$ such that $d_{\mathcal{S}}(\mathcal{O}(g_1), \mathcal{O}(g_2)) = d(g_1, g_2')$. The triangle inequality follows from considering

$$d(g_1, f^*g_3) \leq d(g_1, h^*g_2) + d(h^*g_2, f^*g_3)$$

so that

$$d_{\mathcal{S}}(\mathcal{O}(g_1), \mathcal{O}(g_3)) = \inf_{f \in \mathcal{D}} \{d(g_1, f^*g_3)\} \leq d(g_1, h^*g_2) + d_{\mathcal{S}}(\mathcal{O}(g_2), (g_3))$$

and

$$d_{\mathcal{S}}(\mathcal{O}(g_1), \mathcal{O}(g_3)) \leq d_{\mathcal{S}}(\mathcal{O}(g_1), \mathcal{O}(g_2)) + d_{\mathcal{S}}(\mathcal{O}(g_2), \mathcal{O}(g_3)).$$

Finally we must show that the topology induced from $d_{\mathcal{S}}$ on $\mathcal{S}(M)$ is the quotient topology. Since Π is distance decreasing relative to d and $d_{\mathcal{S}}$, Π is continuous relative to the topologies induced on Riem(M) and $\mathcal{S}(M)$ by d and $d_{\mathcal{S}}$ respectively. Suppose $d_{\mathcal{S}}(\mathcal{O}(g_1), \mathcal{O}(g_2)) < \varepsilon$. Then there is some $g_2' \in \mathcal{O}(g_2)$ such that $d_{\mathcal{S}}(\mathcal{O}(g_1), \mathcal{O}(g_2)) = d(g_1, g_2') < \varepsilon$. Π therefore takes an ε-neighborhood of g_1 (relative to d) onto an ε-neighborhood of $\Pi(g_1)$ (relative to $d_{\mathcal{S}}$), and so is an open mapping relative to the induced metric topologies. But d induces the manifold topology on Riem(M). Since the metric topology induced on $\mathcal{S}(M)$ by $d_{\mathcal{S}}$ makes Π both open and continuous, it coincides with the quotient topology.

IV. The Stratification of Superspace

IV.1 The Manifold Problem

Because of the existence of symmetric geometries, there will be neighborhoods of $\mathcal{S}(M)$ not homeomorphic to neighborhoods of generic geometries (i.e., geometries with no symmetries whatsoever), so that superspace itself cannot be a manifold. If superspace is to be the configuration space for a dynamical analysis (which means differential equations, we are confronted with the Main Problem. We will show that superspace, although itself not a manifold, decomposes into manifolds of geometries, which fit together in a way which will permit a "generalized" dynamical analysis.

The central idea is very simple. We collect together all geometries which have the same symmetry type; these geometries have homeomorphic neighborhoods and therefore a manifold structure. Of these manifolds of geometries, we are naturally lead to ask how they fit together so as to fill out superspace? Conceivably, they could wind around one another, approaching each other infinitely often and infinitely close, or have transversal or self-intersections. Fortunately, pathologies do not occur. The geometries of high symmetry are completely contained in the boundary of those geometries with lesser symmetry. Consequently, a limit of geometries of a particular type must either be of the same type or exhibit higher symmetry. This regularity is characteristic of a stratification and justifies calling a particular manifold of geometries a stratum of $\mathcal{S}(M)$. The Stratification Theorem provides the information that will enable us to pass smoothly from one stratum to another, thereby allowing a "generalized" (or modified) dynamical analysis on the stratified topological space $\mathcal{S}(M)$.

We begin by partitioning Riem(M) by the isometry groups of the metrics. It will be important to note that the isometry group, as a group of diffeomorphisms, describes an action on the underlying manifold, and, moreover, the isometry groups of isometric metrics describe equivalent actions. This coupling of the action of Diff(M) on Riem(M) to $I_g(M)$ on M will be central to the development of Sections IV and V.

Let G be a compact subgroup of Diff(M) and (G) denote all compact subgroups of Diff(M) that are conjugate to G by an element in Diff(M), i.e., $(G) = \{fGf^{-1} | f \in \text{Diff}(M)\}$. If some element of (H) is included in some element of (G), then we say $(H) \leq (G)$. If $(H) \leq (G)$ and $(G) \leq (H)$, a conjugate of H, H', is contained in G, and since H' and G have the same dimension and the same number of components, H' = G and (H) = (G) ($\leq$ is reflexive). Since transitivity is clear, $\leq$ is a partial ordering with associated strong partial ordering < ((H) $\nleq$ (H)). The partially ordered set of conjugacy classes of compact subgroups of Diff(M) will be used to index a partition of $\mathcal{S}(M)$. We first look at this index set a little closer.

If G is a compact (abstract) Lie group, an <u>effective differentiable action of G on M</u> is a continuous isomorphism Φ of G onto a subgroup of Diff(M). An effective action of G on M is therefore represented by a compact group of diffeomorphisms $G^{\text{Diff}} = \Phi(G) \subset \text{Diff}(M)$. Two actions Φ_1 and Φ_2 of G are equivalent if there is a diffeomorphism f of M such that $f \circ \Phi_1(g) = \Phi_2(g) \circ f$ for all $g \in G$. Therefore two actions are equivalent if and only if $fG_1f^{-1} = G_2$ where G_1 and G_2 are the images of G in Diff(M) under Φ_1 and Φ_2. A conjugacy class of Diff(M) therefore determines an equivalence class of actions, and the non-conjugate isomorphic copies of G embedded in Diff(M) are in one to one correspondence with the inequivalent actions of G on M. A conjugacy class (G) can be represented as an (abstract) Lie group G and an equivalence class of actions of G on M.

A few notational words. G^{Diff} and $I_g(M)$ will always be subgroups of Diff(M), whereas G will always be an abstract Lie group (not a subgroup of Diff(M)). (G) will always be a conjugacy class of subgroups of Diff(M) and for emphasis we will sometimes write (G^{Diff}). If $I_g(M)$ is known only as an abstract Lie group (i.e., its action on M is not known) we will write I_g. For example, I_g is equal to an abstract Lie group, $I_g = SO(3)$, say, whereas $I_g(M)$ is isomorphic as a subgroup of Diff(M) to an abstract Lie group, $I_g(M) \cong SO(3)$, say.

Let $\mathscr{M}_{(G)}$ denote those Riemannian metrics such that $I_g(M) \in (G)$, i.e., $\mathscr{M}_{(G)} = \{g \mid g \in \mathscr{M} \text{ and } I_g(M) \in (G)\}$. Each metric g in $\mathscr{M}_{(G)}$ is of <u>symmetry type (G)</u> and <u>G is the symmetry group of g</u>, or <u>g is symmetric by G</u> $(G = G^{\text{Diff}} \in (G))$. The symmetry type of a metric is therefore both an abstract Lie group G (its symmetry group) <u>and</u> an equivalence class of actions of G on M. $\mathscr{M}_G$ will denote those metrics with symmetry group G, $\mathscr{M}_G = \bigcup_\alpha \mathscr{M}_{(G_\alpha)}$, where the union is over all subgroups G_α^{Diff} of Diff(M) isomorphic to G. In favorable cases there is only one action (up to equivalence) of G on M. In these cases $\mathscr{M}_G = \mathscr{M}_{(G)}$. $\mathscr{M}_{(G)}$ might still have an infinite number of components.

$\mathscr{M}_{(G)}$ (respectively, $\mathscr{M}_G$) is an invariant subspace of $\mathscr{M}$ so that the action of $\mathscr{D}$ restricts to $\mathscr{M}_{(G)}$ (respectively, $\mathscr{M}_G$). Define $\mathscr{S}_{(G)}(M) = \dfrac{\mathscr{M}_{(G)}}{\mathscr{D}} = \Pi(\mathscr{M}_{(G)})$ (respectively, $\mathscr{S}_G(M) = \dfrac{\mathscr{M}_{(G)}}{\mathscr{D}} = \Pi(\mathscr{M}_G)$) and let $\mathscr{S}^i_{(G)}(M)$ (respectively, $\mathscr{S}^i_G(M)$) denote its connected components. Geometries in $\mathscr{S}_{(G)}(M)$ (respectively, $\mathscr{S}_G(M)$) are of <u>symmetry type (G)</u> (respectively, have <u>symmetry group G</u>).

IV.2. Stratifications in the Infinite-Dimensional Setting

Stratifications first arose in connection with describing singularities of differentiable maps [50, 51, 56]. The finite-dimensional version of a stratification consists of a partition of a space into manifolds, the strata, indexed by dimension, whose components satisfy the frontier property,

$$X^j_m \cap \overline{X^i_n} \neq \emptyset,\ m \neq n \Rightarrow X^j_m \subset \overline{X^i_n},\ \dim X^j_m < \dim X^i_n .$$

In this case, the boundary $\overline{X^i_n} - X^i_n$ of each stratum is the union of lower dimensional strata.

Perhaps the canonical example of a stratified set is the n × n matrices partitioned by rank [56]. The strata are then matrices of rank r, $M_r(n)$ ($\dim M_r(n) = n^2 - (n - r)^2$). Limit points of $M_r(n)$ not in $M_r(n)$ must lie in lower dimensional strata $M_{r'}(n)$, $r' < r$. Recall the local decreasing property of the isotropy groups (in the compact setting, when a slice exists). Because the dimension of the orbits cannot decrease locally, limit points of higher dimensional orbits must lie in lower dimensional orbits (if the limit orbit itself is not the same high dimension). It is precisely this analogy which leads to the Stratification Theorem. Note that the stratification of superspace is an analog in the infinite-dimensional setting of a stronger regularity, the triangulation of the orbit space, which appears in the theory of compact transformations groups [59].

If X is an infinite-dimensional space (topological dimension) there may be subspaces which should "naturally" be considered as different strata (in any proper definition of stratification) but which have both infinite dimension and codimension. In these cases it is clear that the finite-dimensional definition cannot be maintained. To remedy this situation, the indexing by dimension is replaced by an arbitrary partially ordered set while the crucial frontier property is retained. The resulting definition provides what perhaps might be viewed as minimum requirements for a stratification in the infinite-dimensional setting. The definition of manifold is in the sense of [12].

A partition or decomposition of a second countable Hausdorff topological space X is a set of non-empty subspaces $\{X_\alpha\}$ indexed by a set $\mathscr{A}$ such that

(1) $X = \cup_\alpha X_\alpha$, $\alpha \in \mathscr{A}$

(2) $X_\alpha \cap X_\beta \neq \emptyset \Rightarrow a = \beta$.

The subspace X_α is Hausdorff and second countable whose connected components $\{\{X^i_\alpha\} | i \in C_\alpha\}$ partition X_α (indexed by C_α). If $\{X_\alpha\}$ is a partition for X, $\{X^i_\alpha\}$ is the associated complete partition of X, indexed by $\{(\alpha,i) | \alpha \in \mathscr{A}, i \in C_\alpha\} = \Pi_\alpha C_\alpha$. A partition is a manifold partition (respectively, countable partition; respectively, finite partition) if each X_α is a manifold (respectively, if the index set is countable; respectively, if the index set is finite). If

X_α is a manifold, there are at most a countable number of components X_α^i (because the components of a manifold are open and X_α is second countable), so that if a manifold partition is countable then so is its associated complete partition. A partition is partially ordered if the index set is partially ordered (reflexive and transitive), and < will denote the associated strong partial ordering. The associated complete partition is then partially ordered by $(\alpha,i) < (\beta,j) \Longleftrightarrow \alpha < \beta$. A partially ordered partition $\{X_\alpha\}$ has the frontier property (respectively, inverted frontier property) if $\overline{X_\alpha} \cap X_\beta \neq \emptyset$, $\alpha \neq \beta \Rightarrow X_\alpha \subset \overline{X_\beta}$, $\alpha < \beta$ (respectively, $X_\alpha \cap \overline{X_\beta} \neq \emptyset$, $\alpha \neq \beta \Rightarrow X_\alpha \subset \overline{X_\beta}$, $\beta < \alpha$). A stratification (respectively, inverted stratification) of a connected second countable Hausdorff topological space X is a countable partially ordered manifold partition of X whose associated complete partition has the frontier property (respectively, inverted frontier property), i.e.,

$$X_\alpha^i \cap \overline{X_\beta^j} \neq \emptyset,\ \alpha \neq \beta\ (\Longleftrightarrow (\alpha,i) \neq (\beta,j)) \Rightarrow$$

$$X_\alpha^i \subset \overline{X_\beta^j} \text{ and } \alpha < \beta\ (\Longleftrightarrow (\alpha,i) < (\beta,j))$$

$$(\text{respectively, } \beta < \alpha\ (\Longleftrightarrow (\beta,j) < (\alpha,i))).$$

The manifolds $\{X_\alpha\}$ are the strata of the stratification and the manifolds $\{X_\alpha^i\}$ the connected strata. If $\{X_\alpha\}$ is a stratification and $X_\alpha^i \cap X_\beta^j$, then X_α^i in fact must lie completely in the boundary of X_β^j, $\overline{X_\beta^j} - X_\beta^j$, because $X_\alpha^i \cap X_\beta^j = \emptyset$.

IV.3. The Structure Theorems of Superspace

We now come to the structure theorems for superspace. The first theorem describes a partition of superspace into manifolds of geometries and the second shows how these manifolds are joined in superspace. Together these theorems begin to unravel how the space of geometries of an arbitrary closed manifold is pieced together from its manifold parts. In Section VI this information is used to determine explicitly the finite-dimensional parts of the superspaces of certain "symmetric" manifolds.

Theorem 2 (Decomposition Theorem for Superspace): The decomposition of $\mathcal{S}(M)$ by the subspaces $\{\mathcal{S}_{(G)}(M)\}$ is a countable partially ordered C^∞-Fréchet manifold partition.

Proof: By construction $\{\mathcal{S}_{(G)}\}$ (omitting the manifold variable M) is a partition of $\mathcal{S}(M)$ partially ordered by the conjugacy classes of compact subgroups of Diff(M). Since there are only countable many compact Lie groups (for each dimension n, there are only a finite number of compact Lie groups with m components, and so for each pair (n,m) there is an associated finite set), each of which has at most a countable number of inequivalent differentiable actions on M [42], there is a countable number of conjugacy classes of compact subgroups of Diff(M) (symmetry types). The partition is therefore countable. We will now show that each $\mathcal{S}_{(G)}$ is a manifold, from which it follows that the associated complete partition is also countable.

Let $g^* \in \mathcal{S}_{(G)}$, $g \in \Pi^{-1}(g^*)$, and S a slice at g. S is homeomorphic to an open ball of the origin in

$$\frac{\Gamma^\infty(L_s^2(M))}{\Gamma^\infty(TM)/\mathcal{I}_g(M)}$$

with g taken as the origin. I_g (in this proof and the next, $I_g(M) = I_g \subset \mathrm{Diff}(M)$) acts on S as a compact group of linear operators, whose fixed point set $F(I_g, S)$ is an open ball of some closed linear subspace of

$$\frac{\Gamma^\infty(L_s^2(M))}{\Gamma^\infty(TM)/\mathcal{I}_g(M)}$$

Also, $F(I_g,S) = \mathcal{M}_{(I_g)} \cap S$. For if $\bar{g} \in S \cap \mathcal{M}_{(I_g)}$, $I_{\bar{g}} \subset I_g$ and $I_{\bar{g}} \in (I_g) \Rightarrow I_{\bar{g}} = I_g \Rightarrow \bar{g} \in F(I_g,S)$. Conversely, if $\bar{g} \in F(I_g,S)$, $I_{\bar{g}} \subset I_g$ and $I_{\bar{g}} \supset I_g \Rightarrow I_{\bar{g}} = I_g \Rightarrow \bar{g} \in \mathcal{M}_{(I_g)} \cap S$. By the Slice Theorem, $\Gamma(Q)^*S$ splits into the product $Q \times S$, so that $\Gamma(Q)^*S \cap \mathcal{M}_{(I_g)}$

is homeomorphic to $Q \times (S \cap \mathcal{M}_{(I_g)}) = Q \times F(I_g, S)$. $F(I_g, S)$ is a C^∞-cross-section of $\Gamma(Q)^* S$ and $\Pi(\Gamma(Q)^* S \cap \mathcal{M}_{(I_g)}) \subset \mathcal{S}_{(G)}$ is an open neighborhood of g homeomorphic to the open ball $F(I_g, S)$. If $\bar{g}^* \in \mathcal{S}_{(G)}$, $\bar{g} \in \Pi^{-1}(\bar{g}^*)$, and $\bar{S}$ a slice for $\bar{g}$, then S is an open ball in $\dfrac{\Gamma^\infty(L_s^2(M))}{\Gamma^\infty(TM)/\mathcal{I}_{\bar{g}}(M)}$, which is toplinearly isomorphic to $\dfrac{\Gamma^\infty(L_s^2(M))}{\Gamma^\infty(TM)/\mathcal{I}_g(M)}$. Since the action of $I_{\bar{g}}$ on $\bar{S}$ is equivalent to the action of I_g on S, $F(I_{\bar{g}}, \bar{S})$ is homeomorphic to $F(I_g, S)$. Since $F(I_{\bar{g}}, \bar{S})$ and $F(I_g, S)$ are C^∞cross-sections, $\mathcal{S}_{(G)}(M)$ is a C^∞-differentiable Fréchet manifold modeled on a closed supspace of

$$\frac{\Gamma^\infty(L_s^2(M))}{\Gamma^\infty(TM)/\mathcal{I}_g(M)}. \blacksquare$$

We now come to the main result of the first part of this work.

Theorem 3 (The Stratification of Superspace): The manifold partition $\{\mathcal{S}_{(G)}(M)\}$ of $\mathcal{S}(M)$ is an inverted stratification indexed by symmetry type.

Proof: By the Decomposition Theorem for Superspace, the partition of $\mathcal{S}(M)$ is a countable partially ordered manifold partition. It remains to prove the inverted frontier property for the associated complete partition, i.e.,

$$\mathcal{S}^i_{(G)} \cap \overline{\mathcal{S}^j_{(H)}} \neq \emptyset,\ (G) \neq (H) \Rightarrow \mathcal{S}^i_{(G)} \subset \overline{\mathcal{S}^j_{(H)}} \text{ and } (H) < (G)$$

We first show that if $\mathscr{S}^i_{(G)} \cap \overline{\mathscr{S}^j_{(H)}} \neq \emptyset$, $(G) \neq (H)$, then $(H) < (G)$. Recall that if F is an invariant set of M, $\Pi(\bar{F}) = \overline{\Pi(F)}$. If $F^* = \Pi(F)$, then $\overline{F^*} = \overline{\Pi(F)} = \Pi(\bar{F})$ so that $\Pi^{-1}(\overline{F^*}) = \bar{F}$, since $\bar{F}$ is also invariant. If $g^* \varepsilon\ \mathscr{S}^i_{(G)} \cap \overline{\mathscr{S}^j_{(H)}} \neq \emptyset$, $(H) \neq (G)$, then $\Pi^{-1}(\mathscr{S}^i_{(G)} \cap \overline{\mathscr{S}^j_{(H)}}) = \Pi^{-1}(\mathscr{S}^i_{(G)}) \cap \Pi^{-1}(\overline{\mathscr{S}^j_{(H)}}) = \mathscr{M}_{(G)} \cap \overline{\mathscr{M}_{(H)}} \neq \emptyset$. Let $g\ \varepsilon\ \mathscr{M}_{(G)} \cap \overline{\mathscr{M}_{(H)}}$ and let U_g be a splitting neighborhood of g, $U_g \cong Q \times S$. Since $g\ \varepsilon\ \overline{\mathscr{M}_{(H)}}$, U_g must meet (has non-empty intersection with) $\mathscr{M}_{(H)}$. If $\tilde{g}\ \varepsilon\ \mathscr{M}_{(H)} \cap U_g$, then $(I_{\tilde{g}}) = (H) \leq (G)$ (locally the isotropy groups cannot increase) so that $(H) < (G)$ (because $(H) \neq (G)$).

We will now show that $\mathscr{S}^i_{(G)} \cap \overline{\mathscr{S}^j_{(H)}}$ is both open and closed in $\mathscr{S}^i_{(G)}$, from which it follows $\mathscr{S}^i_{(G)} \cap \overline{\mathscr{S}^j_{(H)}} = \mathscr{S}^i_{(G)}$ and thus $\mathscr{S}^i_{(G)} \subset \overline{\mathscr{S}^j_{(H)}}$. Let $g^*\ \varepsilon\ \mathscr{S}^i_{(G)} - \overline{\mathscr{S}^j_{(H)}}$ (the complement of $\mathscr{S}^i_{(G)} \cap \overline{\mathscr{S}^j_{(H)}}$ in $\mathscr{S}^i_{(G)}$): assume every open neighborhood in $\mathscr{S}^i_{(G)}$, of g^* meets $\overline{\mathscr{S}^j_{(H)}}$. If $\{U^n_{g^*}\}$, $U^n_{g^*} \subset \mathscr{S}^i_{(G)}$, is a nested local base at g^*, we can then choose a sequence $g^*_n\ \varepsilon\ U^n_{g^*} \cap \overline{\mathscr{S}^j_{(H)}}$ which converges to g^*, contrary to assumption. Therefore there must be some open neighborhood $U^n_{g^*}$ of g^* in $\mathscr{S}^i_{(G)}$ such that $U^n_{g^*} \cap \overline{\mathscr{S}^j_{(H)}} = \emptyset$. Thus $\mathscr{S}^i_{(G)} - \overline{\mathscr{S}^j_{(H)}}$ is open in $\mathscr{S}^i_{(G)}$.

We now show that $\mathscr{S}^i_{(G)} \cap \overline{\mathscr{S}^j_{(H)}}$ is also open in $\mathscr{S}^i_{(G)}$, which proves the theorem. Let $g\ \varepsilon\ \mathscr{M}_{(G)} \cap \overline{\mathscr{M}_{(H)}}$ and S a slice at g. $F(I_g,S) = \mathscr{M}_{(G)} \cap S$ is homeomorphic to $F^* = \Pi(F(I_g,S)) \subset \mathscr{S}_{(G)}$, an open neighborhood of $g^* = \Pi(g)$. Let $\{g_n\}$, $g_n\ \varepsilon\ \mathscr{M}_{(H)}$, $g_n \to g$ be a sequence (chosen so that all $g_n\ \varepsilon\ U_g$). By sliding each g_n along its orbit (a C^∞-operation) we can choose another sequence $\{s_n\}$,

$s_n \in S$. By considering S as an open ball in a vector space with origin g, $s_n \to 0$. If $s \in \mathscr{M}_{(G)} \cap S = F(I_g, S)$, we will show that $s \in \overline{\mathscr{M}_{(H)}}$. Hence $F(I_g, S) \subset \overline{\mathscr{M}_{(H)}}$ so that the open set $F^* = \Pi(F(I_g, S)) \subset \mathscr{S}^i_{(G)}$ is also contained in $\overline{\mathscr{S}^i_{(H)}}$. It follows that $\mathscr{S}^i_{(G)} \cap \overline{\mathscr{S}^i_{(H)}}$ is open in $\mathscr{S}^i_{(G)}$. Consider the sequence $\{s - s_n\}$, $s - s_n \to s$ (chosen so that $s - s_n \in S$). We will show that there exists an N such that for $n > N$, $I_{s-s_n} = I_{s_n} \in (H)$ so that $s - s_n \in \mathscr{M}_{(H)} \cap S$ and $s \in \overline{\mathscr{M}_{(H)}}$.

Since $s \in F(I_g, S)$, $I_s = I_g$. If $f \in I_{s_n} \subset I_g = I_s$, $f^*(s-s_n) = f^*s - f^*s_n = s - s_n \Rightarrow f \in I_{s-s_n} \Rightarrow I_{s_n} \subset I_{s-s_n}$ so that $\bigcap_n I_{s-s_n} \supset \bigcap_n I_{s_n}$. If $f \in \bigcap_n I_{s-s_n}$, $f^*(s-s_n) = f^*(s) - f^*(s_n) = s - s_n$ so that $f^*(s) - s = f^*(s_n) - s_n$ for all n. As $n \to \infty$, $s_n \to 0$ and $f^*(s_n) \to 0$, so that $f^*s = s$ ($f^*(s) - s$ can be made arbitrarily small). Therefore $f^*s_n = s_n$ for all $n \Rightarrow f \in \bigcap_n I_{s_n} \Rightarrow \bigcap_n I_{s-s_n} \subset \bigcap_n I_{s_n} \Rightarrow \bigcap_n I_{s-s_n} = \bigcap_n I_{s_n}$. Let $K = \bigcap_n I_{s_n}$, also a compact Lie group. We will show that the $\{I_{s_n}\}$ stabilize at K, i.e., there exists an N such that for $n > N$, $I_{s_n} = K$, from which it follows that $\bigcap_n I_{s-s_n}$ also stabilizes at K, i.e., there exists an N such that for all $n > N$, $I_{s-s_n} = K = I_{s_n}$, and hence the theorem. Define $K_n = \bigcap_1^n I_{s_n}$; then $\lim K_n = K$ and each K_n is a compact Lie subgroup of K_1. There exists an open neighborhood O of K, $K_1 \supset O \supset K$ such that if $O \supset K_n$, then K_n is conjugate to a subgroup of K (i.e., there exists an $f \in K_1$ such that $fK_nf^{-1} \subset K$) [30]. Since $\lim K_n = K$, there exists an N such that if $n > N$, then $O \supset K_n$ and therefore $fK_nf^{-1} \subset K$, so that $K_n = K$ for $n > N$ (since $K_n \supset K$). ■

V. The Topologies of Superspatial Manifolds

V.1. Riemannian Structures and Topology

We are interested in determining the topological implications of a Riemannian structure which exhibits a continuous group of symmetries. In doing so, we complement and globalize the classical work in Riemannian geometry. The setting in classical geometry is as follows: a laboratory metric is found to be invariant under a (local) coordinate Lie group. Due to this invariance, some statements can be made about the metric. If the dimension of the group is high enough, then these local metrics can be classified by the generators of the transformation (Lie algebra).

Here we will be interested only in topological statements about the manifold. The general idea is that a Riemannian metric which has a continuous group of symmetries cannot be supported by an arbitrary manifold; such global rigidity must be passed on to the underlying topology. Conversely, if the topology is "irregular" enough, then no Riemannian structure can exhibit any continuous group of symmetries. There will be many metrics which have local symmetries which do not extend to global symmetries, i.e., the local isometry extends to a global diffeomorphism but not to a global isometry. These metrics are left to the province of the classical physicist, for their presence does not restrict the topology.

The problem which we will investigate is this; first, what Lie groups can occur as isometry groups for Riemannian metrics, and second, what topologies are compatible with a Riemannian metric whose isometry group is isomorphic to a given Lie group? By making the trivial observation that if a Riemannian manifold (M,g) has isometry group $I_g(M)$, then it admits an effective action by $G \cong I_g(M)$, it is possible to apply many results from the theory of compact transformation groups to the above problem. In this setting we ask: What closed 3-dimensional manifolds can support the action of what compact Lie groups? The difficult cases of 1- and 2-dimensional orbits are answered in [32,36 38,46]. We briefly summarize the construction of these manifolds from their orbit structure, and give those manifolds which can support a transitive action. These constructions are used in Section VI, where the analysis is inverted, and the strata over each of these reconstructed manifolds is determined.

The central fact is that $I_g(M)$ is a Lie group of transformations for M [26]. By exploiting the interaction between the coupled actions

$$\mathrm{Diff}(M) \times \mathrm{Riem}(M) \to \mathrm{Riem}(M)$$

$$I_g(M) \times M \to M$$

it is possible to determine the topological types compatible with a Riemannian structure.

If (M,g) is superspatial, then (M,g) has continuous symmetry, finite (or discrete) symmetry, or no symmetry (or is generic) if $\dim I_g(M) > 0$, $\dim I_g(M) = 0$ (and hence $I_g(M)$ is finite) and $I_g(M) \neq \{id_M\}$, or $I_g(M) = \{id_M\}$. If M is superspatial, then M is symmetric, random, or wild, if $\dim I_g(M) > 0$ for some $g \in \mathrm{Riem}(M)$, $\dim I_g(M) = 0$ for all $g \in \mathrm{Riem}(M)$, or $I_g(M) = \{id_M\}$ for all $g \in \mathrm{Riem}(M)$.

If M is random, then $\mathrm{Diff}(M)$ acts almost freely (i.e., with finite isotropy group) on $\mathrm{Riem}(M)$. If M were wild, then $\mathrm{Diff}(M)$ would act freely and $\mathcal{S}(M)$ itself would be a manifold. This is equivalent to $\mathrm{Diff}(M)$ having no finite subgroups. Unfortunately no such manifold is known (in any dimension). In fact there is no manifold for which it can be proved that $\mathrm{Diff}(M)$ does not have a subgroup of order 2 (i.e., an involution) although such cannot always be found.

For any M, the generic metrics $\mathcal{M}_{\{id_M\}}$ form an open dense subset of $\mathrm{Riem}(M)$. Let g_0 and g_1 be generic and let $g_t = (1-t)\, g_0 + t g_1$, $+ \in [0,1]$, be the straight line segment connecting them. The set

$$\sigma = \{ t \mid I_{g_t}(M) \neq \{id_M\} \}$$

is a closed subset of I. Let $t' = \inf \sigma$. The $g_{t'-\varepsilon}$ for $\varepsilon > 0$ is generic. By smoothly attaching a completely unsymmetric bump to $(M, g_{t'-\varepsilon})$, the curve g_t can be deformed so as to avoid the summetric metrics $\{ g_{t''} | t'' \varepsilon\ \sigma\}$. The generic metrics are therefore path connected and connected so that the generic geometries $\mathcal{S}_{\{id_M\}} = \Pi\left(\mathcal{M}_{\{id_M\}}\right)$ form a connected subset of $\mathcal{S}(M)$.

The Classification Theorem lists the symmetric superspatial manifolds according to the possible isometry groups (connected component of the identity) that can occur. Such a list makes precise the intuitive feeling that "most" manifolds are random. A superspatial manifold that does not appear in the table is random (or perhaps even wild) with $I_g(M)$ finite (or trivial) for all $g \in \mathrm{Riem}(M)$. On the other hand, if $I_g(M)$ is "big enough" (i.e., the yet to be constructed M is symmetric "enough"), then M is determined up to diffeomorphism by a single algebraic invariant, $\Pi_1(M)$.

In general, $I_g(M)$ is a compact but not necessarily connected Lie group of diffeomorphisms. If $I_g(M)$ is not connected, the determination of the possible topologies of M is considerably more complicated. Using the idea that "most" of the action of $I_g(M)$ is contained in $I_g^\circ(M)$, the connected component of the identity, the construction of M can be divided into two steps.

First consider the case when $I_g(M)$, isomorphic to the compact non-connected abstract Lie group G, acts transitively on M (which is unknown). $I_g^\circ(M)$ is also a transitive action so that

$$M = \frac{G}{G_m} = \frac{G^\circ}{(G^\circ)_m} \quad .$$

In other words, knowledge of $I_g^\circ(M)$ alone is enough to determine the possible manifolds (topologies) that can support such an action. Moreover, since $(I_g(M))_m$ meets every component of $I_g(M)$ (because M is connected), the full group of isometries can be recovered as a semi-direct product,

$$I_g(M) = I_g^\circ(M) \dot{\oplus} (I_g(M))_m .$$

Suppose now that $I_g(M) = G$, a non-transitive, non-connected effective action on M. Then a fortiori M admits an effective action by the connected normal subgroup G°. If all manifolds which admit such an action can be determined, then those manifolds which can support an action by the properly larger group G will be among them.

Suppose such an M is determined. Then the orbit space $\frac{M}{G^\circ}$ is a Hausdorff topological space (because G is compact) of dimension ≥ 1 (because the action is not transitive). On this space of orbits the finite group

$\Gamma = \frac{G}{G^\circ}$ acts by permuting orbits, i.e., $gG^\circ(G^\circ(m)) = G^\circ(g(m))$.

If $\frac{M}{G^\circ}$ cannot support an action by Γ, then M cannot support the full group G and is ruled out of the class. Unfortunately, since most spaces can support the action of an arbitrary finite group, this usually does not further restrict M, so that knowing G usually does not give any further information about the possible manifolds that can occur than knowing G°.

V.2. The Classification of Superspatial Topologies

We now give a complete determination of those topologies compatible with Riemannian metrics which exhibit continuous groups of symmetries. Moreover, the possible groups which can arise as symmetry groups are given. Note that an arbitrarily large number of wormholes (wormhole model of the universe [53]) can have an SO(2) -symmetric geometry.

Theorem 4 (Classification Theorem of Superspatial Topologies): If (M,g) is a superspatial Riemannian manifold with continuous symmetry (i.e., M is symmetric), then $I_g^\circ(M)$ is isomorphic (as an abstract Lie group) to a group in the first column of the Classification Table and M is diffeomorphic to one of the associated manifolds in the second column.

Corollary: If M is superspatial, symmetric, and prime, then M is either S^3, P^3, T^3, $S^2 \times S^1$, a polyhedral manifold or a lens space. If M is superspatial and symmetric, then M is one of the above, S(p), $P^3 \# P^3$, or a connected sum of lens spaces and an S(p). M is then determined by $\pi_1(M)$ up to arbitrary second integers of lens spaces occurring in connected sums. If dim $I_g(M) > 2$, then M is determined (up to diffeomorphism) by $\pi_1(M)$. ■

In explanation of the table we briefly describe some standard constructions of topological 3-manifolds.

The connected sum $M_1 \# M_2$ of two oriented closed connected 3-manifolds is obtained by removing two open 3-cells E_1 and E_2 in M_1 and M_2 respectively and identifying the boundaries (which are spheres) of $M_1 - E_1$ and $M_2 - E_2$ by an orientation-reversing homeomorphism. $M_1 \# M_2$ is then a closed oriented connected 3-manifold. The connected sum is a commutative associative operation (a semigroup or monoid) on the set of homeomorphism classes of closed orientable connected 3-manifolds with identity element the class of S^3. If one removes two disjoint open 3-cells in M and identifies the resulting boundaries by an orientation-reversing homeomorphism, the result is isomorphic to $M \# (S^1 \times S^2)$. $S^1 \times S^2$ is a handle for M. We will be interested in adding handles only to S^3, and we write S(p) for the sphere with p handles,

$$S(p) = S^3 \# (S^1 \times S^2)_1 \# \ldots \# (S^1 \times S^2)_p .$$

M is prime if it is not the direct sum of any other non-trivial manifolds i.e., $M \neq M_1 \# M_2$, $M_1 \neq S^3$ and $M_2 \neq S^3$.

Let S^3 be viewed as the unit quaternions, that is, the unit sphere in

$$\mathbb{C} \times \mathbb{C},\ Q = \{(z_1,z_2) | (z_1,z_2) \in \mathbb{C} \times \mathbb{C},\ z_1\bar{z}_1 + z_2\bar{z}_2 = 1\}.$$

With this multiplicative structure S^3 is a compact 3-dimensional Lie group isomorphic to SU(2) = Symplectic(1) = Spin(3). Spin(3) is the simply connected universal covering group of SO(3) whose two-fold covering map π has as kernel $\{\pm 1\}$, so that SO(3) is topologically $\frac{S^3}{\{\pm 1\}} = P^3$, real projective 3-space. For p and q positive relatively prime let $\Gamma(p,q) \subset SO(4)$ denote the cyclic group of order p generated by $(e^{2\pi i/p}, e^{2\pi iq/p})$. $\Gamma(p,q)$ can be interpreted as rotating the second factor of $\mathbb{C} \times \mathbb{C}$ q times faster than the first. The lens spaces L(p,q) are the orbit spaces of $\Gamma(p,q)$ acting on S^3 by multiplication, $L(p,q) = \frac{S^3}{\Gamma(p,q)}$ $(L(1,1) = S^3$ and $L(2,1) = P^3)$. The L(p,q) are closed oriented connected 3-manifolds with fundamental group Z_p. For more information see [6,18]. A cyclic subgroup C_p of SO(3) can be viewed as lying in the $x_1 - x_2$ plane. In the embedding of R^3 in $R^4 = \mathbb{C} \times \mathbb{C}$ and SO(3) in SO(4), the action of C_p on S^3 is equivalent to the action of $\Gamma(p,1)$. Hence for

$$C_p \subset SO(3),\ \frac{S^3}{C_p} = \frac{S^3}{\Gamma(p,1)} = L(p,1).$$

To describe the dihedral and the polyhedral manifolds, we refer to [58] where the finite subgroups of SO(3) and SU(2) are classified. The finite subgroups of SO(3) are:

(1) the cyclic group C_p of order p generated by a rotation in the plane by an angle $\frac{2\pi}{p}$,

(2) the dihedral group D_m of order 2m generated by a rotation in the plane by an angle $\frac{2\pi}{m}$ and a flip about a line through the origin.

(3) the polyhedra groups (the symmetry groups of the regular polyhedra in R^3),

- (i) the tetrahedral group T of order 12,
- (ii) the octahedral group O of order 24 which contains T as a normal subgroup of index 2.
- (iii) the icosahedral group I of order 60.

CLASSIFICATION TABLE

of Superspatial Topologies

By Symmetry Groups

	G	M	$\Pi_1(M)$	Number of inequivalent actions	
Full Rotational Symmetry	$SO(4)$	S^3	0	1	Transitive Action
	$SO(3)\times SO(3)$	P^3	Z_2	1	
Axial-Spherical Symmetry	$\frac{T\times SU(3)}{D}$	$L(p,1)$, p odd	Z_p	1	
	$T\times SO(3)$	$L(p,1)$, p even	Z_p	1	
		$S^2\times S^1$	Z	1	
Toroidal Symmetry	T^3	T^3	$Z\oplus Z\oplus Z$	1	
Spherical Symmetry	$SU(2)$	$L(p,1)$, p odd	Z_p	1	
	$SO(3)$	$L(p,1)$ p even	Z_p	1	
		$\frac{S^3}{D_m^*}$	D_m^*	1	
		$\frac{S^3}{T^*},\frac{S^3}{O^*},\frac{S^3}{I^*}$	T^*,O^*,I^*	1,1,1	
		S^3	0	1	2-Dimensional Principal Orbits
		P^3	Z_2	1	
		$P^3 \# P^3$	$Z_2 * Z_2$	1	
		$S^2\times S^1$	Z	1	

CLASSIFICATION TABLE CONT.

	G	M	$\Pi_1(M)$	Number of inequivalent actions	
Reduced Toroidal Symmetry	T^2	T^3	$Z \oplus Z \oplus Z$	1	2-Dimensional Principal Orbits
		$S^2 \times S^1$	Z	1	
		$L(p,q)$	Z_p	1	
Axial or Cylindrical Symmetry	$SO(2)=T^2$	T^3	$Z \oplus Z \oplus Z$	1	1-Dimensional Principal Orbits
		$\frac{S^3}{D_m^*}$	D_m^*	1	
		$\frac{S^3}{T^*};\frac{S^3}{O^*};\frac{S^3}{I^*}$	T^*, O^*, I^*	1,1,1	
		$S(p)$	$Z^*\ldots{}^*Z$ (p factors)	$1+[\frac{p}{2}]$	
		$S(p)\#L(M_1,V_1)\#\ldots\#L(M_n,V_n)$	$Z^*\ldots{}^*Z^*Z_{M_1}\ldots{}^*Z_{M_n}$	$2^n(1+[\frac{p}{2}])$	
		S^3	0	∞	
		$S^2 \times S^1$	Z	∞	
		$L(p,q)$	Z_p	∞	

The other two regular polyhedrons in R^3, the hexahedron (cube) and the dodecahedron reproduce T and I respectively. Moreover, any two isomorphic finite subgroups of SO(3) are conjugate in SO(3).

The finite subgroups of SU(2) are obtained by pulling back the finite subgroups of SO(3) by the covering map π. These pulled back groups are the binary dihedral $D_m^* = \pi^{-1}(D_m)$, the binary polyhedral $T^* = \pi^{-1}(T)$, $O^* = \pi^{-1}(O)$, and $I^* = \pi^{-1}(I)$, and the cyclic groups C_p.

Before beginning the proof, we collect our notation and some facts from the theory of compact transformation groups that we will need in this and the next theorem. Suppose G is a compact Lie group acting differentiably on a connected manifold M with a specified action $\Phi, \Phi: G \times M \to M$. The orbit of m, G(m), is the image of m under the action. As G is compact, the orbits are compact (and hence closed). The isotropy group at m, G_m, is the subgroup of G which leaves m fixed, $G_m = \{g \in G | gm = m\}$. If $\bar{m} \in G(m)$ then $G_{\bar{m}} = gG_m g^{-1}$. G_m is a closed (and hence compact) subgroup of G. The action is free (respectively, almost free) if $G_m = \{id\}$ (respectively, G_m is finite) for all $m \in M$ and is effective if $\bigcap_{m \in M} G_m = G = \{id\}$. If the action is not effective there is at least one element $g \neq id$ which leaves all of M fixed, i.e. $gm = m$ for all $m \in M$, so that all of $\tilde{G}$ is not "really" acting. In this case the kernel of the action, $\tilde{G}$ is a closed normal non-trivial subgroup and $G/\tilde{G}$ is an effective action. We shall only be interested in effective actions; hereinafter we assume (Φ, G) is effective. (Φ, G) induces a continuous isomorphism $\tilde{\Phi} : G \to \mathrm{Diff}(M)$ so that the action (Φ, G) can be represented as a compact subgroup of Diff(M), $G^{\mathrm{Diff}} = \tilde{\Phi}(G)$. The isotropy group at m is then a compact group of diffeomorphisms, G_m^{Diff}, which leave m fixed. The linear isotropy group at m, denoted G_m^*, is the compact group

$$G_m^* = \{T_m f | f \in G_m^{\mathrm{Diff}}\} \subset GL^+(T_m M),$$

where T_m is the tangent functor at m and $GL^+(T_m M)$ is the group of isomorphisms of $T_m M$ with positive determinant. If g is a Riemannian metric for M such that $G^{\mathrm{Diff}} \subset I_g(M)$, then G_m^* is a compact subgroup of $SO(T_m M, g(m))$, the special orthogonal group of the Hilbert space $(T_m M, g(m))$. g(m) will then be said to be G_m^*-invariant.

The action is transitive if there is only one orbit and M is then diffeomorphic to the coset space $\frac{G}{G_m}$. A Riemannian manifold is homogeneous if its isometry group $I_g(M)$ is transitive. If the action is transitive and effective then G_m cannot contain a normal subgroup N because then $G_{\bar{m}} = gG_m g^{-1} \supset N$, so N would be a non-trivial common normal subgroup of the isotropy groups. If dim G(m) = dim M, then G(m) contains an open set of M and is hence open in M. As it is also closed, the action is transitive.

At each point of M there exists a slice [34], from which it follows that of all the isotropy groups that can occur there exists a unique absolute minimal (H) with respect to the partial ordering $\leq$ of conjugacy classes of subgroups of G. (H) is the principal orbit type and the orbits in $M_{(H)}$ are principal orbits. $M_{(H)}$ is an open dense invariant differentiable submanifold of M [28,29]. If the dimension of a principal orbit is r, then the set

$$M_r = \{m | m \in M, \dim G(m) = r\},$$

(those points which lie on orbits of dimension r) are the orbits of highest dimension, and $M_r - M_{(H)}$ are the exceptional orbits of highest dimension. M_r is always connected, and $M_{(H)}$ is connected if G is connected [29].

In the next section we shall be interested in the orbit space of the principal orbits, $P = \frac{M_{(H)}}{G}$. A slice at each point in $M_{(H)}$ is a differentiable local cross-section so that P can be given the structure of a differentiable manifold. The orbit projection map is then differentiable, and $M_{(H)}$ is the total space of a differentiable fiber bundle over P with fiber $\frac{G}{H}$ and structure group $\frac{N(H)}{H}$ (N(H) is the normalizer of H in G) [41]. In favorable cases the bundle is trivial, in which case $M_{(H)} = P \times \frac{G}{H}$ and P can be considered a differentiably embedded submanifold of $M_{(H)}$.

As a final remark we note that an r-dimensional toral Lie group acting effectively and differentiably on M must have trivial isotropy group on the principal orbits. (By the slice theorem G_m is locally constant for a toral action and so is constant on the principal orbits.) It follows that there must be r-dimensional orbits.

We now begin the proof. We examine each case by the dimension of the group. The manifolds which admit a transitive action are considered first. For the case of the 1- and 2-dimensional orbits we refer to the literature. The constructions are briefly summarized as they will be used in the next theorem. We recall that the homeomorphism classes coincide with the diffeomorphism classes for 3-dimensional manifolds.

<u>Proof of Superspatial Topology Theorem</u>: The basic fact is that $I_g(M)$ is a compact effective Lie transformation group for M, and $\dim I_g(M) \leq \frac{1}{2} n(n+1) = 6$ [20,26]. If the highest dimension is attained, then, $M = S^3$ or $P^3 = \frac{S^3}{\{\pm 1\}}$ and $I^o_g(M) \cong SO(4)$ or $\frac{SO(4)}{\{\pm 1\}} = SO(3) \times SO(3)$, respectively [20]. (In this case, M is isometric to these spaces of constant curvature.)

A classical result [13, 15] shows that an n-dimensional Riemannian manifold (n > 2) cannot admit a complete group of motions of dimension $\frac{1}{2}$ n (n+1) - 1. See also [23] for more recent work in this direction.) Hence $\dim I_g(M) \neq 5$.

If $\dim I_g(M) = 4$, then $I^o_g(M)$ is isomorphic to a compact connected 4-dimensional Lie group. T^4 must have a 4-dimensional orbit and so cannot act effectively. I^o_g is therefore a direct product of lower dimensional compact connected Lie groups or a factor group of a direct product by a discrete normal (and hence central) subgroup. The lower dimensional compact connected Lie groups are $SO(2) = T^1$; T^2; T^3, SO(3) and SU(2) = Spin(3) = S^3 (dimensions 1, 2, and 3 respectively) so that I^o_g is either T x SO(3), T x SU(2) or a factor of these by a discrete normal subgroup. If r is the dimension of the highest dimensional orbits, $\dim I_g(M) \leq \frac{1}{2} r(r+1)$ [31] so that r = 3. $I_g(M)$ therefore acts transitively on M (i.e. M is a homogeneous Riemannian manifold; see also [52]) so the isotropy group cannot contain a normal subgroup.

First suppose $I^o_g(M) = T \times SU(2) = G$ and assume that the isotropy group G_m (1-dimensional) lies in the second factor (it cannot lie in the first for an effective action). Then G^o_m is a maximal torus in SU(2). Since the maximal tori completely exhaust SU(2) by conjugation, the central normal subgroup $\{\pm 1\}$of SU(2) must lie in one and hence all maximal tori. In this case $G_m \supset \{\pm 1\}$and so the action is not effective. Hence G_m cannot lie completely in a factor, nor can G_m°. Let π_1 (respectively π_2) denote the projection map onto the first (respectively, second) factor. If G_m is not connected, then $\pi_1(G_m) = \frac{G_m}{\text{Ker } \pi_1}$ is an Abelian factor group and so Ker $\pi_1 \supset SO(2)$, the commutator subgroup of G_m. G_m^o therefore lies in the SU(2) factor, so G_m must be connected. G_m is contained in a maximal torus $T \times T \subset T \times SU(2)$ and is given by

$$U(p,q) = \{(e^{2\pi ipt}, e^{2\pi iqt}) | t \in R\}$$

with p and q non-zero relatively prime integers. $\pi_2(G_m)$ is then a maximal torus of SU(2) so that G_m contains a non-trivial element $(e^{(2\pi i/k)}, -1)$ of the center $T \times \{\pm 1\}$ and the normal subgroup generated by this element, $N = \{(e^{(2\pi in/k)}, (-1)^n) | n \in Z\}$. If k is odd, $N = Z_k \times \{\pm 1\}$ so that the action can be reduced to a $T \times SO(3)$ action. If k is even, then $k = 2h$ and N contains the normal subgroup $H = \{(e^{2\pi in/h}, 1) | n \in Z\}$. The action of G in this case can be reduced to $\frac{G}{N} = \frac{(G/H)}{(N/H)} = \frac{G}{D}$ with $D = \frac{N}{H} = \{(1,1), (-1,-1)\}$, and this is in fact the only reduction. For any discrete normal subgroup must be central, and since it cannot be a direct product it must be generated by $(e^{(2\pi i/k)}, -1)$, k even, and so G is again reduced by D.

To construct the manifolds that can support such an action, we examine the coset space $\frac{(T\times SU(2)/D)}{U(p,q)}$. First consider $\frac{T\times SU(2)}{U(p,q)}$. $(1, SU(2))$ is mapped onto this coset space with kernel $(1,SU(2)) \cap U(p,q) = (1,Z_p)$, so that $\frac{T\times SU(2)}{U(p,q)} = \frac{(1,SU(2))}{(1,Z_p)} = L(p,1)$. If p and q are odd, then $U(p,q) \supset D$ and we can form the quotient space $\frac{U(p,q)}{D}$; then $\frac{(T\times SU(2)/D)}{(U(p,q)/D)} = \frac{T\times SU(2)}{U(p,q)} = L(p,1)$, p odd. If p is odd and q is even, or if p is even and q is odd, $U(p,q) \supset \{(1,1), (-1,1)\}$, or $\{(1,1), (1,-1)\}$ and so the action is not effective (p and q cannot both be even).

If $G = T \times SO(3)$, and G_m lies in the second factor, then $M = \frac{T \times SO(3)}{(1, SO(2))} = S^1 \times S^2$ or $M = \frac{T \times SO(3)}{(1, O(2))} = S^1 \times P^2$, which is not orientable. If G_m does not lie in the second factor and is not connected, then G_m^o lies in the second factor and $M = \frac{T \times SO(3)}{G} = \frac{(T \times SO(3)/G_m^o)}{(G/G_m^o)} = \frac{T \times S^2}{Z_2} = S^1 \times P^2$. G_m is therefore a connected circle group so that $M = \frac{T \times SO(3)}{U(p,q)} = \frac{(1, SO(3))}{(1, Z_p)} = \frac{S^3}{Z_{2p}} = L(2p,1)$.

If $\dim I_g(M) = 3$, then $I_g^o(M)$ is a 3-dimensional compact connected Lie transformation group and hence must be isomorphic to $SU(2)$, $SO(3)$, or T^3. If $I_g^o = T^3$, the action is transitive and free (T^3 has no non-normal subgroups) so M is diffeomorphic to T^3. (M is actually isometric to a flat torus [13, p. 249].) Neither $SU(2)$ nor $SO(3)$ can be reduced by a discrete normal (and hence central) subgroup because $SO(3)$ has no center and $\frac{SU(2)}{\{\pm 1\}} = SO(3)$. Suppose $I_g^o = SU(2)$. $SU(2)$ has no 2-dimensional subgroups (the dimension of maximal tori must have even parity) so that there can be no 1-dimensional orbits. 2-dimensional orbits cannot occur (G_m would contain a maximal torus and hence $\{\pm 1\}$) so that $SU(2)$ must act transitively. M is therefore $\frac{SU(2)}{\Gamma}$ where Γ is a finite subgroup of $SU(2)$. Γ is cyclic, binary dihedral, or binary polyhedral. Since only the cyclic subgroups of odd order do not contain $\{\pm 1\}$, $M = \frac{SU(2)}{C_p} = L(p,1)$, p odd. Up to equivalence, there is only one effective action of $SU(2)$ on $L(p,1)$. If $I_g^o = SO(3)$ and acts transitively, then M is $\frac{SO(3)}{\Gamma}$ where Γ is a finite subgroup of $SO(3)$. Since $SO(3)$ has no non-trivial normal subgroups any finite subgroup of $SO(3)$ can occur as isotropy group. M is therefore one of the following spaces: $\frac{SO(3)}{C_m} = \frac{S^3}{Z_{2m}} = L(2m,1)$; $\frac{SO(3)}{D_m} = \frac{S^3}{D_m^*}$; $\frac{SO(3)}{T} = \frac{S^3}{T^*}$; $\frac{SO(3)}{O} = \frac{S^3}{O^*}$; $\frac{SO(3)}{I} = \frac{S^3}{I^*}$. There is only one effective and transitive action of SO(3) on any of these spaces. One-dimensional orbits cannot occur. The case of the principal orbits being 2-dimensional has been extensively studied in [32,33, 36]. We summarize how M is reconstructed from its orbit space. In the next theorem, these constructions will be used to obtain information about the strata of $\mathscr{S}(M)$.

If G is a compact connected Lie group acting on a compact n-dimensional manifold M with (n-1)-dimensional principal orbits then the orbit space $\frac{M}{G}$ is either the closed interval $\bar{I} = [0,1]$ or S^1. If $\frac{M}{G}$ is S^1, then there is only one orbit type (H) and M is a fiber bundle over S^1 with fiber $\frac{G}{H}$ and structure group $\frac{N(H)}{H}$. Each component of $\frac{N(H)}{H}$ gives a different bundle and action of G. If $\frac{M}{G}$ is $\bar{I}$, then there are two singular (non-principal) orbits of orbit type (H_1) and (H_2) which are mapped onto the end points of $\bar{I}$ by the orbit projection map. M is reconstructed from $\bar{I} \times \frac{G}{H}$ by identifying $\{0\} \times \frac{G}{H}$ to $\{0\} \times \frac{G}{H_1}$ by $(0,g,H) \equiv (0,\bar{g},H)$ if $g^{-1}\bar{g} \in H_1$ and similarly for $\{1\} \times \frac{G}{H}$ and $\{1\} \times \frac{G}{H_2}$. Up to equivalence, there is only one action of G on each manifold so constructed.

Suppose M is superspatial and G = SO(3). The principal orbits are then either $S^2 = \frac{SO(3)}{SO(2)}$ or $P^2 = \frac{SO(3)}{O(2)}$. Since M is orientable the principal orbits are orientable [4, p. 131] so the principal orbit type is SO(2). If the orbit space is S^1, then M is an S^2-bundle over S^1 with structure group $\frac{N(SO(2))}{SO(2)} = \frac{O(2)}{SO(2)} = Z_2$ or {id}. The bundle with structure group Z_2 is not orientable so that M is the trivial bundle $S^2 \times S^1$. If $\frac{M}{G} = \bar{I}$ and both exceptional orbits are fixed points, then $\{0\} \times S^2$ and $\{1\} \times S^2$ are identified to points so that $M = S^3$. When one exceptional orbit is P^2 and the other is a fixed point, then $M = P^3$. In picturing these two cases it is helpful to think of SO(3) acting on a closed solid ball in R^3 by rotation. The orbits are then S^2 and the origin is a fixed point. The boundary sphere of this ball is then identified to either a point (giving S^3) or to P^2 (giving P^3). By considering the connected sum of two closed balls in R^3 it can be seen that if both exceptional orbits are P^2, then $M = P^3 \# P^3$. If we add another closed ball then its boundary must be identified to a point in order that the orbit space be $\bar{I}$, so that $M = P^3 \# S^3 \# P^3 = P^3 \# P^3$.

If $\dim I_g(M) = 2$, then $I_g^o = T^2$ which also acts with 2-dimensional principal orbits and with principal orbit type $(\{1\})$. If the orbit space is S^1, then $\frac{N(T^2)}{T^2} = \{1\}$ so that $M = T^3$. If the orbit space is $\bar{I}$ and the two singular orbits are S^1 (T^2 cannot have fixed points), then M is either S^3, P^3, $S^2 \times S^1$, or a lens space $L(p,q)$. These are the only closed orientable manifolds on which T^2 can act.

If $\dim I_g(M) = 1$, then $I_g^o = SO(2)$. These actions have been recently classified in [38,46]. The results are: M is either S^3, P^3, $S^2 \times S^1$, or a lens space, admitting an infinite number of distinct actions, or $M = S(p) \# L(u_1,v_1) \# \ldots \# L(u_n,v_n)$ with $2^n(1 + [\frac{p}{2}])$ ([ ] means the greatest integer) distinct actions, or $M = T^3$, a polyhedral manifold, or $\frac{S^3}{D_m^*}$, with one distinct SO(2) action. ■

VI. THE STRATA OF SUPERSPACE

VI.1. Coordinates for Superspace

In this section we take a closer look at the manifolds of geometries that wind their way through superspace. An open set in each of these manifolds parameterizes the geometries of a particular symmetry type. Our first theorem shows that some of these manifolds are finite-dimensional. The coordinates of these finite-dimensional strata can be pieced together to give, in a sense to be discussed, coordinates for superspace.

Let $\mathcal{S}^F(M)$ denote the union of the finite-dimensional strata. $\mathcal{S}^F(M)$ is the finite-dimensional part (or region) of $\mathcal{S}(M)$. Let $\mathcal{M}^{Homo} = \{g | g \in Riem(M),\ I_g(M) \text{ is transitive}\}$. $\mathcal{M}^{Homo}$ is then an invariant set (because the equivalent action $f\, I_g(M) f^{-1}$ is transitive if $I_g(M)$ is), and we let $\mathcal{S}^{Homo}(M) = \Pi(\mathcal{M}^{Homo})$; $g^* \in \mathcal{S}^{Homo}(M)$ is a homogeneous geometry.

The Stratum Theorem says that the finite-dimensional part of superspace is precisely the homogeneous geometries, and that, moreover, the finite-dimensional strata are just open cells in a Euclispace. Combined with the Stratification Theorem, we conclude that

$\mathcal{S}^F(M)$ is a complex, "sitting finitely" in $\mathcal{S}(M)$. The finite-dimensional region of superspace is given the "coordinates" of this complex. Note that there might be other finite-dimensional subsets of $\mathcal{S}(M)$ which are not homogeneous, e.g. the Einstein structures [3], but these subsets are not strata.

VI.2. Finite-dimensional Manifolds of Geometries - The Stratum Theorem

The first theorem in our investigation of the strata of superspace gives a necessary and sufficient condition for a stratum to be finite-dimensional, as well as a criterion for an infinite-dimensional stratum to be contractible. The condition is given on the index. In the finite dimensional case, the indices also allow us to write down the strata explicitly. We begin by collecting our notation.

Let V be a real n-dimensional vector space with inner product (,), L(V) the continuous linear operators of V into itself, GL(V) the general linear group (non-singular operators), $L_s^2(V)$ the linear operators symmetric with respect to (,) (i.e., (X,AY) = (AX,Y) for all (X,Y ε V), and $L_s^{2+}(V)$ the positive definite symmetric operators ((X,AX) > 0 for X $\neq$ 0). $L_s^2(V)$ is a closed $\frac{1}{2}$ n(n+1) dimensional subspace of L(V). If A ε $L_s^{2+}(V)$, then $(X,Y)_A$ = (X,AY) is a new inner product for V, and conversely, if [,] is an inner product for V, there exists a unique operator A ε $L_s^{2+}(V)$ such that [X,Y] = (X,AY) (define A by its matrix elements). In other words, $L_s^{2+}(V)$ is isomorphic to the inner products on V. The exponential map on L(V) is a C^∞-diffeomorphism $L_s^2(V)$ onto $L_s^{2+}(V)$ (and of the skew-symmetric operators onto the orthogonal subgroup of GL(V) with respect to the inner product (,)) so that $L_s^{2+}(V)$ is an open $\frac{1}{2}$ n(n+1) dimensional positive cone in $L_s^2(V)$.

Let H be a compact subgroup of GL(V) and $[L_s^{2+}(V),H]$ = $\{A|A \ \varepsilon \ L_s^{2+}(V)$, Ah = hA for all h ε H$\}$, i.e., those positive definite symmetric operators which commute with all of H. If (,) is H-invariant, then $(,)_A$ is H-invariant if and only if A ε $[L_s^{2+}(V),H]$. For if A ε $[L_s^{2+}(V),H]$, then for all X,Y ε V, $(hX,hY)_A$ = (hX,AhY) = (hX,hAY) = (X,AY) = $(X,Y)_A$ for all h ε H. Conversely, if [,] is H-invariant and [,] = $(,)_A$ then [hX,hY] = [X,Y] = (hX,AhY) = (X,AY) = (hX,hAY) for all X,Y ε V, h ε H and so hA and Ah have the same matrix elements. $[L_s^{2+}(V),H]$ is a closed positive cone in $L_s^{2+}(V)$, but an open ball in some lower-dimensional (in general) subspace of $L_s^{2+}(V)$.

Theorem 5 (Stratum Theorem for Superspace): A stratum $\mathscr{S}_{(G)}(M)$ of $\mathscr{S}(M)$ is finite-dimensional (as a manifold) if and only if the action $\Phi : G \times M \to M$ is transitive. If G_o^* is the linear isotropy group at $o \in M$ of a transitive action, then $\mathscr{S}_{(G)}(M) = [L^{2+}(T_oM), G_o^*]$.

Corollary: If the principal orbits are (n-1)-dimensional and $M_{(H)}$ ((H) is the principal orbit type) is a trivial bundle over $\frac{M_{(H)}}{G}$, then $\mathscr{S}_{(G)}(M)$ is contractible.

Proof of Theorem: Suppose the action $G \times M \to M$, represented by G^{Diff}, is transitive. Then M is diffeomorphic to the homogeneous space $\frac{G}{G_m}$; consider G_m as an origin for M denoted o. If g(o) is a G_m^*-invariant inner product for T_oM, then g(o) has a unique extension to a G^{Diff}-invariant Riemannian metric g ($I_g(M) = G^{Diff}$) by translation on M, i.e., $g(m)(X_m,Y_m) = g(o)((f^{-1})^* X_m, (f^{-1})^* Y_m)$ with $X_m, Y_m \in T_mM$, $f \in G^{Diff}$ such that f(o) = m. Since g(o) is G_m^*-invariant, g is independent of the choice of f, and $I_g(M) = G^{Diff}$. Conversely, if g is G^{Diff}-invariant, then g(o) is a G_o^*-invariant inner product on T_oM. Consequently g is determined by its value at a single point, say o. The G^{Diff}-invariant Riemannian metrics are therefore in one to one correspondence with the G_o^*-invariant inner products on T_oM, i.e., with $[L_s^{2+}(T_oM), G_o^*]$.

If $\bar{g}$ is isometric to g, then there exists an $f \in Diff(M)$ such that $f^*g = \bar{g}$. $\bar{g}$ is a $fG^{Diff}f^{-1}$-invariant metric ($I_{f^*g}(M) = fG^{Diff}f^{-1}$) and is therefore also determined by its value at a single point, say $\bar{o} = f^{-1}(o)$. Since g and $\bar{g}$ are isometric, $(T_{\bar{o}}M,\bar{g}(\bar{o}))$ and $(T_oM,g(o))$ are isomorphic as Hilbert spaces (by $T_{\bar{o}}f$). In other words, if $T_{\bar{o}}M$ and T_oM are identified as abstract vector spaces, say to T_oM, by the vector space isomorphism $T_{\bar{o}}f$, then g(o) and $\bar{g}(\bar{o})$ are equivalent G_o^*- and $G_{\bar{o}}^* = (fG_of^{-1})^*$-invariant inner products on T_oM As conjugate groups of operators on T_oM and $T_{\bar{o}}M$, G_o^* and $G_{\bar{o}}^*$ are equivalent actions on T_oM by this definition. Conversely, if $(T_oM, g(o))$ and $(T_{\bar{o}}M, \bar{g}(\bar{o}))$ are isomorphic as Hilbert spaces and $g(o)$ and $\bar{g}(\bar{o})$ are G_o^*- and $G_{\bar{o}}^*$-invariant inner products (with G_o and $G_{\bar{o}}^*$ equivalent actions on T_oM and $T_{\bar{o}}M$ respectively), then the extended metrics g and $\bar{g}$ are isometric. Consequently the G_m^*-invariant (inequivalent) inner products on T_oM (now considered as an abstract vector space, or a general tangent space to M at an unspecified point, not necessarily o) are in one to one correspondence with the isometry classes of the G^{Diff}-invariant metrics, or the (G^{Diff}) -invariant geometries. Hence $\mathscr{S}_{(G)}(M)$ is parameterized by these G_m^*-invariant geometries. Hence $\mathscr{S}_{(G)}(M)$ is parameterized by these G_m^*-invariant products, $\mathscr{S}_{(G)}(M) = [L_s^{2+}(T_oM), G_o^*]$ with T_oM and G_o^* considered abstractly as a vector space and an equivalence class of actions of a compact group of operators.

If the action $G \times M \to M$ is not transitive, we will show that $\mathscr{S}_{(G)}(M)$ is parameterized by the cross-sections of a bundle and is therefore infinite-dimensional. Since G is transitive on each orbit, the dimension of $\mathscr{S}_{(G)}(M)$ is determined, roughly speaking, by the number of orbits. Let (H) be the principal orbit type, $m \in M_{(H)}$, S a slice at m, and $\dim G(m) = r$. Since $M_{(H)}$ is open and (H) is minimal, $\dim S = k = n - r$ and S is a cross-section for the orbits in G(S). If $M_{(H)}$ is not a trivial bundle, we work locally over the slice. This will give the infinite-dimensionality of a "local superspace", from which the result follows. Let $L_s^{2+}(M;S)$ be the bundle $L_s^{2+}(M)$ restricted to S, and $B^\infty(L_{s_\sim}^{2+}(M;S))$ the bounded C^∞-cross-sections (all derivatives bounded). If g is a cross-section of this bundle with $\tilde{g}(s)$ a G_s^*-invariant inner product on $T_sM_{(H)} = T_sM$ for all $s \in S$, then $\tilde{g}$ will be said to be G_s^*-invariant. Since G acts transitively on each orbit every G_s^*-invariant cross-section has a unique extension to a G^{Diff}-invariant metric on G(S). Conversely, the G^{Diff}-invariant metrics on G(S) determine a G_s^*-invariant cross-section of the restricted bundle. Consequently the G^{Diff}-invariant metrics are parameterized by the G_s^*-invariant cross-sections of the restricted bundle.

If F is a diffeomorphism of the open k-cell S onto itself, then f can be extended to a diffeomorphism of M [40]. By this extension, Diff(S) acts on the cross-sections of $B^\infty(L_s^{2+}(M;S))$. The geometries of the invariant set G(S) (local superspace) are thus parameterized by the infinite-dimensional space $\dfrac{B^\infty(L_s^{2+}(M;S))}{\mathrm{Diff}(S)}$. ■

Proof of Corollary: If the dimension of a principal orbit is n - 1 and $M_{(H)}$ is a trivial bundle, then the orbit space of the principal orbits $P = \dfrac{M_{(H)}}{G}$, is a differentiably embedded open interval or circle in M. An orbit G(m) in $M_{(H)}$ is also differentiably embedded so $T_m(G(m))$ can be considered a subspace of T_mM. If g is a G^{Diff}-invariant metric on M, let N(m) denote the orthogonal complement of $T_m(G(m))$ in T_mM. Since N(m) is a 1-dimensional C^∞-distribution (and hence involutive), P can be chosen to be orthogonal to each orbit that it intersects, i.e., $g(p)(T_pP,T_pG(p)) = 0$, $p \in P$. g(p) therefore splits as a direct sum, $g^{\mathrm{transverse}} + g^{\mathrm{longitudinal}}$, with g^t a bounded Riemannian metric for P and g a bounded G_p^*-invariant cross-section of the bundle $L_s^{2+}(T\mathcal{O})$, with $\mathcal{O}$ the bundle of tangents to the orbits, $\mathcal{O} = \bigcup_{p\in P} T_pG(P)$. If $L_s^{2+}(T\mathcal{O};*)$ is the reduced bundle of G_p^*-invariant inner products of $T_pG(P)$, then the G^{Diff}-invariant metrics on the principal orbits are parameterized by $\mathrm{Riem}(P) \times B^\infty(L_s^{2+}(T\mathcal{O};*))$, which then have a unique extension to G^{Diff}-invariant metrics on M (the principal orbits are dense). Reasoning as before, the G_p^*-invariant inner products of $T_pG(P)$ parameterize the (G^{Diff})-invariant geometries of G(P) so that the geometries of M are parameterized by $\dfrac{\mathrm{Riem}(P)}{\mathrm{Diff}(P)} \times B^\infty(L_s^{2+}(T\mathcal{O};*))$, i.e., $\mathcal{S}_{(G)}(M) = \mathcal{S}(P) \times B^\infty(L^{2+}(T\mathcal{O};*))$.

Suppose $P = S^1$. Since any metric on S^1 has O(2) as isometry groups, $\mathrm{Riem}(S^1)$ is fibered by $\mathrm{Diff}(S^1)$ with fiber $\dfrac{\mathrm{Diff}(S^1)}{O(2)}$. Since $\mathrm{Riem}(S^1)$ and $\mathrm{Diff}(S^1)$ are modeled on $C^\infty(S^1,R)$, the manifold $\mathcal{S}(S^1)$ is modeled on $\frac{C^\infty(S^1,R)}{C^\infty(S^1,R)/R} = R$. Hence $\mathcal{S}(S^1) = R^+$. Similarly for P = I (using bounded metrics). Since each fiber of $L_s^{2+}(T\mathcal{O};*)$ is a cell (and hence contractible), $\mathcal{S}_{(G)}(M)$ is contractible.■

VI.3. The Classification of Superspatial Strata

In this section we apply the results of the Stratum Theorem to the strata of superspatial manifolds, the superspatial strata. The Superspatial Strata Theorem determines explicitly the finite dimensional and contractible superspatial strata, thereby giving parameters (coordinates) for these geometries. The other infinite-dimensional strata are themselves quotient spaces.

Theorem 6 (Superspatial Strata Theorem): If M heads a column in the Superspatial Strata Table, then $\mathcal{S}_{(G)}(M)$ is given in the slot corresponding to G.

Remark: For dim G > 1, G ≠ SO(3), there is at most one action of G on a superspatial manifold, i.e., $\mathcal{S}_{(G)}(M) = \mathcal{S}_G(M)$. If G = SO(3) then P^3 admits either a transitive action or an action with 2-dimensional orbits. This is the only ambiguity.

Proof: We first determine the finite dimensional strata. If G = SO(4) or SO(3) x SO(3), then M is S^3 or P^3 respectively and G_m = SO(3). SO(3) acts effectively on $R^3 = T_mM$, and is therefore topologically equivalent to the action of all proper rotations around the origin [31, p. 260]. Any inner product on R^3 left invariant by this action is a positive scalar multiple of the usual Euclidean one, so that $\mathcal{S}_{SO(4)}(S^3) = R^+$ and $\mathcal{S}_{SO(3)xSO(3)}(P^3) = R^+$. (These geometries are spheres and projective spaces of constant curvature parameterized by their radii.)

If $G = \frac{T \times SU(2)}{D}$ or T x SO(3), then G_m = SO(2) or O(2) and acts on R^3 as a circle group of rotations (and a reflection if G_m = O(2)) around some fixed axis [31, p. 260]. SO(2) or O(2) acts on the plane perpendicular to this fixed axis so the SO(2)- or O(2)-invariant inner products on R^2 are parameterized by R^+. Any dilation of distances (change of scale) on the fixed axis is also SO(2)-or O(2)-invariant, so that

$\mathscr{S}_{\frac{T\times SU(2)}{D}}(L(p,1)) = R^+ \times R^+ = (R^+)^2$, p odd; $\mathscr{S}_{T\times SO(3)}(L(p,1)) =$

$(R^+)^2$, p even; and $\mathscr{S}_{T\times SO(3)}(S^2 \times S^1) = (R^+)^2$.

If $G = T^3$, then $G_m = \{e\}$ and $M = T^3$, so there is no restriction on the inner products on $T_mM = R^3$. Each inner product of R^3 induces a geometry on R^3 with isometry group T^3. Therefore $\mathscr{S}_{T^3}(T^3) = L_s^{2+}(R^3) = (R^+)^6$. (These geometries are the isometry classes of the flat tori [58, p. 123].)

If $G = SU(2)$, G_m is odd cyclic. If $G_m = \{e\}$, $M = S^3$ and each inner product on R^3 determines a geometry on S^3, i.e., $\mathscr{S}_{SU(2)}(S^3) = (R^+)^6$. If $G_m = C_p$, p odd, > 2, C_p acts as a finite group of rotations of R^3, leaving an axis fixed. R^+ parameterizes the Z_p-invariant distances on the fixed axis. An inner product on R^n deforms the unit sphere (in the standard inner product) into an ellipse. If we parameterize the inner products on R^2 by the major and minor axes of these ellipses and their orientation in the plane, an inner product is G_m-invariant if and only if G_m is a symmetry group for the corresponding ellipse. Any ellipse is Z_2-invariant, but only the circle is Z_p-invariant, $p > 2$. Consequently, a Z_p-invariant inner product, $p > 2$, is SO(2)-invariant and is therefore parameterized by R^+. Hence $\mathscr{S}_{SU(2)}(L(p,1)) = (R^+)^2$, p odd, > 2.

If $G = SO(3)$ and acts transitively with $G_m = \{e\}$, then $M = P^3$ and $\mathscr{S}_{SO(3)}(P^3) = (R^+)^6$. If $G_m = Z_2$, then $\mathscr{S}_{SO(3)}(L(4,1)) = (R^+)^4$ (any ellipse is Z_2-invariant). If $G_m = Z_p$, $p > 2$, then the G_m-invariant inner products on R^2 are SO(2)-invariant, so

$\mathscr{S}_{SO(3)}(L(2p,1)) = (R^+)^2$, $p > 2$. If $G_m = D_2$, then any ellipse with major and minor axes oriented along the x and y axes is invariant by D_2 so that $R^+ \times R^+$ (the major and minor axes) parameterizes the D_2-invariant inner products. Hence $\mathscr{S}_{SO(3)}(\frac{S^3}{D_2^*}) = (R^+)^3$.

If $G_m = D_m$, $m > 2$, the D_m-invariant inner products on R^2 are parameterized by R^+ because D_m contains Z_m as a cyclic subgroup. Z_2 acts on the fixed axis but any change of scale on this axis is Z_2-invariant. Therefore $\mathscr{S}_{SO(3)}(\frac{S^3}{D_m^*}) = R^+ \times R^+ = (R^+)^2$, $m > 2$.

The inner products on R^3 are isomorphic to the ellipsoids of revolution parameterized by the lengths of their three axes and the three Euler angles of rotation. If any ellipsoid is left invariant by T, I or O, then the ellipsoid is a sphere and is therefore left invariant by all of SO(3). Hence the inner products on R^3 left invariant by T, I or O are parameterized by R^+ so that

$$\mathscr{S}_{SO(3)}(\frac{S^3}{T^*}) = \mathscr{S}_{SO(3)}(\frac{S^3}{I^*}) = \mathscr{S}_{SO(3)}(\frac{S^3}{O^*}) = R^+.$$

If SO(3) acts non-transitively, the principal orbits are topologically S^2 and the principal orbit space P is either S^1 or the open interval I. In either case the principal orbits are a trivial bundle over P so that if g is an SO(3)-invariant metric on M, P can be considered an orthogonally embedded submanifold. g is determined by $g|P$, which splits at each point into an inner product on T_pP and an SO(2)-invariant inner product on $T_pG(P)$. Since the SO(2)-invariant inner products on R^2 are parameterized by R^+, $L_s^{2+}(T\mathcal{O};*)$ is a 1-dimensional sub-bundle of $L_s^{2+}(T\mathcal{O})$, which we denote (P,R^+). The SO(3)-invariant metrics on M are therefore parameterized by Riem(P) x $B^\infty(P,R^+)$. Metrically, the principal orbits are of constant curvature [13, p. 245] so that $B^\infty(P,R^+)$ is interpreted as a C^∞-assignment of radii to each sphere which intersects P.

If $P = S^1$ then $M = S^2 \times S^1$ and the longitudinal cross-sections are parameterized by $C^\infty(S^1,R^+)$. If P = I and the singular orbits are fixed points then $M = S^3$. Since M is reconstructed by identifying the spheres to points at the endpoints of P, we further require that $\lim_{p\to 0} g^\ell(p) = \lim_{p\to 1} g^\ell(p) = 0$. These boundary conditions essentially compactify P so that g^ℓ is again parameterized by $C^\infty(S^1,R^+)$. If the non-principal orbits are P^2 then $M = P^3 \# P^3$ and we only require that $\lim_{\substack{p\to 0\\ p\to 1}} g^\ell(p)$ is finite (i.e., that g^ℓ is bounded)

SUPERSPATIAL STRATA TABLE

G \ M	S^3	P^3	$S^2 \times S^1$	L(4,1)	L(p,1), p odd, >2	L(p,1), p even, >4
SO(4)	R^+					
SO(3) × SO(3)		R^+				
$\frac{T \times SU(2)}{D}$	$(R^+)^2$				$(R+)^2$	
T × SO(3)		$(R^+)^2$	$(R^+)^2$	$(R^+)^2$		$(R^+)^2$
T^3						
SU(2)	$(R^+)^6$				$(R^+)^2$	
SO(3)	$R^+ \times C^\infty(S^1,R^+)$	$(R^+)^6$; $R^+ \times B^\infty(I,R^+)$; $\lim_{p \to 0} g^\ell(p) = 0$	$R^+ \times C^\infty(S^1,R^+)$	$(R^+)^4$		$(R^+)^2$
T^2	$R^+ \times B^\infty(I,(R^+)^3)$	$R^+ \times B^\infty(I,(R^+)^3)$	$R^+ \times B^\infty(I,(R^+)^3)$			

SUPERSPATIAL STRATA TABLE CONT.

G \ M	T^3	$\frac{S^3}{D_2^*}$	$\frac{S^3}{D_m^*}$, m>2	$\frac{S^3}{T^*}, \frac{S^3}{O^*}, \frac{S^3}{I^*}$	$P^3 \# P^3$	L(p,q), p > 2
SO(4)						
SO(3) × SO(3)						
$\frac{T \times SU(2)}{D}$						
T × SO(3)						
T^3	$(R^+)^6$					
SU(2)						
SO(3)		$(R^+)^3$	$(R^+)^2$	R^+	$R^+ \times B^\infty(R^+)$	
T^2	$R^+ \times C^\infty(S^1, (R^+)^3)$					$R^+ \times B^\infty(I, (R^+)^3)$

so that g^ℓ is parameterized by $B^\infty(I,R^+)$. If the singular orbits are P^2 and a fixed point, $M = P^3$ and g^ℓ is parameterized by $B^\infty(I,R^+)$, together with the boundary condition $\lim_{p\to 0} g(p) = 0$.

If $G = T^2$, the principal orbit type is {id} and the principal orbits are T^2. If $P = S^1$, then $M = T^3$ and g^ℓ is parameterized by $C^\infty(S^1, (R^+)^3)$. In all other cases $P = I$ and g^ℓ is parameterized by $B^\infty(I,(R^+)^3)$, so $\mathcal{S}_{T^2}(M) = R^+ \times B^\infty(I,(R^+)^3)$ or $\mathcal{S}_{T^2}(T^3) = R^+ \times C^\infty(S\ (R^+)^3)$. Metrically, the orbits are flat tori, so that $B^\infty(I,(R^+)^3)$ can be interpreted as a bounded C^∞-choice of flat tori along the cross-section I. Note that the flat tori are parameterized by a 3-dimensional space [58, p. 79]. In each case the geometries of M are parameterized by the product of $\mathcal{S}(P) = R^+$ with the appropriate space of cross-sections.

VII. THE EXTENSION OF SUPERSPACE TO A PROPER MANIFOLD

When we begin to cast our eyes toward the dynamical situation it is important to know how the strata influence the geometric motion. For classical motions we expect a reflection to occur at the boundaries of the strata; for the quantum case, a scattering, either back into $\mathcal{S}(M)$ or into some other $\mathcal{S}(M')$. (This is not to say that a generic geometry cannot also be scattered into some $\mathcal{S}(M')$.) In the finite-dimensional regions of superspace (i.e. for the homogeneous geometries), it has already been demonstrated by Misner's mixmaster model of the universe [see Misner, these proceedings] that the boundaries of the strata influence the geometric motion by reflections. The special situation adds weight to the feeling that the strata play a fundamental role and must be taken into account to describe fully the motion in superspace.

Nonetheless, for the purposes of writing differential equations, it would be helpful if superspace could be extended to a proper manifold. By projecting down the geometric motion on this manifold to superspace, the influence of the strata could then be taken into account. In this section we show that such an extension, denoted $\mathcal{S}^{ext}(M)$, always exists.

This extended superspace comes from Riem(M) by the action of a subgroup of Diff(M). By choosing the subgroup judiciously, the resulting orbit space is a manifold. Roughly speaking, $\mathcal{S}^{ext}(M)$ lies between Riem(M) and $\mathcal{S}(M)$, or, equivalently, comes from Riem(M) by fewer identifications than were made for $\mathcal{S}(M)$. In turn $\mathcal{S}(M)$

can be recovered from $\mathcal{S}^{ext}(M)$ by completing the identifications by means of a projection map $\Pi^{ext} : \mathcal{S}^{ext}(M) \to \mathcal{S}(M)$.

We would like some way to measure "how far" $\mathcal{S}^{ext}(M)$ is from $\mathcal{S}(M)$, or, in other words, to measure how many identifications have been "left out." Otherwise, for example, we could leave out all the identifications. The extended proper manifold in this case would just be Riem(M).

If $\mathcal{H}$ is a Lie subgroup of $\mathcal{D}$, then the action of $\mathcal{D}$ on $\mathcal{M}$ restricts to an action of $\mathcal{H}$ on $\mathcal{M}$, $\mathcal{H} \times \mathcal{M} \to \mathcal{M}$, whose orbit space we denote $\mathcal{S}_{\mathcal{H}}(M)$. $\mathcal{H}$ defines a projection map $\Pi_{\mathcal{H}}: \mathcal{S}_{\mathcal{H}}(M) \to \mathcal{S}(M)$ by $\Pi_{\mathcal{H}}(\mathcal{H}g) = \mathcal{O}(g)$. We will take the <u>codimension of the extension</u> (or the <u>codimension of $\mathcal{S}(M)$ relative to $\mathcal{S}_{\mathcal{H}}(M)$</u>), denoted $\dim(\mathcal{S}_{\mathcal{H}}(M)/\mathcal{S}(M))$, and defined as the codimension of $\mathcal{H}$ in $\mathcal{D}$, as a measure of the closeness of $\mathcal{S}_{\mathcal{H}}(M)$ to $\mathcal{S}(M)$. Roughly speaking, the codimension of the extension is the "dimension of the identifications" that is needed to recover $\mathcal{S}(M)$ from $\mathcal{S}_{\mathcal{H}}(M)$ under the projection $\Pi_{\mathcal{H}}$.

<u>Theorem 7 (Extension Theorem for Superspace)</u>: For every superspatial M, $\mathcal{S}(M)$ can be extended to a proper manifold, $\mathcal{S}^{ext}(M)$, such that $\dim(\mathcal{S}^{ext}(M)/\mathcal{S}(M)) = n(n+1)$ ($n = \dim M$).

<u>Proof</u>: Let $m \in M$ and let $\mathcal{D}_m^{id}$ denote the subgroup $\mathcal{D}_m^{id} = \{f \in \mathcal{D} | f(m) = m \text{ and } T_m f = id_{T_m M}\}$.

In other words, $f \in \mathcal{D}_m^{id}$ iff f is tangent to id_M at m (i.e. $f(m) = id_M(m)$ and $T_m f = T_m id_M = id_{T_m M}$). $\mathcal{D}_m^{id}$ is a Lie subgroup of $\mathcal{D}$ which acts on $\mathcal{M}$ by restricting the action of $\mathcal{D}$,

$$\mathcal{D}_m^{id} \times \mathcal{M} \to \mathcal{M}.$$

We now recall that if two isometries are tangent at a point they must be equal [17, P. 62]. Consequently if $f \in \mathcal{D}_m^{id}$ is an isometry for some g, i.e. $f^*g = g$, $f = id_M$. Therefore $\mathcal{D}_m^{id}$ has a trivial isotropy at each point of $\mathcal{M}$ and so acts freely. The slice for each $g \in \mathcal{M}$ reduces to a local cross-section and the action of $\mathcal{D}_m^{id}$ defines a principal fiber bundle. The orbit space of this action, $\mathcal{S}^{ext}(M)$, is the base space for this bundle and is a manifold.

To show that $\mathscr{D}_m^{id}$ has finite codimension in $\mathscr{D}$ we consider the subgroup

$$\mathscr{D}_m = \{f \in \mathscr{D} \mid f(m) = m\}.$$

$\mathscr{D}$ acts on M transitively with isotropy group at m equal to $\mathscr{D}_m$. Therefore $\mathscr{D}/\mathscr{D}_m$ is diffeomorphic to M. $\mathscr{D}_m^{id}$ is a normal subgroup of $\mathscr{D}_m$ and $\mathscr{D}_m/\mathscr{D}_m^{id} = GL^+(T_m M)$. Since

$$\dim \frac{\mathscr{D}}{\mathscr{D}_m^{id}} = \dim \frac{\mathscr{D}}{\mathscr{D}_m} + \dim \frac{\mathscr{D}_m}{\mathscr{D}_m^{id}} ,$$

we have

$$\dim \frac{\mathscr{S}^{ext}(M)}{\mathscr{S}(M)} = n(n+1). \blacksquare$$

VIII. ACKNOWLEDGEMENTS

I would like to take this opportunity to express my thanks to Professors R. Abraham, G. Bredon, W. Browder, P. Chernoff, B. Dewitt, J. Marsden, L. Michel, P. Orlik, C. Papakyriakopoulos, J. Stasheff, N. Steenrod, and to the Princeton graduate students Charles Fefferman, Dick Graff, Ralph Greenberg, Arthur Greenspoon, Harsh Pittie, and Frank Quinn for their conversations, interest and encouragement.

I would especially like to thank Professors Richard Palais and John Wheeler whose guidance and ideas were a continued source of inspiration. At several dark moments during the preparation of this work their fortitude prevented despair. Lastly, I would like to thank Gwen Townsend of Smith College for the original preparation of the tables.

IX REFERENCES

[1] Abraham, R., Lectures of Smale on differential topology, Notes at Columbia University, New York, 1962.

[2] Abraham, R. and J. Robbin, Transversal Mappings and Flows, Benjamin, New York (1967).

[3] Berger, M. and D. Ebin, Some decompositions of the space of symmetric tensors on a Riemannian manifold, Journ. of Diff. Geom. (to appear).

[4] Borel, A., Seminar on Transformation Groups, Princeton University Press, Princeton, N.J., 1960.

[5] Bourbaki, N., Topologie Génerale, (Actualitiés Scientifiques et Industrielles, No. 1142) 2nd ed. Paris: Herman et Cie, 1951. Ch. I, Sections 9 and 10.
[6] Cooke, G., and R. Finney, Homology of Cell Complexes, Princeton University Press, Princeton (1967).
[7] DeWitt, B., Quantum theory of gravity. I. The canonical theory, Phys. Rev. 160 (1967), 1113-1148.
[8] Ebin, D.G., On the space of Riemannian metrics, Doctoral Thesis, M.I.T., Cambridge, Mass., 1967.
[9] _________, On the space of Riemannian metrics, A.M.S. Bulletin, 74 (1968), 1001-1003
[10] _________, The manifold of Riemannian metrics, Proceedings of the 1968 A.M.S. Summer Institute on Global Analysis (to appear).
[11] Eells, J., On the geometry of function spaces, Symp. Inter. de Topologia Alg., Mexico, (1956); 1958, 303-308.
[12] _________, A setting for global analysis, Bull. Amer. Math. Soc. 72 (1966), 751-807.
[13] Eisenhart, L., Riemannian Geometry, Princeton University Press, Princeton, 1949.
[14] Freifeld, C., One-parameter subgroups do not fill a neighborhood of the identity in an infinite-dimensional Lie (pseudo-) group, Battelle Rencontres, Benjamin, New York, 1968. (Ed. J. A. Wheeler and C. DeWitt.)
[15] Fubini, G., Sugli Spaziche amettono ungruppo continuo dimovimenti; Annali di Mat., 8 (1903), 39-81.
[16] Geroch, R. P., Topology in general relativity, J. Math. Phys. 8 (1967), 782-786.
[17] Helgason, S., Differential Geometry and Symmetric Spaces, Academic Press, New York, 1962.
[18] Hilton, P.J., and S. Wylie, Homology Theory, Cambridge University Press, Cambridge, 1960.
[19] Kelley, J., General Topology, Van Nostrand, New York, 1955.
[20] Kobayashi, S., and K. Nomizu, Foundations of Differential Geometry I, Interscience Publishers. New York. 1963.
[21] Lang, S., Introductions to Differentiable Manifolds, Interscience Publishers, New York, 1962.
[22] Leslie, J., On a differential structure for the group of diffeomorphisms, Topology 6 (1967), 263-271.
[23] Mann, L.N., Gaps in the dimensions of compact transformation groups, Conference on Transformation Groups, Springer-Verlag, New York, 1968, 293-296.
[24] Markov, A., Insolubility of the problem of homeomorphy, Proceedings Intern. Congress of Math. 1958, Cambridge University Press.
[25] McMillan, D., Some contractable open 3-manifolds, Trans. Am. Math. Soc. 102 (1962), 373-382.
[26] Meyers, S. and N. Steenrod, The group of isometries of a Riemannian manifold, Ann. of Math. 40 (1939), 400-416.

[27] Moise, E., Affine structures in 3-manifolds, V: The triangulation theorem and Hauptvermutung, Ann. Math. 56 (1952), 92-114.

[28] Montgomery, D., H. Samelson, and L. Zippin, Singular points of a compact transformation group, Ann. of Math., 63 (1956), 1-9.

[29] Montgomery, D., Exceptional orbits of highest dimension, Ann. of Math., 64 (1956), 131-141.

[30] Montgomery, D. and L. Zippin, A theorem on Lie groups, Bull. Amer. Math. Soc., 48 (1942), 448-452.

[31] ________________, Topological Transformation Groups, Interscience, New York, 1955.

[32] Mostert, P., On compact Lie groups acting on a manifold, Ann. of Math., 65 (1957), 447-455.

[33] ________, Errata, Ann. Math. 66 (1957), 589.

[34] Mostow, G. D., Equivariant embeddings in Euclidean space, Ann. of Math., 65 (1957), 432-446.

[35] Munkres, J.R., Elementary Differential Topology, Ann. of Math. Studies, No 54, Princeton University Press, Princeton 1963.

[36] Neumann, W.D., 3-dimensional G-manifolds with 2-dimensional orbits, Conference on Transformation Groups, Springer-Verlag, New York, 1968, 220-222. (ed. P. Mostert.)

[37] Omori, H., On the group of diffeomorphisms of a compact manifold, Proceedings of the 1968 A.M.S. Summer Institute on Global Analysis (to appear).

[38] Orlik, P., and F. Raymond, Actions of SO(2) on 3-manifolds, Conference on Transformation Groups, Springer-Verlag, New York, 1968, 297-318. (Ed. P. Mostert.)

[39] Palais, R., Natural operations on differential forms, Trans. Amer. Math. Soc., 92 (1959), 125-141.

[40] ________, Extending diffeomorphisms, Proc. Amer. Math. Soc., 11 (1960), 274-277.

[41] ________, The Classification of G-Spaces, Memoirs A.M.S., No. 36 (1960).

[42] ________, Equivalence of nearby differentiable actions of a compact group, Bull. Amer. Math. Soc., 67 (1961), 362-364.

[43] ________, On the existence of slices for actions of non-compact Lie groups, Ann. of Math., 73 (1961), 295-322.

[44] ________, Homotopy theory of infinite dimensional manifolds, Topology 5 (1966), 1-16.

[45] Papakyriakopoulos, C. D., Some problems on 3-dimensional manifolds, Bull. Amer. Math. Soc., 64 (1958), 317-335.

[46] Raymond, F., Classification of the actions of the circle on 3-manifolds, A.M.S. Trans. 131 (1968), 51-78.

[47] Seifert, H. and W. Threlfall, Lehrbuch der Topologie, Cheldea, New York, 1947.

[48] Stern, M., Investigations of the Topology of Superspace, A.B. thesis, Princeton University, Princeton, 1967 (Unpublished.)

[49] Stone, A., Paracompactness and product spaces, Bull. A.M.S. 54 (1948), 977-982.

[50] Thom, R., Les singularitiés des applications differentiables, Seminaire Bourbaki, Paris, May, 1956.

[51] __________, Les singularitiés des applications differentiables, Annales de l'Institut Fourier, VI (1956), 43-87.

[52] Wang, H.C., On Finsler spaces with completely Integrable equations of Killing, Journ. of the London Math Soc., 22 (1947), 5-9.

[53] Wheeler, J. A., Geometrodynamics, Academic Press, New York, 1962.

[54] ____________________, Superspace and the nature of quantum geometrodynamics, Battelle Rencontres, Benjamin, New York, 1968. (Ed. J. A. Wheeler and C. DeWitt.)

[55] ____________________, Geometrodynamics and the issue of the final state, Relativity, Groups and Topology, Gordon and Breach, New York, 1964. (Ed. C. DeWitt. and B. DeWitt.)

[56] Whitney, H., Elementary structures of real algebraic varieties, Ann. of Math. 66 (1957), 545-556.

[57] __________, Tangents to an analytic variety, Ann. of Math. 81 (1965), 496-549.

[58] Wolf, J., Spaces of Constant Curvature, McGraw Hill, New York, 1967.

[59] Yang, C. T., The triangulability of the orbit space of a differentiable transformation group, Bull. Amer. Math. Soc., 69 (1963), 405-408.

[60] Yano, K., On n-dimensional Riemannian spaces admitting a group of motions of order $\frac{1}{2}n(n-1)+1$, Trans. Amer. Math. Soc., 74 (1953), 260-279.

SPACETIME AS A SHEAF OF GEODESICS IN SUPERSPACE*

Bryce S. DeWitt

University of North Carolina at Chapel Hill

Superspace - the space of C^∞ geometries on a compact Hausdorff 3-dimensional manifold M without boundary - has been defined and analysed by Arthur Fischer in an accompanying paper. The results of his analysis show that superspace is not a manifold but is a stratified union of manifolds. It is the purpose of the present paper to give some nonrigorous arguments which suggest that superspace can nevertheless be extended in such a way that it becomes a manifold, in fact a Riemannian manifold, and that a spacetime which satisfies Einstein's equations can be regarded as a sheaf of geodesics in this extended manifold.

We briefly review Fischer's elementary definitions, with some slight changes of notation. Let Riem(M) be the space of C^∞ Riemannian metrics on M and Diff(M) the group of C^∞ diffeomorphisms of M. Riem(M) is an open subspace of the space of all C^∞ symmetric 3 x 3 matrix functions on M and possesses a natural topology induced therefrom. In coordinate language a "point" of Riem(M) is determined by giving six functions $\gamma_{ij}(\underset{\sim}{x})$ of three (coordinate) variables $(x^1, x^2, x^3) = (\underset{\sim}{x})$, these functions being subject to the constraints $\gamma_{ij} = \gamma_{ji} (i,j = 1,2,3)$ and $\gamma \equiv \det(\gamma_{ij}) > 0$. We shall use the generic symbol $\underset{\sim}{\gamma}$ to denote a point of Riem(M).

Each element f of Diff(M) maps Riem(M) into itself by the transformation law for covariant symmetric tensors, which may be written symbolically $\underset{\sim}{\gamma}' = f^*\gamma$, and hence Diff(M) acts as a transformation group on Riem(M). The <u>orbit</u> of each point $\underset{\sim}{\gamma}$ of Riem(M)

*Work supported by the National Science Foundation.

under the action of Diff(M) is defined by

$$\text{orb}\ \underset{\sim}{\gamma} \equiv \{\underset{\sim}{\gamma}' | \underset{\sim}{\gamma}' \varepsilon\ \text{Riem}(M),\ \underset{\sim}{\gamma}' = f^{*}\underset{\sim}{\gamma}\ \text{for some}\ f\ \varepsilon\ \text{Diff}(M)\}. \tag{1}$$

There is a one-to-one correspondence between the orbits in Riem(M) and the points, often denoted by ${}^{(3)}\mathscr{G}$ (3-geometry), of superspace. In fact, the most efficient mathematical procedure is simply to identify them:

$$\text{orb}\ \underset{\sim}{\gamma} = {}^{(3)}\mathscr{G}. \tag{2}$$

A formal definition of superspace is then given by

$$\mathscr{S}(M) \equiv \{A | A\ \varepsilon\ P(\text{Riem}(M)),\ A = \text{orb}\ \underset{\sim}{\gamma}\ \text{for some}\ \underset{\sim}{\gamma}\ \varepsilon\ \text{Riem}(M)\} \tag{3a}$$

where P(Riem(M)) denotes the power set of Riem(M). Another formal representation of $\mathscr{S}(M)$ is

$$\mathscr{S}(M) = \frac{\text{Riem}(M)}{\text{Diff}(M)}\ . \tag{3b}$$

A given 3-geometry may be singled out by displaying an explicit covariant metric tensor $\gamma_{ij}(\underset{\sim}{x})$ in a particular coordinate system. It can equally well be specified by displaying the cooresponding contravariant metric tensor $\underset{\sim}{\gamma}^{ij}(\underset{\sim}{x})$ in the same coordinate system:

$$\gamma_{ik}\gamma^{kj} = \delta^{j}_{i}\ . \tag{4}$$

In fact, any invertible set of <u>functionals</u> of the $\gamma_{ij}(\underset{\sim}{x})$ (e.g., harmonic transforms) can serve as well. This leads us to try to consider Riem(M) as a kind of <u>functionally differentiable</u> manifold having a number of dimensions which may be characterized by the mathematically meaningless but heuristically useful symbol $6 \times \infty^3$. From this point of view the specification of a point in Riem(M) by means of the functions $\gamma_{ij}(\underset{\sim}{x})$ corresponds to the adoption of a particular coordinate system in Riem(M). Moreover, if f ε Diff(M) then $f^{*}\gamma$ is a linear functional of γ, and hence the actions of Diff(M) on Riem(M) are <u>functional diffeomorphisms</u> of Riem(M).

If Riem(M) can be regarded as a (functionally) differentiable manifold then why not try to regard it as a Riemannian manifold? Is it, in fact, possible to endow Riem(M) with a Riemannian metric g? Suppose we have a 3-metric $\gamma_{ij}(\underset{\sim}{x})$ on M, and suppose we subject this 3-metric to an infinitesimal C^{∞} variation $\delta\gamma_{ij}(\underset{\sim}{x})$. What kind of an expression would we write down for the arc length $\delta \mathfrak{s}$ between the points γ_{ij} and $\gamma_{ij} + \delta\gamma_{ij}$ in Riem(M)? The obvious answer is

$$\delta \mathfrak{s}^2 = \int d^3\underset{\sim}{x} \int d^3\underset{\sim}{x}' \mathcal{G}^{ijk'l'} \delta\gamma_{ij}\delta\gamma_{k'l'} \tag{5}$$

where

$$\delta\gamma_{ij} = \delta\gamma_{ij}(\underset{\sim}{x}), \quad \delta\gamma_{k'l'} = \delta\gamma_{kl}(\underset{\sim}{x}'), \tag{6}$$

and the $\mathcal{G}^{ijk'l'}$ are a set of 21 tempered distributions on M x M, constructed out of (and hence functionals of) the functions $\gamma_{ij}(\underset{\sim}{x})$ and subjected solely to the restriction that they be invertible, i.e. that there exist another (unique) set of tempered distributions $\mathcal{G}_{ijk'l'}$ satisfying

$$\int \mathcal{G}_{ijm''n''} \mathcal{G}^{m''n''k'l'} d^3\underset{\sim}{x}'' = \delta_{ij}^{\ \ k'l'} \equiv \delta_{ij}^{\ \ kl}\delta(x,x') \tag{7}$$

where $\delta(\underset{\sim}{x},\underset{\sim}{x}')$ is the delta function on M x M and

$$\delta_{ij}^{\ \ kl} \equiv \frac{1}{2}(\delta_i^{\ k}\delta_j^{\ l} + \delta_i^{\ l}\delta_j^{\ k}). \tag{8}$$

Distributions having these properties are not at all difficult to find, and hence we certainly <u>can</u> make Riem(M) into a Riemannian manifold.

The following question now arises: Is it possible to choose $\mathcal{G}^{ijk'l'}$ in such a way that it defines, by a natural projection, a Riemannian structure on superspace? Suppose $\delta_{\parallel}\gamma_{ij}$ and $\delta_{\perp}\gamma_{ij}$ are two infinitesimal C^∞ displacements from the point γ_{ij} in Riem(M), of which the first lies in the orbit through γ_{ij} and the second is perpendicular to the orbit. Then

$$\delta_{\parallel}\gamma_{ij} = \xi_{i.j} + \xi_{j.i} \text{ for some infinitesimal functions } \xi_i \tag{9}$$

and

$$\int \mathcal{G}^{ijk'l'}{}_{.j}\,\delta_{\perp}\gamma_{k'l'}\,d^3\underset{\sim}{x}' = 0 \tag{10}$$

where the dots denote "covariant derivatives" based on γ_{ij}, calculated by regarding the ξ_i as the components of a covariant vector and the unprimed indices of $\mathcal{G}^{ijk'l'}$ as those of a contravariant tensor density of unit weight.

The points γ_{ij} and $\gamma_{ij} + \delta_{||}\gamma_{ij}$ lie on the same orbit, whereas the points γ_{ij} and $\gamma_{ij} + \delta_{\perp}\gamma_{ij}$ lie on two different orbits. The square of the perpendicular distance, at γ_{ij}, between the two orbits in the latter case is

$$\delta_{\perp}\mathscr{S}^2 = \int d^3\underset{\sim}{x}\int d^3\underset{\sim}{x}'\,\mathscr{G}^{ijk'l'}\delta_{\perp}\gamma_{ij}\delta_{\perp}\gamma_{k'l'} \tag{11}$$

If it were possible to choose $\mathscr{G}^{ijk'l'}$ in such a way that this perpendicular distance remained constant as we moved to other positions on the two orbits we would then have a Riemannian structure defined naturally on $\mathscr{S}(M)$. We would simply define the distance between two infinitesimally differing 3-geometries in $\mathscr{S}(M)$ as the (constant) perpendicular distance between the corresponding orbits in Riem(M).

How can the constancy of $\delta_{\perp}\mathscr{S}$ be guaranteed? Let f be an element of Diff(M). Then the action of f on the points γ_{ij} and $\gamma_{ij} + \delta_{\perp}\gamma_{ij}$ is to shift them to other points on the two orbits, namely $f^*\gamma_{ij}$ and $f^*\gamma_{ij} + f^*\delta_{\perp}\gamma_{ij}$ respectively. The new displacement $f^*\delta_{\perp}\gamma_{ij}$ leads to a new perpendicular distance given by

$$(f^*\delta_{\perp}\mathscr{S})^2 = \int d^3\underset{\sim}{x}\int d^3\underset{\sim}{x}'(f^*\mathscr{G}^{ijk'l'})(f^*\delta_{\perp}\gamma_{ij})(f^*\delta_{\perp}\gamma_{k'l'}) \tag{12}$$

If $\mathscr{G}^{ijk'l'}$ could be chosen in such a way that the actions of Diff(M) were isometries of Riem(M) then this perpendicular distance, and indeed all arc lengths, would remain invariant under the actions of Diff(M).

What is the necessary and sufficient condition on $\mathscr{G}^{ijk'l'}$ for the actions of Diff(M) to be isometries of Riem(M)? It is that the arc length (5) be invariant under coordinate transformations. Since $\delta\gamma_{ij}$ transforms like a covariant tensor it follows that $\mathscr{G}^{ijk'l'}$ must transform, under the actions of Diff(M), like a contravariant bitensor density of unit weight at both $\underset{\sim}{x}$ and $\underset{\sim}{x}'$. When this condition holds $\mathscr{S}(M)$ becomes endowed with a Riemannian structure.

Unfortunately, this does not convert $\mathscr{S}(M)$ into a Riemannian manifold since, as Fischer has shown, $\mathscr{S}(M)$ is not even a manifold. What goes wrong may be illustrated by considering a much simpler example. Suppose we replace Riem(M) by Euclidean 3-space and Diff(M) by the group U(1) of rotations about an axis. Choose Cartesian axis x,y,z and let the rotations take place about the z axis. Then the necessary and sufficient condition that the rotations act isometrically on the 3-space is that its metric depend only on z and $r(\equiv \sqrt{x^2 + y^2})$. The Euclidean metric, of course, satisfies this condition, although it is not the only one which does. But let us choose it for simplicity. The orbits are then circles perpendicular to and centered on the z axis, and the space of orbits is the Euclidean half plane (endowed with the Euclidean metric) together with its boundary. The boundary points are the points of the z axis. It is their presence in the orbit space which prevents it from being a manifold. The boundary points have neighborhoods which are structurally different from those of interior points.

It will be noted that the boundary points are precisely those points which remain invariant under the actions of the group. More generally, if any point of a manifold on which a group acts remains invariant under the actions of a nontrivial subgroup, then the corresponding orbit differs homeomorphically from the generic orbits (typically it is of lower dimension), and its neighborhoods in the orbit space differ structurally from those of generic orbits. The maximal subgroup which leaves a given point invariant is known as the isotropy group at that point. The isotropy group is typically either a Lie group or a discrete group. If it is a nontrivial Lie group, of dimensionality $n(>0)$ say, then the corresponding orbit lies on a "boundary" having n fewer dimensions than that of the orbit itself.

As a slightly more complicated example which illustrates these remarks, consider the group $U(1) \times U(1)$ of rotations about the subspaces $x_1 = x_2 = 0$ and $x_3 = x_4 = 0$ in Euclidean 5-space (coordinates x_1, x_2, x_3, x_4, x_5). The orbits are tori ($S^1 \times S^1$) and the orbit space is the Euclidean quarter 3-space having coordinates $u \equiv \sqrt{x_1^2 + x_2^2}$, $v \equiv \sqrt{x_3^2 + x_4^2}$, and x_5. The boundary points comprise the two half planes $u = 0, v > 0$ and $u > 0, v = 0$, together with their common boundary $u = v = 0$. The isotropy group associated with the points on the two half planes is $U(1)$, while that associated with the points on their common boundary is $U(1) \times U(1)$ itself. The orbit space is seen to be the union of a partially ordered set of nonintersecting (open) manifolds $\bigcup_n M_n$. The partial ordering is defined by: $M_m < M_n$ if and only if $M_m \subset \text{bd } M_n$, where "bd M_n" denotes the boundary of M_n. It is possible to show that any orbit space can be stratified by decomposition into such a union of open sets related by a partial ordering based on inclusion in boundaries.

In particular, superspace can be so stratified.

Fischer has classified the stratifications of $\mathcal{S}(M)$ for all M which can support Riemannian geometries having isometry groups of dimensionality greater than zero. What do isometry groups have to do with the problem? Let $\underset{\sim}{\gamma}$ be a point of Riem(M). Then the isotropy group at $\underset{\sim}{\gamma}$ is just the isometry group of $\underset{\sim}{\gamma}$ regarded as a 3-metric. It is the subgroup of Diff(M) which leaves $\underset{\sim}{\gamma}$ invariant:

$$I_{\underset{\sim}{\gamma}}(M) \equiv \{f | f \in \mathrm{Diff}(M),\ f^{*}\underset{\sim}{\gamma} = \underset{\sim}{\gamma}\} \tag{13}$$

A natural topology can be imposed on Diff(M) which defines a natural topology on the coset space $\mathrm{Diff}(M)/I_{\underset{\sim}{\gamma}}(M)$, and it can be shown that

$$\frac{\mathrm{Diff}(M)}{I_{\underset{\sim}{\gamma}}(M)} \approxeq \mathrm{orb}\ \underset{\sim}{\gamma} \tag{14}$$

where "$\approxeq$" denotes homeomorphism. If $I_{\underset{\sim}{\gamma}}(M)$ is nontrivial then the neighborhoods of orb $\underset{\sim}{\gamma}$ in $\mathcal{S}(M)$ differ structurally from those of generic orbits, and orb $\underset{\sim}{\gamma}$ is a "boundary" point of $\mathcal{S}(M)$. That is to say, the boundary points of $\mathcal{S}(M)$ are those 3-geometries which possess some kind of symmetry, i.e., which admit nontrivial groups of motions.

Most 3-manifolds are <u>wild</u>, in the sense that no metrics can be imposed upon them which admit isometry groups of dimensionality greater than zero. However, even wild manifolds may support nontrivial discrete isometry groups. It is not yet known for which 3-manifolds (if any) discrete isometry groups <u>cannot</u> occur. Thus it may be that almost all 3-manifolds yield superspaces which are themselves not manifolds.

We now ask the question: Is it possible to extend superspace in such a way that it becomes a Riemannian manifold? We can certainly reduce superspace to a Riemannian manifold by simply removing its boundary points. But quite apart from the fact that this would amount to taking a prejudiced stand against symmetrical 3-geometries, for which there is no apparent physical motivation, such a reduced manifold would be trivially geodesically incomplete.

In order to see what happens to geodesics in an orbit space let us return to the simplified example of rotation about an axis in Euclidean 3-space. Let L be a straight line (geodesic) in this space. If L is skew to the rotation axis then the orbits which it intersects constitute a hyperbola in the orbit space (Euclidean half plane). If L intersects the rotation axis then it is orthogonal to every orbit which it intersects, and these orbits comprise a straight line (geodesic) in the orbit space. This straight line, however, is broken. It undergoes specular reflection at the boundary of the half plane. In our other previous example, in which U(1) x U(1) acts on Euclidean 5-space, similar geodesics also undergo specular reflection at the orbit-space boundary.

These examples illustrate a general rule which applies to all orbit spaces, and to superspace in particular. Suppose Riem(M) has been endowed with a metric which admits Diff(M) as an isometry group, so that $\mathcal{S}(M)$ becomes endowed with a corresponding Riemannian structure. Let L be a geodesic in Riem(M) and let orb L be the collection of orbits (3-geometries) which it intersects. Orb L will not generally be a geodesic in $\mathcal{S}(M)$. However, if L is orthogonal to one of the orbits which it intersects then it is orthogonal to them all, and orb L <u>is</u> a geodesic in $\mathcal{S}(M)$. Moreover, if this geodesic strikes a boundary point of $\mathcal{S}(M)$ it must suffer a break by undergoing some kind of reflection.

In the two simple examples above it is obvious how the breaks in the geodesics may be mended. In the first example one simply extends the half plane to the full plane, and in the second example one extends the quarter 3-space to the whole 3-space. The extended orbit spaces are true Riemannian manifolds, possessing unbroken geodesics.

It is very likely that superspace can be similarly extended, so likely in fact that we shall give its extension a symbol: "$\mathcal{S}^{ex}(M)$". A natural mapping can be expected to exist from $\mathcal{S}^{ex}(M)$ to $\mathcal{S}(M)$, in which several points in $\mathcal{S}^{ex}(M)$ (a finite number?) correspond to the same point in $\mathcal{S}(M)$. Figure 1 illustrates the many-one character of this mapping, and at the same time shows how $\mathcal{S}^{ex}(M)$ resolves the problem of broken geodesics.

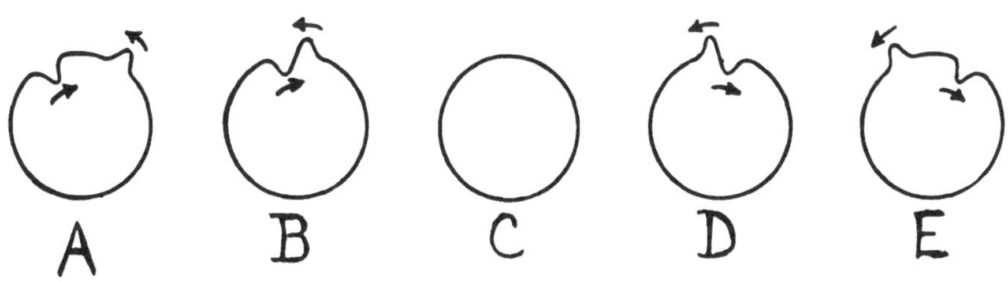

Fig. 1. 3-geometries lying on a geodesic in $\mathcal{S}^{ex}(M)$. A and E are identical, and so are B and D. C is a symmetric 3-geometry which in $\mathcal{S}(M)$ is a boundary point.

Diagrams A, B, C, D, E are schematic representations of five representative points (3-geometries) taken in order along a geodesic in $\mathscr{S}^{ex}(M)$. Geometries A and E (also B and D) correspond to the same point in $\mathscr{S}(M)$* but to distinct points in $\mathscr{S}^{ex}(M)$. The geodesic on which these points are situated passes through the symmetric 3-geometry C. Although in $\mathscr{S}^{ex}(M)$ there is nothing exceptional about C, in $\mathscr{S}(M)$ it is a boundary point, and the geodesic suffers a break here.

We now come to the question of the specific choice of metric to be imposed on superspace. There are infinitely many bitensor densities $\mathscr{G}^{ijk'l'}$ which can be imposed as metrics on Riem(M) and which yield corresponding geometries for $\mathscr{S}(M)$ (and hence for $\mathscr{S}^{ex}(M)$). Which choice is "right"? This depends on what we are trying to do. If our interest in superspace is primarily that of a mathematician, we may wish to make $\mathscr{S}(M)$ into a proper metric space, so that two 3-geometries are identical if and only if the distance between them is zero. One of the simplest ways of achieving this is to choose for Riem(M) the following positive definite metric.

$$\mathscr{G}^{ijk'l'} \equiv \frac{1}{2}\gamma^{1/2}(\gamma^{ik}\gamma^{jl} + \gamma^{il}\gamma^{jk})\,\delta(\underset{\sim}{x},\underset{\sim}{x}')$$

If, on the other hand, our interest is canonical general relativity then a different and slightly more complicated choice suggests itself. 3-geometries in general relativity are slices through spacetime, and the aim of the canonical theory is to describe the dynamics of these slices. The dynamics may be summed up in the Hamilton-Jacobi equation [1]

$$G_{ijkl}\frac{\delta\mathscr{W}}{\delta\gamma_{ij}}\frac{\delta\mathscr{W}}{\delta\gamma_{kl}} = \gamma^{1/2}\,{}^{(3)}R, \tag{15}$$

$$G_{ijkl} \equiv \frac{1}{2}\gamma^{-1/2}(\gamma_{ik}\gamma_{jl} + \gamma_{il}\gamma_{jk} - \gamma_{ij}\gamma_{kl}), \tag{16}$$

together with the supplementary constraints

$$\left(\frac{\delta\mathscr{W}}{\delta\gamma_{ij}}\right)_{.j} = 0. \tag{17}$$

$\mathscr{W}$ is known as the Hamilton-Jacobi functional. It is a real function over Riem(M). The constraints (17) show that it has also the property of being invariant under Diff(M):

---

*In the figure they are oriented oppositely.

$$f^*\mathscr{W} = \mathscr{W} \text{ for all } f \in \mathrm{Diff}(M). \tag{18}$$

It is often said that $\mathscr{W}$ may consequently be regarded as a function over $\mathscr{S}(M)$. I shall argue in this paper that $\mathscr{W}$ should rather be regarded as a function over $\mathscr{S}^{ex}(M)$.

The method of obtaining solutions of Einstein's (empty space) equations from the Hamilton-Jacobi functional has been described in references [2] and [3]. The 1 + 3 decomposition of spacetime, which is characteristic of every canonical field theory, leads one to introduce a corresponding decomposition of the 4-metric of spacetime

$$(g_{\mu\nu}) = \begin{pmatrix} -\alpha^2 + \gamma^{kl}\beta_k\beta_l & \beta_j \\ \beta_i & \gamma_{ij} \end{pmatrix}, \quad \alpha > 0. \tag{19}$$

The functional derivative $\delta\mathscr{W}/\delta\gamma_{ij}$ may then be identified with $-G^{ijkl}K_{kl}$ where

$$G^{ijkl} \equiv \frac{1}{2}\gamma^{1/2}(\gamma^{ik}\gamma^{jl} + \gamma^{il}\gamma^{jk} - 2\gamma^{ij}\gamma^{kl}), \tag{20}$$

$$G_{ijmn}G^{mnkl} = \delta_{ij}{}^{kl}, \tag{21}$$

$$K_{ij} \equiv -\frac{1}{2}\alpha^{-1}(\gamma_{ij,0} - \beta_{i\cdot j} - \beta_{j\cdot i}). \tag{22}$$

This permits one to write

$$\gamma_{ij,0} = 2\alpha\, G_{ijkl}\frac{\delta\mathscr{W}}{\delta\gamma_{kl}} + \beta_{i\cdot j} + \beta_{j\cdot i}. \tag{23}$$

If now this last equation is differentiated with respect to x^0, and all reference to $\mathscr{W}$ in the resulting expression is eliminated by using the equation itself again, together with the result of functionally differentiating Eq. (15) with respect to $\gamma_{k'l'}$, one obtains six of the ten Einstein equations. The other four are simply Eqs. (15) and (17) with $\delta\mathscr{W}/\delta\gamma_{ij}$ replaced by $-G^{ijkl}K_{kl}$.

It is helpful at this point to indulge in a bit of transfinite numerology. Equations (15) and (17) together constitute a set of $4 \times \infty^3$ simultaneous first order (functional) differential equations*

---

*The integrability of these equations is guaranteed by the consistency of the corresponding constraints in the canonical theory. See, for example, reference [2].

in the $6 \times \infty^3$ independent variables $\gamma_{ij}(\underset{\sim}{x})$ and the single dependent variable $\mathscr{W}$. Their general solution involves $2 \times \infty^3 + 1$ constants of integration. One of these constants is a trivial additive constant. Equations (15) and (17) therefore possess a $2 \times \infty^3$ parameter family of essentially distinct solutions.

Suppose we are given one of these solutions, and suppose we adopt coordinate conditions which fix α and the β_i as certain functionals of the γ_{ij}, transforming, under the actions of Diff(M), as a scalar and the components of a covariant vector respectively. Then each point of Riem(M) defines a (one-dimensional) curve in Riem(M), which may be obtained by starting at that point and integrating Eqs. (23). Since all the points on any one of these curves define the same curve, it follows that each solution of Eqs. (15) and (17) (together with a set of coordinate conditions) defines a $6 \times \infty^3 - 1$ parameter congruence of curves in Riem(M). Since Eqs. (23) transform covariantly under the actions of Diff(M) this congruence immediately projects onto a $3 \times \infty^3 - 1$ parameter congruence in $\mathscr{S}^{ex}$(M). (We introduce $\mathscr{S}^{ex}$(M) here rather than $\mathscr{S}$(M) since we want to avoid broken curves and the nuisance of boundary points.) Each curve in the latter congruence corresponds (a) to a spacetime satisfying Einstein's equations and (b) to a particular slicing of that spacetime. Choose one of these curves and denote by Ω the spacetime which it generates. Let U be the subset of $\mathscr{S}^{ex}$(M) consisting of all the 3-dimensional spacelike slices through Ω. Then <u>every</u> curve in the congruence which intersects U also generates Ω and lies in U. This follows from the complete arbitrariness of α and the β_i in the derivation (outlined above) of Einstein's equations from (15), (17) and (23), and from the fact that every sequential slicing of a given spacetime can be achieved by appropriate choice of α for given β_i. Of course, once α and the β_i are fixed the congruence is fixed, and the class of slice sequences to which its curves give rise is correspondingly restricted. Nevertheless, every slice through a given spacetime can be found on at least one of the curves which generate that spacetime.

These considerations lead to the idea of decomposing a given congruence into "sheaves", each sheaf consisting of all the curves which generate a given spacetime. For example, all the curves which intersect the subset U above comprise a single sheaf. The dimensionality of U is ∞^3 since this is the number of free choices involved in selecting a scalar function (e.g.,α) on M and hence of slicing spacetime.* Since each curve which intersects U lies in U, U contains an $\infty^3 - 1$ parameter family of curves. Every other sheaf contains a similar $\infty^3 - 1$ parameter family. Since each sheaf in the congruence corresponds to a distinct spacetime and since the congruence itself is a $3 \times \infty^3 - 1$ parameter family, it follows that each solution of Eqs. (15) and (17) defines a $2 \times \infty^3$ parameter family of distinct spacetimes. Finally, since (15) and (17) themselves possess a $2 \times \infty^3$ parameter family of distinct solutions, we see that Einstein's equations possess a $4 \times \infty^3$ parameter family of physically distinct solutions. This number corresponds precisely to the $2 \times \infty^3$ degrees of freedom which the gravitational field is always said to possess.

We now show how the metric of Riem(M) can be chosen so that the congruence of curves in $\mathcal{S}^{ex}(M)$, defined by a given $\mathcal{W}$ and a given choice of α and β_i, becomes a congruence of geodesics in $\mathcal{S}^{ex}(M)$. That this is possible should perhaps not be surprising since dynamical problems can often be reformulated as problems in the theory of geodesics. In the present case no generality is lost if we set

$$\beta_i = 0 \tag{24}$$

since other choices for the β_i merely correspond to coordinate transformations in M. Moreover, no generality is lost if we impose an over-all normalization on α, since changes in its normalization correspond merely to changes in the parameter x^0. It will be convenient to adopt the following normalization:†

$$2\int \alpha\gamma^{1/2}\ {}^{(3)}R\ d^3\underset{\sim}{x} = \pm 1 \tag{25}$$

We then multiply Eq. (15) by 2α and integrate over M, obtaining

$$\int d^3\underset{\sim}{x} \int d^3\underset{\sim}{x}'\, \mathcal{G}_{ijk'l'} \frac{\delta \mathcal{W}}{\delta\gamma_{ij}} \frac{\delta \mathcal{W}}{\delta\gamma_{k'l'}} = \pm 1 \tag{26}$$

where

$$\mathcal{G}_{ijk'l'} \equiv 2\alpha\, G_{ijkl}\ (\underset{\sim}{x},\underset{\sim}{x}') \,. \tag{27}$$

---

*We are here excluding topological changes in M.

†In order not to have to worry about units at this point, it is also convenient to assume that the absolute units $\hbar = c = 16\pi G = 1$ are being employed.

If $\mathscr{G}_{ijk'l'}$ is chosen as the metric of Riem(M) then equation (26) becomes the eikonal equation for Riem(M). This choice of metric guarantees that the gradient field of $\mathscr{W}$ defines a congruence of geodesics in Riem(M). The tangent to the geodesic through any point is given by

$$\frac{d\gamma_{ij}}{d\varsigma} = \int \mathscr{G}_{ijk'l'} \frac{\delta \mathscr{W}}{\delta\gamma_{k'l'}} d^3\underset{\sim}{x}' = 2\alpha\, G_{ijkl} \frac{\delta \mathscr{W}}{\delta\gamma_{kl}}, \tag{28}$$

which, on comparison with Eq. (23) allows one to identify x^0 with the arc length ς. Moreover, equations (17) may be rewritten in the form

$$\int \mathscr{G}^{ijk'l'}{}_{\cdot j} \frac{d\gamma_{k'l'}}{d\varsigma} d^3\underset{\sim}{x}' = 0, \quad \mathscr{G}^{ijk'l'} \equiv \frac{1}{2}\alpha^{-1} G^{ijkl}\delta(\underset{\sim}{x},\underset{\sim}{x}'), \tag{29}$$

which, in view of Eq. (10), implies that the geodesics are constrained to be orthogonal to all orbits, so that the gradient field of $\mathscr{W}$ in fact defines a congruence of geodesics in $\mathscr{S}^{ex}(M)$. From our previous discussion it follows that this congruence decomposes into a $2 \times \infty^3$ parameter family of sheaves, one sheaf for each distinct spacetime which $\mathscr{W}$ generates.

We note that the above choice of metric for Riem(M), and hence for $\mathscr{S}^{ex}(M)$, is not unique. For each choice of α, subject to the normalization condition (25), a metric can be introduced which permits every solution of Einstein's equations to be regarded as a sheaf of geodesics in $\mathscr{S}^{ex}(M)$. This flexibility in the choice of α is an asset. With a careful choice of α, "kinks" which develop in the geometry of M as one moves along a geodesic may be smoothed out to the maximum extent, so that spurious singularities in the 3-geometry may be avoided. A possible choice for α might be

$$\alpha = A \exp(-{}^{(3)}R_{ij}\,{}^{(3)}R^{ij}) \tag{30}$$

with

$$A^{-1} = 2\int \gamma^{1/2}\,{}^{(3)}R \exp(-{}^{(3)}R_{ij}\,{}^{(3)}R^{ij})\, d^3\underset{\sim}{x}, \tag{31}$$

The reader may easily invent better (and more complicated) choices.*

---

*For some geodesics (e.g., those which pass through a flat 3-geometry in the case $M = S^1 \times S^1 \times S^1$) it may also be necessary to change the normalization condition on α to

$$\int \alpha\gamma^{1/2}\,{}^{(3)}R\, d^3\underset{\sim}{x} = 0.$$

These geodesics then become null geodesics in $\mathscr{S}^{ex}(M)$ and ς becomes an affine parameter.

If, no matter what choice is made for α, the 3-geometry ultimately develops a singularity as one moves along a given geodesic, then one may be fairly certain that a genuine singularity in spacetime has been encountered. Such singularities are of rather frequent occurrence and may, in fact, be almost universal. The latter would be true if it could be shown that $\mathcal{S}^{ex}(M)$ itself is not geodesically complete but possesses a frontier of infinite curvature (for all choices of α). This possibility is strongly suggested by the simpler analysis [2] of the 6-dimensional manifold having G^{ijkl} as its metric, which indeed possesses such a frontier.

It is tempting to believe that the representation of spacetime as a sheaf of geodesics in superspace will prove to be a powerful tool in the study of the singularity problem. This, however, is probably being too optimistic. Advances in our knowledge of the structure of superspace and its geodesics are likely to come from advances in our understanding of spacetime singularities and their classification, rather than the other way around. Nevertheless, the geodesic theory should prove to be a useful framework for ideas. If nothing else, it makes the concept of superspace more accessible by placing it in a familiar setting.

Some final remarks are now in order regarding the use of $\mathcal{S}^{ex}(M)$ rather than $\mathcal{S}(M)$ as the basic manifold of the theory. In classical general relativity my preference for $\mathcal{S}^{ex}(M)$ is mainly a matter of convenience; it avoids troublesome boundary points and broken geodesics. In the quantum theory, however, it has a deeper significance. Consider the situation shown in Fig. 2. This is a highly schematic and oversimplified representation of superspace. Lines OX and OY

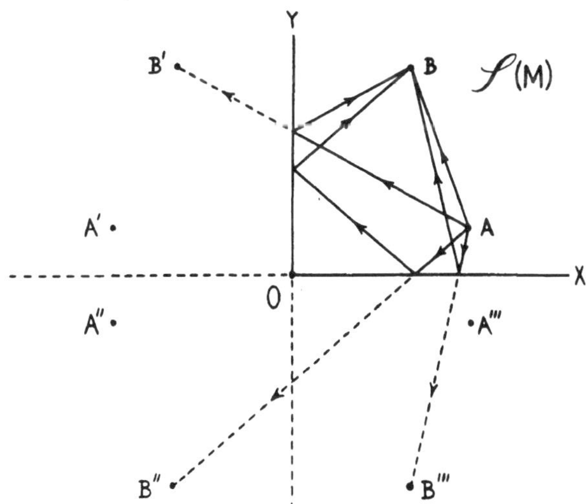

Fig. 2. A schematic representation of superspace. The 3-geometries A and B can be connected by four geodesics (three of them broken) in $\mathcal{S}(M)$ but by only a single geodesic in $\mathcal{S}^{ex}(M)$.

are boundaries of $\mathscr{S}(M)$, on which lie 3-geometries having some kind of symmetry. $\mathscr{S}^{ex}(M)$ is obtained by reflection in these two lines. The points A and B represent two 3-geometries, in $\mathscr{S}(M)$, and the points A', A'', A''', and B', B'', B''' are respectively their images in $\mathscr{S}^{ex}(M)$.

The straight line (geodesic) joining A and B represents a portion of a spacetime which has A and B among its infinity of cross sections. In the language of the Sandwich Problem [4] it is the "filling" of the sandwich. The diagram shows, however, that the Sandwich Problem does not have a unique solution. Three other geodesics also connect A and B and can serve as acceptable sandwich fillings. These, however, are broken; each of them passes through a 3-geometry with partial symmetry. If the Hamilton-Jacobi functional $\mathscr{W}$ is to be able to describe all of these geodesics it must necessarily be multiple valued. In $\mathscr{S}^{ex}(M)$, on the other hand, $\mathscr{W}$ can be single valued. If A and B are close enough together so that no curvature-induced caustic intervenes, there is only a single geodesic between them. The three broken geodesics which connect A and B in $\mathscr{S}(M)$ become unbroken geodesics in $\mathscr{S}^{ex}(M)$ and no longer connect A with B but with three other points B', B'', and B''' instead.

Let $\mathscr{W}(B|A)$ be the Hamilton-Jacobi functional (over $\mathscr{S}^{ex}(M)$) which generates all spacetimes having A as a cross section. Here A is fixed and B is to be understood as the variable. Alternatively B may be regarded as fixed while A ranges over $\mathscr{S}^{ex}(M)$, in which case $\mathscr{W}(B|A)$ generates all spacetimes having B as a cross section. From the eikonal equation (26) one may easily show that $\mathscr{W}(B|A)$ is, apart from a trivial additive constant, just the geodesic distance between A and B (times i if the geodesic is "timelike"). It therefore has a characteristic branch point behavior at A = B, which stems from the fact that the specification of A and B implies nothing about time orientation along the geodesic. It is also revealed in the fact that since the Hamilton-Jacobi equation is quadratic in $\mathscr{W}$, if $\mathscr{W}$ is a solution then so is $-\mathscr{W}$. When passing the branch point we shall always take that branch which permits us to write

$$\mathscr{W}(A|B) = -\mathscr{W}(B|A). \tag{32}$$

In the quantum theory the analog of $\mathscr{W}(B|A)$ is the <u>amplitude</u> $\psi(B|A)$ for A and B both to occur as cross sections of spacetime. $\mathscr{W}$ and ψ are related by the WKB approximation

$$\psi(B|A) \simeq \mathscr{A}(B|A) \exp [i\mathscr{W}(B|A)] \tag{33}$$

Where $\mathscr{A}$ satisfies the probability conservation law [2]

$$\delta(\mathscr{A}^2 G_{ijkl}\,\delta\mathscr{W}/\delta\gamma_{kl})\,/\,\delta\gamma_{ij} = 0. \tag{34}$$

and the conjugation rule

$$\mathscr{A}(B|A)^* = \mathscr{A}(A|B). \tag{35}$$

Any other "wave function" for spacetime may be constructed as a superposition of the $\Psi(B|A)$.

The domain of $\Psi(B|A)$ is $\mathscr{S}^{ex}(M) \times \mathscr{S}^{ex}(M)$. Suppose, however, that the real physical superspace were not $\mathscr{S}^{ex}(M)$ but $\mathscr{S}(M)$. Then the amplitude for A and B both to occur as cross sections of spacetime would not be $\Psi(B|A)$ but rather a sum of the amplitudes $\Psi(B|A)$, $\Psi(B'|A)$, $\Psi(B''|A)$ and $\Psi(B'''|A)$, i.e., a "Feynman sum" over the possible classical trajectories (in $\mathscr{S}(M)$!) between A and B. Wave packets emanating from A and propagating along these separate trajectories could interfere with one another at B. However, there is no a priori principle which tells us what phase factors to assign to such packets after they have been reflected at the boundary. More precisely, we do not know what phase factors to assign to the various contributions to the Feynman sum above. For example, we might choose the combination

$$\Psi(B|A) - \Psi(B'|A) + \Psi(B''|A) - \Psi(B'''|A)$$

which vanishes when either A or B is a boundary point of $\mathscr{S}(M)$, or we might choose

$$\Psi(B|A) + \Psi(B'|A) + \Psi(B''|A) + \Psi(B'''|A)$$

which has vanishing normal derivative at the boundary. Since there seems to be no physical motivation for making the partially symmetric 3-geometries play the role either of nodal or of antinodal points, nor of singling them out in any other way, I believe that $\mathscr{S}^{ex}(M)$ should be adopted as the physical superspace, and not $\mathscr{S}(M)$. Of course, only some kind of fantastic experiment or deduction from observation can ultimately determine whether I am right.*

---

*To those who still insist that $\mathscr{S}(M)$ is the real superspace, the following point should be stressed: It will not do to argue that, after all, 3-space is 3-space, and we only inhabit one universe which, at any instant, is completely specified by its 3-geometry. An object like $\Psi(B|A)$ is quite plainly a wave function which embraces all possible observers, and which can only be interpreted on the basis of an Everett-Wheeler-type metaphysics [5]. Such a wave function describes not one universe but a tremendous number of universes, each one following its own track through a maze of quantum-probability-branch-points. There is no more reason to suppose that the tracks AB and AB' in Fig. 2 will interfere with one another than there is to suppose that the various Everett-Wheeler tracks will interfere.

References

[1] A. Peres, Nuovo Cimento 26, 53 (1962).

[2] B. S. DeWitt, Phys. Rev. 160, 1113 (1967).

[3] U. H. Gerlach, Phys. Rev. 177, 1929 (1969).

[4] J. A. Wheeler, in Relativity Groups and Topology, 1963 Les Houches Lectures (Gordon and Breach Science Publishers, Inc., New York, 1964).

[5] H. Everett, III Rev. Mod. Phys. 29, 454 (1957); J. A. Wheeler, Rev. Mod. Phys. 29, 463 (1957); B. S. DeWitt, in Battelle Rencontres, 1967 Lecture in Mathematics and Physics, C. M. DeWitt and J. A. Wheeler eds. (W. A. Benjamin, Inc. New York, 1968).

AUTHOR INDEX

SUBJECT INDEX

MIX
Papier aus verantwortungsvollen Quellen
Paper from responsible sources
FSC® C105338

If you have any concerns about our products,
you can contact us on
ProductSafety@springernature.com

In case Publisher is established outside the EU,
the EU authorized representative is:
Springer Nature Customer Service Center GmbH
Europaplatz 3, 69115 Heidelberg, Germany

Printed by Libri Plureos GmbH
in Hamburg, Germany